HERMETICITY OF ELECTRONIC PACKAGES

HERMETICITY OF ELECTRONIC PACKAGES

by

Hal Greenhouse

Material Scientist Emeritus
AlliedSignal Corporation
Baltimore, Maryland

NOYES PUBLICATIONS
Park Ridge, New Jersey, U.S.A.

WILLIAM ANDREW PUBLISHING, LLC
Norwich, New York, U.S.A.

Library of Congress Catalog Card Number: 99-29617
ISBN: 0-8155-1435-2
Printed in the United States

Published in the United States of America by
Noyes Publications / William Andrew Publishing, LLC
Norwich, New York, U.S.A.

10 9 8 7 6 5 4 3 2 1

Library of Congress Cataloging-in-Publication Data

Greenhouse, Hal..
Hermeticity of electronic packages / by Hal Greenhouse.
p. cm.
Includes bibliographical references and index.
ISBN 0-8155-1435-2
1. Electronic packaging. 2. Electronics--Materials--Permeability. 3. Sealing (Technology) I. Title.
TK7870.15.G71 1999
621.381'046--dc21 99-29617
CIP

MATERIALS SCIENCE AND PROCESS TECHNOLOGY SERIES

Series Editors

Rointan F. Bunshah, University of California, Los Angeles
Gary E. McGuire, Microelectronics Center of North Carolina
Stephen M. Rossnagel, IBM Thomas J. Watson Research Center

Electronic Materials and Process Technology

CHARACTERIZATION OF SEMICONDUCTOR MATERIALS, Volume 1: edited by Gary E. McGuire

CHEMICAL VAPOR DEPOSITION FOR MICROELECTRONICS: by Arthur Sherman

CHEMICAL VAPOR DEPOSITION OF TUNGSTEN AND TUNGSTEN SILICIDES: by John E. J. Schmitz

CHEMISTRY OF SUPERCONDUCTOR MATERIALS: edited by Terrell A. Vanderah

CONTACTS TO SEMICONDUCTORS: edited by Leonard J. Brillson

DIAMOND CHEMICAL VAPOR DEPOSITION: by Huimin Liu and David S. Dandy

DIAMOND FILMS AND COATINGS: edited by Robert F. Davis

DIFFUSION PHENOMENA IN THIN FILMS AND MICROELECTRONIC MATERIALS: edited by Devendra Gupta and Paul S. Ho

ELECTROCHEMISTRY OF SEMICONDUCTORS AND ELECTRONICS: edited by John McHardy and Frank Ludwig

ELECTRODEPOSITION: by Jack W. Dini

HANDBOOK OF CARBON, GRAPHITE, DIAMONDS AND FULLERENES: by Hugh O. Pierson

HANDBOOK OF CHEMICAL VAPOR DEPOSITION, Second Edition: by Hugh O. Pierson

HANDBOOK OF COMPOUND SEMICONDUCTORS: edited by Paul H. Holloway and Gary E. McGuire

HANDBOOK OF CONTAMINATION CONTROL IN MICROELECTRONICS: edited by Donald L. Tolliver

HANDBOOK OF DEPOSITION TECHNOLOGIES FOR FILMS AND COATINGS, Second Edition: edited by Rointan F. Bunshah

HANDBOOK OF ION BEAM PROCESSING TECHNOLOGY: edited by Jerome J. Cuomo, Stephen M. Rossnagel, and Harold R. Kaufman

HANDBOOK OF MAGNETO-OPTICAL DATA RECORDING: edited by Terry McDaniel and Randall H. Victora

HANDBOOK OF MULTILEVEL METALLIZATION FOR INTEGRATED CIRCUITS: edited by Syd R. Wilson, Clarence J. Tracy, and John L. Freeman, Jr.

HANDBOOK OF PLASMA PROCESSING TECHNOLOGY: edited by Stephen M. Rossnagel, Jerome J. Cuomo, and William D. Westwood

HANDBOOK OF POLYMER COATINGS FOR ELECTRONICS, 2nd Edition: by James Licari and Laura A. Hughes

HANDBOOK OF REFRACTORY CARBIDES AND NITRIDES: by Hugh O. Pierson

HANDBOOK OF SEMICONDUCTOR SILICON TECHNOLOGY: edited by William C. O'Mara, Robert B. Herring, and Lee P. Hunt

HANDBOOK OF SEMICONDUCTOR WAFER CLEANING TECHNOLOGY: edited by Werner Kern

HANDBOOK OF SPUTTER DEPOSITION TECHNOLOGY: by Kiyotaka Wasa and Shigeru Hayakawa

HANDBOOK OF THIN FILM DEPOSITION PROCESSES AND TECHNIQUES: edited by Klaus K. Schuegraf

HANDBOOK OF VACUUM ARC SCIENCE AND TECHNOLOGY: edited by Raymond L. Boxman, Philip J. Martin, and David M. Sanders

HANDBOOK OF VLSI MICROLITHOGRAPHY: edited by William B. Glendinning and John N. Helbert

HIGH DENSITY PLASMA SOURCES: edited by Oleg A. Popov

HYBRID MICROCIRCUIT TECHNOLOGY HANDBOOK, Second Edition: by James J. Licari and Leonard R. Enlow

IONIZED-CLUSTER BEAM DEPOSITION AND EPITAXY: by Toshinori Takagi

MOLECULAR BEAM EPITAXY: edited by Robin F. C. Farrow

SEMICONDUCTOR MATERIALS AND PROCESS TECHNOLOGY HANDBOOK: edited by Gary E. McGuire

ULTRA-FINE PARTICLES: edited by Chikara Hayashi, R. Ueda and A. Tasaki

Ceramic and Other Materials—Processing and Technology

ADVANCED CERAMIC PROCESSING AND TECHNOLOGY, Volume 1: edited by Jon G. P. Binner

CEMENTED TUNGSTEN CARBIDES: by Gopal S. Upadhyaya

CERAMIC CUTTING TOOLS: edited by E. Dow Whitney

CERAMIC FILMS AND COATINGS: edited by John B. Wachtman and Richard A. Haber

CORROSION OF GLASS, CERAMICS AND CERAMIC SUPERCONDUCTORS: edited by David E. Clark and Bruce K. Zoitos

FIBER REINFORCED CERAMIC COMPOSITES: edited by K. S. Mazdiyasni

FRICTION AND WEAR TRANSITIONS OF MATERIALS: by Peter J. Blau

HANDBOOK OF CERAMIC GRINDING AND POLISHING: edited by Ioan D. Mavinescu, Hans K. Tonshoff, and Ichiro Inasaki

HANDBOOK OF INDUSTRIAL REFRACTORIES TECHNOLOGY: by Stephen C. Carniglia and Gordon L. Barna

SHOCK WAVES FOR INDUSTRIAL APPLICATIONS: edited by Lawrence E. Murr

SOL-GEL TECHNOLOGY FOR THIN FILMS, FIBERS, PREFORMS, ELECTRONICS AND SPECIALTY SHAPES: edited by Lisa C. Klein

SOL-GEL SILICA: by Larry L. Hench

SPECIAL MELTING AND PROCESSING TECHNOLOGIES: edited by G. K. Bhat

SUPERCRITICAL FLUID CLEANING: edited by John McHardy and Samuel P. Sawan

DEDICATION

This book is dedicated to the memory of Bill Vergara, a pioneer in the development of the Hybrid Microcircuit Technology, and a friend and colleague for twenty-two years.

Preface

The technology of *hermeticity* addresses the transfer of fluids in and out of sealed enclosures. This technology is based on physics and chemistry, and (like many such technologies) is difficult to grasp when the exposure is brief or infrequent. One's first exposure to this technology usually involves an application related problem. The understanding of, and particularly the solution to, the problem requires a considerable specific background. Not having such a background, the physical concept of the problem is just out of one's grasp and its solution is nowhere in sight. Subsequent exposure to this technology only helps a little, as the background is still missing and the new application is often slightly different.

The purpose of this monograph is to provide the necessary background and problem solving examples, so that packaging engineers and other specialists can apply this knowledge to solving their problems. Ninety nine problems and their solutions are presented. These problems are representative of the type of problems occurring in industry. Many of the included problems are those that the author has experienced.

The technology of hermeticity is an offshoot from *vacuum science.* Vacuum science has a long history, going back to two Italians: Gasparo Berti in 1640, and Evangelista Torricelli in 1644. During the next three hundred and some years, scientists have tried to produce better and better vacuums. They realized that the degree of vacuum achieved, not only depends upon how much and how fast the gas can be removed from the vessel, but also upon the amount and rate of gas leaking into the vessel. This lack of an hermetic vessel eventually led to the technology of hermeticity.

One method of finding leaks in a vacuum system was to connect the system to a mass spectrometer which was tuned to the gas; helium. Helium was selected because the amount of helium in the atmosphere is only 1 part in 200,000 (the rate of its diffusion through a leak is greater than any other gas except hydrogen), and that no other gas can be mistaken for helium by a mass spectrometer. Helium was then sprayed at various parts of the system and if there was a leak, the mass spectrometer would so indicate. This technique, slightly modified, would eventually be used to detect leaks in sealed packages when they contained helium.

The leak testing of sealed packages, when the initial atmosphere in the enclosure had some helium, became a common practice by the early nineteen sixties. In 1965 D. A. Howl and C. A. Mann reported on a leak testing method for enclosures which were not sealed in an atmosphere containing helium. This new method forced helium under pressure through the leakage path into the enclosure. A helium mass spectrometer then detected the helium escaping the enclosure. Subsequently, MIL-STD 883 adopted a leak test method based on this work.

Bibliographies at the end of chapters will lead the reader to areas beyond the present scope of this monograph.

Baltimore, Maryland
November, 1999

Hal Greenhouse

Contents

1

Gas Kinetics

1.0 GENERAL CONSIDERATIONS

This book is primarily about the movement of gases between a sealed package and its external environment. The properties of gases are many, depending upon the gas pressure, the temperature, and the type of gas.

The father of the *Kinetic Theory of Gases* is usually considered to be Gassendi.[1] In 1658, he proposed that gases were composed of rigid atoms made from a similar substance, different in size and form. Gassendi was able to explain many physical phenomena based on these simple assumptions. In 1678, Hooke[2] took Gassendi's theory and expanded it to explain the elasticity of a gas. The elasticity was said to be due to the impact of hard, independent particles on the material which surrounded it. The next advance was due to Newton[3] in 1718. Newton proposed that the pressure of a gas was the result of the repulsive forces between molecules. This was later proven to be only a secondary effect. The first major improvement to the theory was due to Daniel Bernoulli[4] in 1738.

The Kinetic Theory of Gases was well established by the year 1870. It was based on classical physics such as Newtonian mechanics. Some of the proofs of the theory did not occur until the beginning of the twentieth century but they did not alter the classical theory. The classical theory accounts for the motion and energy of the gases, and only has deviations when the pressure of the gas is greatly different from one atmospheric pressure.

Although the Kinetic Theory of Gases explained the gases macroscopically, the theory did not include their vibration, rotation, and the effect of these motions on translation. These later microscopic properties were explained by quantum and statistical mechanics. The classical theory suffices for the purpose of this book, except for the topic of *Residual Gas Analysis* that is based on mass spectrometry. The flow of gases between a container and the container's environment can be completely understood by the classical theory.

Molecules are moving very fast, about 500 meters per second. This rapid motion results in a very large radius of curvature so that the molecular motion can be considered to be in a straight line. The effect of gravity is negligible. The classical theory assumes perfect elasticity of a completely rigid molecule whose size is negligible.

Consider a gas in a sealed container. The fast moving molecules strike the walls of the container as well as colliding with each other. The molecular collisions with the walls exert a pressure on the walls in proportion to the number of impacts. If we now decrease the size of the container in half, keeping the number of molecules in the container the same, the density of the molecules doubles. The number of molecules striking the walls in a given time has now doubled, so the pressure at the wall has also doubled.

The *Mean Free Path* (mfp) of a gas is the average distance a molecule travels before it collides with another molecule. The mfp is therefore inversely proportional to the density (or pressure) of the gas as well as the size of the molecule. If a gas is not confined, the mfp is proportional to the absolute temperature. The mfp of a mixture of gases considers the differences in diameters of each gas. The different diameters of the gases cause a greater variation from the mean than in the case where only one kind of gas is present.

The opportunity of gases to have different densities and different mfp, give rise to three different kinds of gas flow. They are viscous, molecular, and transitional (diffusive).

The flow is viscous when the collisions between molecules dominate (compared to collisions with the walls of the gas container) and there is a difference in the total pressure across the interface. Molecular flow takes place when the predominate collisions are with the walls of the container. The transitional flow is when both types of collisions are common and extends over two magnitudes of pressure. For example: viscous flow would be applicable at pressures greater than 10^{-3} atmospheres, molecular flow at pressures less than 10^{-5} atmospheres, and transitional flow between the pressures 10^{-3} and 10^{-5} atmospheres. The type of flow can be defined in terms of the mfp and the diameter of the channel.

- When the mfp/D is less than 0.01, the flow is viscous
- When the mfp/D is greater than 1.0, the flow is molecular
- When the mfp/D is between 0.01 and 1.0, the flow is transitional

Another condition for defining molecular flow is the relationship between the smallest dimension of the leak channel and the average pressure.

- When $DP_{ave} > 600$ μ-cm, the flow is viscous
- When $DP_{ave} < 10$ μ-cm, the flow is molecular
- Where P_{ave} is the average pressure in μ (microns of Hg)
- Where D is the diameter or smallest dimension of a channel or tube in cm.

It is seen from the latter condition that molecular flow is possible even at a pressure of one atmosphere, if the minimum leak dimension is small enough. This is the typical condition for fine leaks in packages where the collisions with the walls predominate.

The above is a general description of molecular motion. The specific laws by which the molecules behave are attributed to several scientists.

1.1 Boyle's Law

Robert Boyle in 1662, and E. Mariotte in 1676, showed experimentally that at a constant temperature, the volume of a definite mass of gas is inversely proportional to the pressure. There is a deviation from this law at high pressures. The deviation is related to the temperature at which the gas can be liquefied. Boyle's Law at 50 atm has a 3.3% error for hydrogen, and an error of 26% for carbon dioxide. At 10 atm, the error for carbon dioxide drops to less than 10% and hydrogen to less than 1%. The pressure under consideration in the study of hermeticity is seldom greater than 5 atm, so that Boyle's Law is adequate.

1.2 Charles's Law (1787) or Gay-Lussac's Law (1802)

The pressure of any gas in a fixed volume is directly proportional to the absolute temperature of the gas. This law implies that the cubical coefficient of expansion is the same for all gases. This is not true for high pressure (similar to the deviations to Boyle's Law).

1.3 Dalton's Law (1801)

The pressure exerted by a mixture of gases is the sum of the pressures exerted by the constituents (partial pressures) separately. Here again, we have deviations at high pressures that will be of no concern to the present topic.

1.4 Avogadro's Law (1811)

The number of molecules in a gas of a specific volume, temperature and pressure is the same for all gases, and is therefore independent of the nature of the gas.

1.5 Avogadro's Number

The number of molecules in a gram molecular weight (the weight equal to the molecular weight, or 1 mole) for any substance is 6.023×10^{23}.

1.6 Loschmidt's Number

The number of molecules per cubic centimeter needed to produce one atmosphere pressure at the standard temperature of 0°C is the same for all gases and equals 2.685×10^{19}.

2.0 MATHEMATICAL RELATIONSHIPS

Boyle's Law in mathematical terms is $V \propto 1/P$ or $P \propto 1/V$. From the law of Charles and Gay-Lussac, $P \propto T$. Combining these two equations, gives: $P \propto T/V$ or $PV = \text{constant} \times T$.

Applying Avogadro's law for one molecular weight (1 mole), leads to the conclusion that the constant is independent of the type of gas. Designating this constant as R_o, the equation is:

Eq. (1-1) $PV = R_o T$

For n moles:

Eq. (1-2) $PV = nR_o T$

where: P = the pressure

V = the volume

R_o = the gas constant

T = the absolute temperature in °K

R_o = the gas constant for one mole and is standardized at one atmosphere and at 0°C (273.16°K)

One method of verifying Avogadro's law was to measure the volume of different gases when they consisted of one mole at 0°C. This volume was determined to be 22.414 liters.

From Eq. (1-1):

$$R_o = \frac{1\ \text{atm}}{\text{mole}} \frac{22.144\ \text{liters}}{273.16°\text{K}} = 0.08205 \frac{\text{liter atm}}{\text{mole°K}}$$

and since 1 atm = 760 torr,

$$R_o = 0.08205 \times 760 = 62.358\ \text{torr-liters deg}^{-1}\ \text{mole}^{-1}$$

R_o can also be evaluated in ergs, where 1 erg is 1 dyne cm, and 1 dyne is 1 gram cm sec^{-2}. Here the pressure is 76 cm of mercury and the volume is in cubic centimeters.

$$R_o = \frac{8.314 \times 10^7\ \text{ergs}}{\text{mole deg}}$$

Since 4.184×10^7 ergs = 1 calorie:

$$R_o = \frac{8.314 \times 10^7 \text{ ergs}}{\dfrac{4.184 \times 10^7 \text{ ergs mole deg}}{\text{cal}}} = \frac{1.987 \text{ cal}}{\text{mole deg}}$$

The pressure of a homogeneous gas is:

Eq. (1-3) $$P = \frac{1}{3} v_s^{\,2} \rho$$

where: v_s^2 = the root mean square of the average velocity

ρ = the density of the gas

The derivation of Eq. (3) is lengthy but can be found in Glasstone[5] and in Jeans.[6]

The density of a gas is:

Eq. (1-4) $$\rho = \frac{\gamma}{V}$$

where γ is the mass of the gas in volume V.

Combining Eq. (3) with Eq. (4):

$$P = \frac{v_s^{\,2} \rho}{3} = \frac{v_s^{\,2} \dfrac{\gamma}{V}}{3}$$

Note that $3PV = v_s^2\, \gamma$, and if we divide by 2:

Eq. (1-5) $$\frac{3}{2} PV = \frac{1}{2} v_s^{\,2} \gamma$$

The last term is the well known equation for kinetic energy.

The velocity of a molecule is related to the Boltzmann constant.

Eq. (1-6) $$v_s^{\,2} = \frac{3kT}{m}$$

where: v_s^2 = the root mean square of the velocity

k = Boltzmann constant = 1.38×10^{-16} ergs deg^{-1}, and is the gas constant for a single molecule

T = absolute temperature

m = the mass of one molecule

Putting Eq. (1-6) into Eq. (1-5),

$$(3/2)\, PV = (1/2)\, \gamma 3kT/m, \text{ or } PV = \gamma kT/m$$

Now $\gamma/m = N$, the number of molecules, so that:

Eq. (1-7) $$PV = NkT$$

If we consider only one mole, then N becomes N_A, Avogadro's number, and V is the volume for one mole = 22.41 liters. Therefore:

Eq. (1-8) $$PV = N_A kT$$

The average velocity is v_a.

Eq. (1-9) $$v_a = \sqrt{\frac{8kT}{\pi m}}$$

or

Eq. (1-10) $$v_a = \sqrt{\frac{8R_oT}{\pi M}} \text{ in cm/sec}$$

where: $$R_o = 8.3146 \times 10^7 \frac{\text{ergs}}{(\text{degree K})(\text{mole})}$$

1 erg = 1 dyne cm

1 dyne = gram cm/sec^2

so that

$$R_o = 8.3146 \times 10^7 \frac{(\text{g})(\text{cm}^2)}{(\sec^2)(\deg \text{K})(\text{mole})}$$

$$v_a = \sqrt{\frac{(8)(8.3146\times10^7)(g)(\text{cm}^2)(T)(\deg \text{K})}{(\pi)(\sec^2)(\deg \text{K})(\text{mole})\frac{(\text{M})(g)}{\text{mole}}}}$$

$$v_a = \sqrt{\frac{(8)(8.3146\times10^7)(T)(cm^2)}{(\pi)(\sec^2)(\text{M})}}$$

$$v_a = \sqrt{\frac{(6.65168\times10^8)(cm^2)}{(\pi)(\sec^2)(\text{M})}}\sqrt{\frac{T}{\text{M}}}$$

$$v_a = 14{,}551\sqrt{\frac{T}{\text{M}}} \text{ in cm/sec}$$

where T is in degrees K and M is the molecular weight

In any gas, there is a distribution of velocities (Maxwellian). The mfp for multiple velocities is:

Eq. (1-11) $$\text{mfp} = \frac{1}{\sqrt{2}\cdot\pi\frac{N}{V}\sigma^2}$$

where: N = the number of molecules in volume V

σ = the diameter of the molecule

From Eq. (1-7), $N/V = P/kT$, so that

Eq. (1-12)
$$\text{mfp} = \frac{1}{\sqrt{2}\,\pi \frac{P}{kT}\sigma^2}$$

$$\text{mfp} = \frac{kT}{\sqrt{2}\,\pi P\sigma^2}$$

where: k is the Boltzman constant

T is the absolute temperature

P is the pressure in cgs unit (dynes/cm^2)

one atm = 760 mm of mercury = 1.0133×10^6 dynes/cm^2

The equation for viscous flow is:

Eq. (1-13)
$$\eta = \frac{\sqrt{mkT}}{\sqrt{\pi^3}\,\sigma^2}$$

where: η = viscosity

m = mass of the molecule

k is the Boltzmann constant

T is the absolute temperature

σ is the molecular diameter

This equation is valid from one atmosphere to a pressure where the mfp is approximately equal to the size of the gas's container.

3.0 PROBLEMS AND THEIR SOLUTIONS

Problem 1. A vessel is perfectly sealed so that gases can neither enter nor leave the vessel. The vessel contains a partition separating the vessel into two equal volumes. One of the volumes contains air at atmospheric pressure, the other volume contains no gas. The partition is now punctured producing a hole 1 mm in diameter. What is the pressure in each volume?

Solution. Boyle's Law states that the volume is inversely proportional to the pressure. After the puncture, the volume of the gas has doubled so that the pressure has halved, and equals 1/2 atm.

Problem 2. A microcircuit package having an internal volume of 1 cc, is sealed in an atmosphere of 90% nitrogen and 10% helium. What is the number of molecules by type in the package?

Solution. The number of molecules in one molecular weight is Avogadro's number, 6.023×10^{23}. The volume of any gas that has a weight equal to its molecular weight is one molar volume = 22.414 liters. The difference between one cc and one ml is approximately one part in 22,000 (the units can be used interchangeably).

Considering the nitrogen: the number of molecules in 0.9 ml = the fraction of 1 molar volume $\times\ 6.023 \times 10^{23} = [0.9\ \text{ml}/22414\ \text{ml}] \times 6.023 \times 10^{23}$ molecules $= 2.418 \times 10^{19}$ molecules.

For the helium: the number of molecules in 0.1 ml is:

$$\frac{0.1\,\text{ml}}{22414\,\text{ml}} \times 6.023 \times 10^{23} = 2.687 \times 10^{18}$$

The type of gas does not determine the number of molecules as Avogadro's number is true for all gases. The molecular weight does not enter into this problem. If the weight of the gas is used in approaching the problem, the number of molecules = [weight/molecular weight] $\times\ 6.023 \times 10^{23}$. But the weight = [volume × molecular weight/22414 ml] so that:

$$\text{the number of molecules} = \frac{\dfrac{\text{volume} \times \text{molecular weight}}{22414\,\text{ml}}}{\text{molecular weight}} \times 6.023 \times 10^{23}$$

The molecular weights cancel, so we are left with: (volume/22414) $\times\ 6.023 \times 10^{23}$.

Problem 3. What is the mean free path (mfp) of nitrogen at one atmosphere of pressure? The diameter of the nitrogen molecule is 3.8 × 10^{-8} cm.

Solution. Using Eq. (1-11):

$$\text{mfp} = \frac{1}{\sqrt{2} \cdot \pi \frac{N}{V} \sigma^2}$$

$$\text{mfp} = \frac{1}{4.44 \times \frac{6.023 \times 10^{23}}{22414\,\text{cm}^3} \times \left(3.8 \times 10^{-8}\right)^2 \text{cm}^2}$$

$$\text{mfp} = \frac{22414\,\text{cm}^3}{4.44 \times 6.023 \times 10^{23} \times 14.44 \times 10^{-16}\ \text{cm}^2}$$

$$\text{mfp} = 58 \times 10^{-7}\ \text{cm} = 5.8 \times 10^{-6}\ \text{cm}$$

Equation (1-12) can also be used:

$$\text{mfp} = \frac{\text{kT}}{\sqrt{2} \cdot \pi P \sigma^2}$$

Here the pressure is in cgs unit, i.e. dynes/ cm^2:

one atm = 1.013 × 10^6 dynes/cm^2

assume $T = 0^{\circ}\text{C} = 273^{\circ}\text{K}$

$k = 1.38 \times 10^{-16}$ dyne-cm/deg

$$\text{mfp} = \frac{\frac{1.38 \times 10^{-16}\ \text{dyne cm}}{\text{deg}} \times 273\ \text{deg}}{4.44 \times 1.013 \times 10^6\ \frac{\text{dynes}}{\text{cm}^2} \times \left(3.8 \times 10^{-8}\ \text{cm}\right)^2}$$

$$\text{mfp} = 376.7 \times 10^{-16}\,\text{cm}/64.8 \times 10^{-16} \times 10^{6}$$

$$\text{mfp} = 5.8 \times 10^{-6}\,\text{cm}.$$

Problem 4. What is the mean free path of nitrogen when the temperature is 0°C and the pressure is 1×10^{-5} torr?

Solution. 1 torr = 1 mm Hg, and 760 mm = 1.0133×10^{6} dynes/cm^2, so that the pressure (P) in dynes/cm^2 =

$$\frac{1\times10^{-5}\,\text{mm}\times1.0133\times10^{6}\,\dfrac{\text{dynes}}{\text{cm}^2}}{760\,\text{mm}} = 0.01333\,\frac{\text{dynes}}{\text{cm}^2}$$

Using Eq. (1-12):

$$\text{mfp} = \frac{1.38\times10^{-16}\times273}{4.44\times0.0133\times14.4\times10^{-16}}$$

$$\text{mfp} = 443\,\text{cm}$$

Problem 5. A Multi-Chip Module (MCM) is sealed with one atmosphere of nitrogen. The temperature of the nitrogen in the MCM during sealing is 30°C. The cover is 0.010" thick Kovar, and the size of the inside cavity is 2 in^2. When the MCM is tested at –55°C, how much does the cover deflect into the MCM package? Assume that the deflection can be calculated using Roark's equation number thirty.[7] This equation for a kovar cover reduces to:

$$y = -\frac{1.524\times10^{-9}\times\text{psi}\times\text{D}^2}{t^3}$$

where:

psi = the pressure difference in psi

D = the width of the non-supported cover in inches = 2

t = the thickness of the cover in inches = 0.01

y = the deflection in inches

Solution. If the psi is known, the deflection can be calculated. When the MCM is at –55°C, the density or pressure of the gas inside the package will be less than when it was sealed and less than the ambient atmospheric pressure of one atmosphere. The volume of the package remains the same when the temperature is changed. Since the pressure inside the package varies inversely with the absolute temperature, we can write:

$$P_i / P_c = T_i / T_c$$

where:

- P_i = the initial pressure of 1 atm
- P_c = the pressure cold
- T_i = the initial temperature of 30°C = 293°K
- T_c = the cold temperature of –55°C = 218°K
- P_c = 1 atm. × 218°K/293°K = 0.744 atm
- 1 atm = 14.7 psi

Therefore, the pressure in the package is 0.744 × 14.7 = 10.9 psi

Placing the values into the deflection equation:

$$y = -\frac{1.524 \times 10^9 \times (14.7 - 10.9) \times 4}{0.01^3}$$

$$y = -6.10 \times 10^{-3} \text{ in}$$

The center of the cover will deflect inward 6.1 mils.

REFERENCES

1. Gassrendi, *Syntagma Philosphicum,* Lugduni (1658)
2. Hooke, Robert, *Lectiones Cutlerianae,* London (Burndy Library) (1674–1679)
3. Newton, I., *Principia,* 1726 Edition, Univ. California Press (1999)

4. Bernoulli, D., "De affectionibus atque motibus fluidorum elasticorum, praecipue autem aeris," *Hydrodynamica, Argentoria*, Sectio decima (1738)
5. Glasstone, S., *Text Book of Physical Chemistry*, second edition, D. Van Nostrand Company, pp. 249–251 (1946)
6. Jeans, J., *An Introduction to the Kinetic Theory of Gases*, Cambridge Univ. Press, pp. 51–53 (1940)
7. Roark, R. J., *Formulas for Stress and Strain*, McGraw-Hill (1988)

2

Viscous and Molecular Conductance of Gases

1.0 CONDUCTION OF GASES

Imagine a chamber with a partition down the center so that the volumes are equal on each side. Assume the pressure on each side is also equal. Considering the volume on the left side of the partition, there are a finite number of molecules striking one square centimeter of the partition. The identical number of molecules are striking the same square centimeter on the other side of the partition. If the square centimeter becomes an orifice, the molecules which were striking the partition keep traveling into the other half of the chamber. The same number of molecules are traveling in opposite directions through the orifice at the same rate, so there is no net change in the number of molecules on either side.

Now, imagine a similar chamber with a partition down the center, but the pressure in the left side is greater than the pressure in the right side. The number of molecules striking a square centimeter of the partition from the left side is greater than the number of molecules striking the partition from the right side. If the square centimeter becomes an orifice, the molecules continue in their path, and there will be a net change in the number of molecules on each side. The number will increase on the right side and decrease on the left side. This exchange will continue until the number (pressure) is the same on each side.

This flow in opposing directions exists, unless there are no molecules on one side. The opposing flow also holds true for gases traveling through a constriction such as a cylinder.

Consider two chambers of equal volume, connected to each other by a cylinder having a cross-section of one square centimeter. Assume the same type of gas is in each chamber, but the chamber on the left has a pressure greater than the one on the right. Gas from the left chamber will travel through the cylinder into the right chamber until the pressures are equal. The rate of this flow is less when there is a connecting cylinder than when there was just an orifice. The cylinder presents an impedance to the flow because of the pressure drop along the cylinder. In the case of viscous flow, there is an additional impedance because of the friction between the gas and the wall of the cylinder. In the case of molecular flow, some molecules striking the wall will be reflected backwards because the surface irregularities of the wall are many times larger than the molecule.

A detailed discussion concerning the flow of gases through orifices and tubes of various cross-sections can be found in texts on Vacuum Technology, such as Gutherie and Wakerling[1] and Dushman.[2]

The quantity of gas that flows per unit of time is usually designated as Q and is analogous to electrical current. It may be written as:

Eq. (2-1) $$Q = F(P_1 - P_2)$$

where:
Q = gas flow per second
F = the conductance (analogous to electrical conductance)
P_1 = the pressure on the high pressure side
P_2 = the pressure on the low pressure side

where the pressures are analogous to voltages.

The conductance F is the reciprocal of the impedance Z. The conductance F, consists of two parts: the conductance of the orifice and the conductance of the channel attached to the orifice, such as a tube. The impedances add, so that:

Eq. (2-2) $$Z = Z_o + Z_t$$

or

$$\frac{1}{F} = \frac{1}{F_o} + \frac{1}{F_t}$$

where: F_o = the conductance of the orifice

F_t = the conductance of the tube

2.0 VISCOUS CONDUCTION

Poiseuille[3] derived an equation for viscous flow through a circular pipe of uniform cross-section:

Eq. (2-3) $$F_{vc} = \frac{\pi D^4 P_a}{128\eta\ell} \text{ in cgs units}$$

where: F_{vc} = the viscous conductance of the cylinder in $cm^3\ sec^{-1}$

D = the diameter of the cylinder in cm

P_a = the average pressure in the tube in dynes/cm^2 = $(P_1+P_2)/2$

where: P_1 = the pressure on one side and P_2 the pressure on the other side

η = the viscosity of the gas in poise, its value is found in Table 2-2

ℓ = the length of the tube in cm

Davey[4] has developed an equation for viscous flow that contains a correction for slip and for the end effect. The slip correction no longer assumes that the velocity of the molecules at the wall is zero. The molecules can "slip" along the surface of the walls. The end correction includes the conduction of the orifice. This modified equation is:

Eq. (2-4) $$F_{vc} = \frac{\pi D^4 P_a}{128\eta}\left[\frac{Y}{\ell + \left(\frac{Y}{Z}\right)D}\right]$$

where $\left[\ell + \left(\dfrac{Y}{Z}\right)D\right]$ is substituted for ℓ

where
$$Y = 1 + \left[\frac{4.4445}{\left(\dfrac{D}{\text{mfp}}\right) + 1.1949}\right]$$

The term within the square bracket is the end correction.

where
$$Z = 1.6977 + \left[\frac{4.6742}{\left(\dfrac{D}{\text{mfp}}\right) + 2.0444}\right]$$

The terms within the square bracket are the slip correction. The values of Y and Z are based on data from Lund and Berman.[5]

Two variations of the basic viscous flow Eq. (2-3) have been developed by Guthrie and Wakerling:[6] the first for a cylinder and the second for a rectangular duct. The viscous conductance of a cylinder at 20°C for air is:

Eq. (2-5)
$$F_{vc} = \frac{0.182 D^4 P_{ave\mu}}{\ell} \text{ in liters per second}$$

where:

F_{vc} = the viscous conductance of a cylinder in liters/second the constant 0.182 is the coefficient for air at room temperature and has the dimensions of 1/micron-seconds

D = the inside diameter of the cylinder in cm

$P_{ave\mu}$ = the average pressure in microns of mercury

ℓ = the length in cm

The viscous conductance for a rectangular cross-section is:

Eq. (2-6) $$F_{vr} = \frac{0.26Ga^2b^2P_{ave\mu}}{\ell} \text{ in liters per second}$$

where: the 0.26 coefficient is for air at room temperature

the coefficient for helium at room temperature is 0.24, and they have the dimensions of 1/micron-seconds

a and b = the sides of the rectangle in cm

$P_{ave\mu}$ is the average pressure in microns of mercury

G is a function of the a/b ratio and is given in Table 2-1*(a)* and *(b)*

Table 2-1*(b)* is a log-log extrapolation of *(a)*

Table 2-1. *(a)* The Value of G for Viscous Flow Through a Rectangular Leak Channel *(b)* Log-Log Extrapolation from *(a)*

(a)

a/b	1.0	0.9	0.8	0.7	0.6	0.5	0.4	0.3	0.2	0.1
G	1.00	0.99	0.98	0.95	0.90	0.82	0.71	0.58	0.42	0.23

(b)

a/b	0.05	0.02	0.01	0.005	0.002	0.001
G	0.13	0.065	0.038	0.021	0.01	0.006

Viscous flow does not exist when the total pressure inside the package equals the total pressure out side the package, and therefore is not applicable to packages in normal use at an ambient of one atm. Viscous flow could be applicable when bombing or leak testing packages, or if a package has a pressure other than one atm.

3.0 MOLECULAR CONDUCTION

The flow of gases through a fine leak is molecular because of the small dimensions of the leak channel. The number of molecules striking a unit area of surface is proportional to the pressure of the gas and inversely proportional to the square root of its molecular weight. The molecular weight is easily found in several sources, but molecular sizes are not. Table 2-2 gives the molecular weight, the diameter, the viscosity and the mass of molecules of interest.

Table 2-2. Properties of Molecular Species

Molecular Species	Molecular Weight in Grams	Diameter $\times 10^{-8}$ cm #	Viscosity in Micro-Poise @ 20°C &	Molecular Mass $\times 10^{-24}$ Grams*
Helium	4.00	2.2	194	6.64
Neon	20.2	2.6	311	33.5
Argon	40.0	3.7	222	66.2
Nitrogen	28.0	3.8	177	46.5
Oxygen	32.0	3.6	202	53.1
Air	28.7	3.7 Ave	184	47.6
Water	18.0	3.2	125 @ 100°C	29.9
Carbon Dioxide	44.0	4.6	148	73.0

#Values have been rounded to accommodate several sources

& From the Handbook of Chemistry and Physics, CPC

*From Guthrie and Wakerling[1]

The volume of gas that strikes an area A inside a vessel is given by Dushman[7] and is equal to the conductance of an orifice, F_o:

Eq. (2-7) $$F_o = 3638\sqrt{\frac{T}{M}}A \quad \text{in cc/sec per cm}^2$$

where: the constant 3638 has the dimensions of 1/cm-sec

T = the absolute temperature

M = the molecular weight of the gas

A = the area in cm^2

It is assumed that the orifice dimension is small, compared to the dimensions of the vessel. The above equation is the *volume* of gas, not the number of molecules. Note that the volume of gas is independent of the pressure. The number of molecules striking the unit area is a function of the pressure and can be derived as follows:

Rewriting Eq. (1-7),

$N = PV/kT$ and when $V = 1$ cc,

Eq. (2-8) $$N = \frac{P}{kT}$$

where: N = the number of molecules

P = the pressure in micro-bars

k = Boltzman constant = 1.38×10^{-16} dyne - cm/deg

T = the absolute temperature

$$N = \frac{P_{\mu bars}\ \text{in}\ \dfrac{\text{dynes}}{\text{cm}^2}}{1.38\times10^{-16}\ \text{in}\ \dfrac{\text{dyne}-\text{cm}}{\text{deg}}(T)\text{deg}}$$

1 torr = 1 mm = 1.333×10^3 micro-bars, so that:

$$N = \frac{P_{\text{torr}} \text{ in } \dfrac{\text{dynes}}{\text{cm}^2}}{\left(1.38\times10^{-16}\right)\left(1.333\times10^{3}\right)\text{in } \dfrac{\text{dyne}-\text{cm}}{\text{deg}}(T)\text{deg}}$$

Eq. (2-9) $$N = \frac{5.44\times10^{12}\,P_{\text{torr}}}{T} \text{ molecules/cc}$$

Combining Eq. (2-7) with Eq. (2-9), we arrive at the number of molecules striking a unit area (cm^2), n_s. This equals the volume of gas that strikes a square centimeter in a second times the number of molecules in that volume.

$$n_s = F_o \times N$$

$$n_s = \left(3638\sqrt{\frac{T}{M}}\cdot A \text{ in } \frac{\text{cc}}{\text{sec}-\text{cm}^2}\right)\cdot\left(\frac{5.44\times10^{12}}{T}\,\frac{P_{torr}}{\text{cc}}\right)$$

Eq. (2-10) $$n_s = \frac{1.98\times10^{16}\,P_{torr}}{\sqrt{MT}}A \text{ in molecules/cm}^2\text{-sec}$$

where A = area in cm^2.

The conductance of an orifice is equal to the volume of gas passing through the orifice and is therefore equal to Eq. (2-7).

An investigation into the cause of fine leaks in sealed microcircuit packages led to a microscopic visual examination of the glass to metal seals, after each process and stress test, in the manufacture of over 100 hybrids. There was no correlation between fine leakers and visual examination. In many instances, perfect looking seals leaked and those with questionable visual criteria did not leak. The conclusion reached from this investigation was that the fine leaks were so small that they could not be seen under a microscope. The resolution of an optical microscope is 40,000 lines per centimeter which corresponds to a dimension of 2.5×10^{-5} cm (0.01 mils).

The general equation for molecular *flow* comes from the kinetic theory of gases, and is:

Eq. (2-11) $$Q = \sqrt{\frac{kT}{2\pi m}} \cdot (P_1 - P_2) A$$

where: Q = the molecular gas flow in atm. cc/sec

k = Boltzman constant

T = the absolute temperature

m = the mass of the molecule

P_1 = the pressure on the high pressure side, in dynes/cm^2

P_2 = the pressure on the low pressure side

A = the area in square cm, and is assumed to be small compared to the size of the vessel

The molecular conductance can be derived from Eq. (2-11), by referring to Eq. (2-1), $Q = F(P_1 - P_2)$, so that

Eq. (2-12) $$F_m = \sqrt{\frac{kT}{2\pi m}} \cdot A \quad \text{in cgs units}$$

The general equation for the *molecular* conductance for a *long tube* of uniform cross-section is:[8]

Eq. (2-13) $$F_m = \frac{4A^2 v_a}{3H\ell} \quad \text{in cc/sec}$$

where: F_m = the molecular conductance

A = the cross-sectional area in cm^2

v_a = the average velocity in cm/sec

H= the perimeter of the cross-section of the tube in cm

ℓ = the length in cm

and v_a from Eq. (1-10) is $\sqrt{\dfrac{8R_oT}{\pi M}}$ so that

$$F_m = \frac{4A^2 \sqrt{\dfrac{8R_o T}{\pi M}}}{3H\ell}$$

where $R_o = 8.314 \times 10^7$ ergs/deg K-mole

Eq. (2-14) $$F_m = \frac{19.40A^2 \sqrt{\dfrac{T}{M}}}{H\ell} \text{ in liters/sec}$$

where $\sqrt{\dfrac{T}{M}}$ is dimensionless

The terms $H\ell$ = the internal surface area of the tube. The greater this area, the more molecules will be reflected backwards, thus decreasing the molecular conductance.

If the tube is a cylinder, the equation becomes:

Eq. (2-15) $$F_m = \frac{3.81D^3 \sqrt{\dfrac{T}{M}}}{\ell} \text{ in liters/sec}$$

where the constant 3.81 is in units of cm/sec.

Equation (2-15) assumes the length of the cylinder to be much greater than the diameter, which is the usual case for fine leaks. This is not always the case and a modification to Eq. (2-15) by Clausing[9] includes a factor "*K*" which is a function of the length to diameter ratio. The value of *K* was an empirical value. Dushman[10] arrived at a factor *K′* which is very close to Clausing's *K*, but could be expressed in an equation:

Eq. (2-16) $$K' = \frac{1}{1 + \dfrac{3\ell}{4D}}$$

The value of K' has a maximum error with respect to Clausing's K, of about 12%, at ℓ/D values between 1.0 and 1.5. Outside this range, the error progressively gets smaller. The complete range for K or K' is from 1 to $4D/3\ell$. Some values of K' for ℓ/D values of 5, 10, 20, 40, and 500 are 0.210, 0.117, 0.0625, 0.0322, and 0.00266 respectively. The equation for the molecular conductance of a cylinder with the K' correction is:

Eq. (2-17)
$$F_{mcK'} = \frac{3.81 D^3 \sqrt{\frac{T}{M}}}{\ell} \frac{1}{1 + \frac{3\ell}{4D}} \text{ in liter/sec}$$

Equations (2-16) and (2-17) are generally not applicable to fine leaks, as the correction factors apply to small values of ℓ/D.

The molecular conductance for a rectangular tube, F_{mr}, of uniform cross-section can be derived from Eq. (2-14):

where $A = a \times b$

$H = 2(a + b)$

and a, b, and ℓ are in cm

Eq. (2-18)
$$F_{mr} = \frac{9.7 a^2 b^2}{(a+b)\ell} \sqrt{\frac{T}{M}} \text{ in liters per second}$$

where the constant 9.7 is in units of cm/sec.

Clausing[9] has a correction factor for molecular flow through tubes with a rectangular cross-section as shown in Table 2-3.

Applying this correction factor, Eq. (2-18) becomes

Eq. (2-19)
$$F_{mrk} = \frac{9.7 a^2 b^2}{(a+b)\ell} \sqrt{\frac{T}{M}} \cdot K_R \text{ in liters per second}$$

This equation is also not applicable to fine leaks.

Table 2-3. Clausing Correction Factors for Molecular Flow Through Rectangular Tubes

ℓ/b	0.1	0.2	0.4	0.8	1.0	2.0	4.0	10.0	Infinity
K	0.953	0.907	0.836	0.727	0.685	0.542	0.400	0.246	$\frac{b}{\ell}\ln\frac{\ell}{b}$

4.0 CONDUCTION IN THE TRANSITIONAL RANGE[11]

There is no acceptable theory that can lead to equations for conductance in the transitional range. There is however empirical data by Knudsen which leads to the empirical equation:

Eq. (2-20) $$F = F_v + U_m F_m$$

where: F = the total conductance of a cylinder

F_v = the viscous conductance of the cylinder

F_m = the molecular conductance of the cylinder

U_m = a correction factor for the molecular conductance, and has values between 0.81 and 1.0

This equation states that the molecular conductance decreases in the transitional region. Experiments show that the conductance in this region actually increases and is greater than the viscous and molecular conductance equation would predict.[12] The conductance (F) has greater values than the viscous and molecular straight line extrapolations in the transitional region. This region is between values of D/mfp of 1–100. To account for this increase in conductance, the viscous conductance must be increasing in the transitional region. A new equation can be written for the total conductance:

Eq. (2-21) $$F = U_v F_v + U_m F_m$$

where: U_v = the correction factor for viscous conductance

U_m = the correction factor for molecular conductance

5.0 COMPOSITE CONDUCTANCE EQUATIONS

Composite conductance equations include conductances for both viscous and molecular flow, as well as a correction factor for the transitional region.

For a cylinder whose length >> the diameter ($F_t \ll F_o$), F_o can be neglected. An equation for the conductance of such a tube was experimentally verified by Knudsen.[12] It includes derived equations for the viscous and molecular conductance and an empirical modification for the transitional pressure range. The equation is:

Eq. (2-22)

$$F_C = \left(\frac{\pi D^4 PA}{128\eta\ell}\right) + \frac{1}{6}\left(\frac{\sqrt{\frac{2\pi KT}{m}}}{\ell} D^3\right)\left(\frac{1+\sqrt{\frac{m}{kT}}\frac{DPa}{\eta}}{1.24+\sqrt{\frac{m}{kT}}\frac{DPa}{\eta}}\right) \text{ in cgs units}$$

The first term to the right of the = sign is the conductance for viscous flow, and is Eq. (3). The second and third terms are for the molecular conductance modified for the transitional pressure range.

where: F_c = the total conductance of the cylinder in cm^3 sec^{-1}

D = the diameter of the cylinder in cm

P_a = the average pressure in the tube in dynes/ cm^2 = $(P_1+P_2)/2$,

where: P_1 = the pressure on one side and P_2 the pressure on the other side

η = the viscosity of the gas in poise

ℓ = the length of the tube in cm

k = Boltzmann constant = 1.38×10^{-16} ergs/degree K

m = mass of the gas molecule

T = the absolute temperature

Another composite equation for a cylinder has been derived by Davey.[4] This equation includes Eq. (2-4) for viscous conduction as well as a modified form of the molecular portion of Knudsen's Eq. (2-22). The equation in cgs units is:

Eq. (2-23)

$$F_C = \frac{\pi P_a D^4}{128\eta}\left[\frac{Y}{\ell + \left(\frac{Y}{Z}\right)D}\right] + \frac{\pi v_{ave}}{6}\left[\frac{D^3}{\left(\ell + \frac{4D}{3}\right)\left(\frac{D}{\text{mfp}} + 1.509\right)}\right]$$

Assuming the viscosity of helium and the average velocity, and converting the pressure from dynes per square centimeter to atmospheres:

Eq. (2-24)

$$F_C = 1.282 \times 10^8 P_{ave} D^4 \left[\frac{Y}{\ell + \left(\frac{Y}{Z}\right)D}\right] + 4.96 \times 10^4 \left[\frac{D^3}{\left(\ell + \frac{4D}{3}\right)\left(\frac{D}{\text{mfp}} + 1.509\right)}\right] \quad \text{cc/sec}$$

where P_{ave} is in atmospheres.

Several molecular conductances equations have been presented. With the exception of the equations by Davey, the equations have been derived from the science of vacuum technology. Davey derived his equations considering the leakage of gases from packages. It is instructive to compare these equations to the smallest theoretical leak.

6.0 SMALLEST THEORETICAL LEAK

When the opening of the leak orifice is smaller than the size of the molecule, the molecule cannot go through the orifice. If we assume an orifice with a diameter the size of the molecule, the resulting conductance calculations are the lower limit of a leak. Applying Eq. (2-7), Table 2- 4 is for a circular orifice at 20°C with its diameter equal to the diameter of the molecule, and for a rectangular orifice where the smallest side is equal to the diameter of the molecule and equal to one tenth the larger size of the opening. F_{oc} is the conductance of a circular orifice, and F_{or} is the conductance for a rectangular opening.

Table 2-4. Theoretical Minimum Leak Rates

Gas	Mol. Wt. in Grams	$\sqrt{\frac{T}{M}}$	σ $\times 10^{-8}$	s $\times 10^{-15}$	Area $\times 10^{-15}$	F_{oc} $\times 10^{-11}$	F_{or} $\times 10^{-10}$
He	4	8.56	2.2	0.484	0.38	1.18	1.50
Ne	20.2	3.81	2.6	0.675	0.531	0.74	0.94
Ar	40	2.71	3.7	1.37	1.075	1.06	1.35
N_2	28	3.23	3.8	1.44	1.13	1.33	1.69
O_2	32	3.02	3.6	1.30	1.02	1.12	1.42
H_2O	18	4.03	3.3	1.02	0.804	1.18	1.50
CO_2	44	2.58	4.6	2.12	1.66	1.56	1.99

We can calculate the diameter of an opening which would produce the smallest limiting leak for a leak channel 0.1 cm long. Assume a package filled with one atm of helium leaking into air which also is at one atm. The flow is molecular and Eq. (2-15), rewritten so that the answer is in cc/sec, is:

Eq. (2-25)

$$F_{mc} = \frac{3810D^3}{\ell}\sqrt{\frac{T}{M}} \text{ in cc/sec}$$

$$D^3 = \frac{F_{MC}\ell}{3810\sqrt{\frac{T}{M}}} \text{ in cm}$$

$$D^3 = \frac{1.18 \times 10^{-11}(.1)}{(3810)(8.56)} \text{ in cm}$$

$$D^3 = 3.62 \times 10^{-17}$$

$$D = 3.3 \times 10^{-6} \text{ cm}$$

This is the diameter of a 0.1 centimeter long channel to allow the minimum true helium leak rate of 1.18×10^{-11} atm -cc/sec.

If we assume an aspect ratio for a rectangular opening, similar size openings can be calculated. Equation (2-18) in cc/sec is:

Eq. (2-26) $$F_{mr} = \frac{(9710)(a^2b^2)}{(a+b)\ell}\sqrt{\frac{T}{M}}$$

assume $a = 10b$, $\ell = 0.1$ then

$$F_{mr} = \frac{(9700)(10b)^2(b)^2(8.56)}{(b+10b)(0.1)}$$

$F_{mr} = F_{or}$ in Table 2-4 for helium $= 1.5 \times 10^{-10}$

$$1.5\times10^{-10} = \frac{(9700)(100)(b)^4(8.56)}{(1\ 1\text{b})(0.1)}$$

$$\frac{100b^4}{11b} = \frac{(1.50\times10^{-10})(0.1)}{(9700)(8.56)} = 1.81 \times 10^{-16}$$

$$b^3 = 1.99 \times 10^{-17}$$

$$b = 2.71 \times 10^{-6} \text{ cm}$$

$$a = 2.71 \times 10^{-5} \text{ cm}$$

Table 2-5 summarizes the opening sizes for the minimum helium leak rate calculated above.

Table 2-5. Dimensions of Orifices for the Minimum Possible Leak Rate of Helium

Leak Rate (atm.-cc/s)	ℓ	Diameter (cm)	a (cm)	b (cm)
$F_{oc} = 1.18 \times 10^{-11}$	0	2.2×10^{-8}	–	–
$F_{or} = 1.50 \times 10^{-10}$	0	–	2.2×10^{-7}	2.2×10^{-8}
$F_{mc} = 1.18 \times 10^{-11}$	0.1	3.3×10^{-6}	–	–
$F_{mcr} = 1.50 \times 10^{-10}$	0.1	–	2.71×10^{-5}	2.71×10^{-6}

Another illuminating set of calculations can be made, comparing conductances for an equal area, but having a different geometry.

Assume a circular orifice whose diameter is 2×10^{-4} cm. Using Eq. (2-7),

$$F_o = 3638\sqrt{\frac{T}{M}} \text{ Area in cc/sec per cm}^2$$

The area is $\frac{\pi D^2}{4} = \frac{\pi(2 \times 10^{-4})^2}{4} = 3.14 \times 10^{-8}$ cm

$$F_o = (3638)(8.56)(3.14 \times 10^{-8}) = 9.78 \times 10^{-4} \text{ cc/sec}$$

Again assigning 0.1 cm as the length of the leak channel, Eq. (2-25) calculates the conductance:

$$F_{mc} = \frac{(3810)D^3}{\ell}\sqrt{\frac{T}{M}} \quad \text{in cc/sec}$$

$$F_{mc} = \frac{(3810)(2\times10^{-4})^3(8.56)}{0.1} \quad \text{in cc/sec}$$

$$F_{mc} = 2.61 \times 10^{-6} \text{ cc/sec}$$

Selecting different length to width ratios for rectangular channels, while maintaining the same cross-sectional area (3.14×10^{-8} cm^2), yields the following calculations:

$$F_{mr} = \frac{(9700)a^2b^2}{(a+b)\ell}\sqrt{\frac{T}{M}}$$

For $a = 3.14 \times 10^{-4}$ cm, and $b = 1 \times 10^{-4}$ cm

$$F_{mr} = \frac{(9700)(3.14\times10^{-4})^2(10^{-4})^2(8.56)}{[(3.14\times10^{-4})+(1\times10^{-4})]\times0.1}$$

$$F_{mr} = 1.97 \times 10^{-6} \text{ cc/sec}$$

For $a = 3.14 \times 10^{-3}$, $b = 10^{-5}$

$$F_{mr} = \frac{(9700)(3.14\times10^{-3})^2(10^{-5})^2(8.56)}{(3.14\times10^{-3}+10^{-5})\times(0.1)}$$

$$F_{mr} = 2.61 \times 10^{-7} \text{ cc/sec}$$

For $a = 3.14 \times 10^{-2}$, and $b = 10^{-6}$,

$$F_{mr} = \frac{(9700)(3.14\times10^{-2})^2(10^{-6})^2(8.56)}{[(3.14\times10^{-2})+10^{-6}](0.1)}$$

$$F_{mr} = 2.61 \times 10^{-8} \text{ cc/sec}$$

Table 2-6 compares the conductances of channels having the same cross-sectional area but having a different geometry.

Table 2-6. Conductances of Uniform Channels of Equal Cross-Sectional Areas but Different Shape Orifices*

Conductance (cc/s)	Diameter (cm)	*a* (cm)	*b* (cm)
$F_{mc} = 2.61 \times 10^{-6}$	2×10^{-4}	–	–
$F_{mr} = 1.97 \times 10^{-6}$	–	3.14×10^{-4}	1×10^{-4}
$F_{mr} = 2.61 \times 10^{-7}$	–	3.14×10^{-3}	1×10^{-5}
$F_{mr} = 2.61 \times 10^{-8}$	–	3.14×10^{-2}	1×10^{-6}

*Cross-sectional area = 3.14×10^{-8} cm^2, length of channel = 0.1 cm

7.0 DISCUSSION

The correction factors for molecular flow in Eq. (2-16) and in Table 2-3 are either empirical or derived from experiments at low pressures and with cross-sectional areas which are large compared to leak channels. There is no experimental data for very small cross-sections because of the difficulty in forming such a channel. It is not likely that these correction factors apply to channels with a very small cross-section.

The smallest theoretical helium leak was calculated to be 1.18×10^{-11} atm cc/sec. This is for an orifice the size of the helium molecule with a zero length leak channel. The diameter of a channel when it is 0.1 cm long to yield the same leak rate is given in Table 2-5 as 3.3×10^{-6} cm. The minimum leak for the oxygen molecule, whose size is 3.6×10^{-8} cm, is 1.12×10^{-11} atm cc/sec., again for a zero length channel. For a channel 0.1 cm long, the diameter corresponding to this leak rate is 4.60×10^{-6} cm. When the diameter is smaller (for example, 4.6×10^{-7} cm) the leak rate will be less than the minimum.

$$F_{mc} = \frac{(3810)(4.6\times10^{-7})^3}{0.1}\sqrt{\frac{T}{M}} = \frac{(3810)(4.6\times10^{-7})^3}{0.1}\sqrt{\frac{293}{32}}$$

$$F_{mc} = \frac{(3810)(4.6\times10^{-7})^3(3.03)}{0.1}\sqrt{\frac{T}{M}}$$

$$F_{mc} = 1.12 \times 10^{-14} \text{ cc/sec for oxygen}$$

This is magnitudes smaller than the theoretical minimum leak. When the orifice of the 0.1 cm long channel is this large, no oxygen will flow through the channel. Equations to calculate the quantify of a gas leaking in or out of a package will be derived in Ch. 3. The results of these equations are based on the reliable minimum limit of detectability by Residual Gas Analysis (RGA) of 100 ppm. The time for oxygen to leak into a one cc package is: for the minimum oxygen leak rate of 1.12×10^{-11} atm cc/sec,

$$\text{time} = 4.46 \times 10^{8} \text{ sec} = 14.2 \text{ years}$$

The minimum leak corresponding to 14.2 years can be considered hermetic.

Correction factors to conductance equations which yield conductances smaller than the minimum theoretical leak should not be used. Eqs. (2-16), (2-17), and (2-19) are not applicable to fine leak channels.

8.0 PROBLEMS AND THEIR SOLUTIONS

Problem 1. What is the viscous conductance of helium at 20°C through a slit shaped channel with dimensions of 1.0 cm by 0.1 cm, and 0.1 cm in length? The pressure on one side of the slit is two atm and on the other side it is one atm.

Solution. The viscous conductance for a rectangular channel is found by using Eq. (2-6):

$$F_{vr} = \frac{0.26 G a^2 b^2 P_{ave\mu}}{\ell} \text{ in liters/sec}$$

$$P_{ave} = (2 \text{ atm} + 1 \text{ atm})/2 = 1.5 \text{ atm}$$

$$P_{ave\mu} = 1.5 \times 760 \times 10^3 = 1140 \times 10^3\ \mu$$

$$(a/b) = (0.1/1.0) = 0.1$$

From Table 2-1, $G = 0.23$

$$F_{vr} = (0.26)(0.23)(1)^2(0.1)^2(1140 \times 10^3)/0.1$$

$$F_{vr} = 6{,}817.2 \text{ liters/sec}$$

Problem 2. A microcircuit package, having a volume of 5.0 cc is sealed with 1.0 atm of helium. It is placed into a leak detector for 1.0 minute at 20°C. It is noticed during removal, that a pin was missing. The diameter of the pin is 0.4 mm, and the glass depth which surrounded the pin is 1.5 mm. How much helium escaped during the one minute?

Solution. We must first determine if the flow is viscous, molecular or transitional. The criteria was given in Ch. 1 and is:

- When the mfp/D is less than 0.01, the flow is viscous
- When the mfp/D is greater than 1.0, the flow is molecular
- When the mfp/D is between 0.01 and 1.0, the flow is transitional

The mfp is found using Eq. (1-12):

$$\text{mfp} = \frac{kT}{\sqrt{2}.\pi P\sigma^2}$$

where: $k = 1.38 \times 10^{-16}$ ergs/°K

$T = 293$°K

P = the average pressure=(1 atm)/2 = $(1.0133 \times 10^6)/2$

$\sigma^2 = 0.485 \times 10^{-15}$ (see Table 4)

$$\text{mfp} = \frac{1.38 \times 10^{-16}(293)}{\pi\sqrt{2}\left(\frac{1.0133 \times 10^6}{2}\right)\left(0.485 \times 10^{-15}\right)}$$

$$\text{mfp} = \frac{80.868}{1.54 \times 10^6} = 37 \times 10^{-6}\ \text{cm}$$

$$\frac{\text{mfp}}{D} = \frac{37 \times 10^{-6}}{0.4} = 9.2 \times 10^{-5}$$

Therefore the flow is viscous.

Using Eq. (2-3), the value for the viscosity of helium can be used:

$$F_{vc} = \frac{\pi D^4 Pa}{128\eta\ell} \quad \text{in cgs units}$$

$$P_a = 0.5\ \text{atm} = (1.0133 \times 10^6)/2$$

$$F_{vc} = \frac{\pi(0.04)^4\left(1.0133 \times 10^6\right)}{2(128)\left(194 \times 10^{-6}\right)(0.15)}$$

where 194×10^{-6} is the viscosity of helium at 20°C (taken from Table 2-2)

$$F_{vc} = 1093\ \text{cc/sec}$$

Correcting for the slip and the end effect, Eq. (2-3) becomes Eq. (2-4):

$$F_{vc} = \frac{\pi D^4 Pa}{128\eta}\left[\frac{Y}{\ell + \left(\frac{Y}{Z}\right)D}\right]$$

where
$$Y = 1 + \left[\frac{4.4445}{\left(\frac{D}{\text{mfp}}\right) + 1.1949}\right]$$

where
$$Z = 1.6977 + \left[\frac{4.6742}{\left(\frac{D}{\text{mfp}}\right) + 2.0444}\right]$$

$$Y = 1 + \frac{4.4445}{\frac{0.04}{37 \times 10^{-6}} + 1.1949} = 1 + 0.004 \cong 1$$

$$Z = 1.6977 + \frac{4.6742}{\frac{0.04}{37 \times 10^{-6}} + 2.0444} = 1.69774 + 0.004 = 1.6981$$

$$\frac{Y}{Z} = \frac{1}{1.6981} = 0.589$$

$$\left(Y+\frac{Y}{Z}D\right)=0.15+[(0.589)(0.04)]=0.17$$

$$F_{vc}=1093\times 0.15/0.17=966 \text{ cc/sec}$$

All the helium in the 5 cc package escaped in a fraction of a second.

Problem 3. A sealed package has a circular hole with a diameter of 5×10^{-4} cm., and a length of 0.1 cm. What is the molecular conductance of nitrogen passing through this hole when the temperature is 25°C?

Solution. The molecular conductance of a cylinder in liters per second is calculated using Eq. (2-15):

$$F_{mc}=\frac{3.81D^3\sqrt{\frac{T}{M}}}{\ell}$$

The molecular weight of nitrogen is 28 and the temperature is 298°K.

$$F_{mc}=\frac{3.81\left(5\times 10^{-4}\right)^3\sqrt{\frac{298}{28}}}{0.1}$$

$$F_{mc}=(3.81)(1.25\times 10^{-10})(3.26)/0.1$$

$$F_{mc}=1.552\times 10^{-9} \text{ liters/sec}=1.552\times 10^{-6} \text{ cc/sec}$$

Problem 4. What is the total molecular conductance of nitrogen at 25°C for a package having two cylindrical leak channels. The first leak channel has the dimensions of Problem 3. The second channel has a diameter of 0.005 cm and a length of 0.1 cm.

Solution. The total conductance is the two in parallel; using Eq. (2-2):

$$\frac{1}{F_{\text{total}}}=\frac{1}{F_1}+\frac{1}{F_2}$$

F_1 is the conductance of Problem 3 = 1.552×10^{-6} cc/sec

$$F_2 = \frac{3810(0.005)^3(3.26)}{0.1}$$

$$F_2 = 0.001552 \text{ cc/sec}$$

$$\frac{1}{F_{\text{total}}} = \frac{1}{1.552 \times 10^{-6}} + \frac{1}{0.001552}$$

$$\frac{1}{F_{\text{total}}} = \frac{(1.552 \times 10^{-6})(0.001552)}{(1.552 \times 10^{-6}) + (0.001552)}$$

$$F_{\text{total}} = 1.552 \times 10^{-6} \text{ cc/sec}$$

Problem 5. *Part A:* The high vacuum valve (assume no thickness to the valve) of a leak detector opens automatically when the pressure on the high side is 20 microns of mercury. The diameter of the open valve is 2 in. What is the molecular conductance of this opening for nitrogen at 20°C? *Part B:* The valve is connected to an eight inch long circular pipe of the same diameter. What is the molecular conductance of this cylinder? *Part C:* How many molecules are entering this pipe per second?

Solution. *Part A:* 2 inches = 5.08 cm. The conductance of the open valve is the conductance of an orifice, and Eq. (2-7) is applicable.

$$F_o = 3638A\sqrt{\frac{T}{M}} \text{ in cc/sec per cm}^2$$

$$F_o = 3638\pi\frac{5.08}{4}\sqrt{\frac{293}{28}}$$

$$F = 238{,}545 \text{ cc/sec} = 238 \text{ l/sec}$$

Part B: 8 inches = 20.32 cm. Equation (2-15) calculates the molecular conduction of a cylinder.

$$F_{mc} = \frac{3.81 D^3 \sqrt{\dfrac{T}{M}}}{\ell} \text{ in l/sec}$$

$$F_{mc} = \frac{(3.81)(5.08)^3(3.23)}{20.32}$$

$$F_{mc} = 79.39 \text{ l/sec}$$

The total conductance considers the conductance of the orifice as well as the pipe.

$$\frac{1}{F_{\text{total}}} = \frac{1}{F_o} + \frac{1}{F_{mc}}$$

$$\frac{1}{F_{\text{total}}} = \frac{1}{238} + \frac{1}{79.39}$$

$$F_{\text{total}} = 59.53 \text{ l/sec}$$

Part C: The number of molecules entering the pipe is found using Eq. (2-10):

$$n_s = \frac{1.98 \times 10^{16} P_{\text{torr}}}{\sqrt{\text{MT}}} A \text{ in molecules/sec}$$

$$n_s = \frac{(1.98 \times 10^{16})(20 \times 10^{-3})\pi(5.08)^2}{4(\sqrt{28 \times 293})}$$

$$n_s = 8.86 \times 10^{13} \text{ molecules/sec}$$

Problem 6. What is the total conductance of a leak in a package filled with 100% helium at one atm, when it is being leak tested in a helium leak detector? The leak channel is 0.001 cm in diameter and 0.1 cm long.

Solution. There is an absolute total pressure difference between the inside and outside of the package, so that viscous flow is possible. Equations (2-23) and (2-24) consider both viscous and molecular flow:

$$F_c = 1.278\times10^8 P_{ave}\left[\frac{YD^4}{\ell+\left(\dfrac{Y}{Z}\right)D}\right] + 4.961\times10^4\left[\frac{D^3}{\left(\ell+\dfrac{4D}{3}\right)\left(\dfrac{D}{\text{mfp}}+1.509\right)}\right]\ \text{cc/sec}$$

where P_{ave} is in atm.

$$Y = 1+\left[\frac{4.4445}{\left(\dfrac{D}{\text{mfp}}\right)+1.1949}\right]$$

$$Z = 1.6977+\left[\frac{4.6742}{\left(\dfrac{D}{\text{mfp}}\right)+2.0444}\right]$$

$$\text{mfp} = \frac{\text{kT}}{\sqrt{2}\pi P\sigma^2}$$

where: $k = 1.38 \times 10^{-16}$ ergs/°K

$T = 293$ °K

P = the average pressure=(1 atm)/2 = $(1.0133 \times 10^{6})/2$

$\sigma^2 = 0.485 \times 10^{-15}$ (see Table 2-4)

$$\text{mfp} = \frac{1.38 \times 10^{-16}(293)}{\pi\sqrt{2}\left(\frac{1.0133 \times 10^{6}}{2}\right)\left(0.485 \times 10^{-15}\right)}$$

$$\text{mfp} = \frac{80.868}{1.54 \times 10^{6}} = 37 \times 10^{-6} \text{ cm}$$

$$\frac{D}{\text{mfp}} = \frac{0.001}{37 \times 10^{-6}} = 27.02$$

$$Y = 1 + \frac{4.4445}{27.02 + 1.1949} = 1.15$$

$$Z = 1.6977 + \frac{4.6742}{27.02 + 2.0444} = 1.86$$

$$\frac{Y}{Z} = 0.62$$

$$F_c = 1.282 \times 10^{8}(0.5)\left[\frac{(1.15)(0.001)^4}{(0.1) + (0.62 \times 0.001)}\right] + 4.961 \times 10^{4}\left[\frac{(0.001)^3}{\left(0.1 + \frac{4 \times 0.001}{3}\right)(27.02 + 1.509)}\right]$$

$$F_c = 0.641 \times 10^8 \left(\frac{1.15 \times 10^{-12}}{0.1} \right) + 4.961 \times 10^4 \left[\frac{1 \times 10^{-9}}{(0.1)(28.529)} \right]$$

$$F_c = 7.37 \times 10^{-4} + 1.739 \times 10^{-5}$$

$$F_c = 7.54 \times 10^{-4} \text{ cc/sec}$$

Problem 7. A 5 cc package filled with one atm of helium has a leak channel of uniform circular cross-section and 0.1 cm long. What is the diameter of the channel if the molecular leak rate of helium at 20°C is 1×10^{-7} atm-cc/sec? The package is in an ambient of air at one atm.

Solution. From Eq. (2-1),

$$Q = F(P_1 - P_2) = 1 \times 10^{-7} \text{ atm-cc/sec}$$

where P_1 is the helium pressure on the inside of the package = 1 atm, P_2 is the helium pressure outside the package = 0, therefore, $F = Q = 1 \times 10^{-7}$ cc/sec.

There is no total pressure difference between the inside and the outside (the pressure inside and outside is one atm, so there is no viscous flow).

Equation (2-25), the equation for molecular conductance of a cylinder, can be used to calculate the diameter:

$$D^3 = \frac{F_{mc}(\ell)}{3810\sqrt{\frac{T}{M}}}$$

$$D^3 = \frac{(1 \times 10^{-7})(0.1)}{3810\sqrt{\frac{293}{4}}}$$

$$D^3 = \frac{1\times 10^{-8}}{(3810)(8.56)} = 3.066\times 10^{-13}$$

$$D = 6.74 \times 10^{-5}\text{ cm}$$

The second half of Eq. (2-24) could also be used. This equation has the correction for the end effect and for additional conductance in the transitional range.

$$F_{mc} = 4.961\times 10^4\left[\frac{D^3}{\left(\ell + \frac{4D}{3}\right)\left(\frac{D}{\text{mfp}} + 1.509\right)}\right]\text{ cc/sec}$$

$$\text{mfp} = \frac{kT}{\sqrt{2}\pi P\sigma^2}$$

where: $P = 1\text{ atm} = 1.0133 \times 10^6\text{ dynes/cm}^2$

$\sigma^2 = 0.484 \times 10^{-15}$ for helium

$$\text{mfp} = \frac{\left(1.38\times 10^{-16}\right)(293)}{\pi\sqrt{2}\left(1.0133\times 10^6\right)\left(0.484\times 10^{-15}\right)}$$

$$\text{mfp} = 1.856 \times 10^{-5}\text{ cm}$$

$$F_{mc} = 4.961\times 10^4\left[\frac{D^3}{\left(0.1 + \frac{4D}{3}\right)\left(\frac{D}{1.856\times 10^{-6}} + 1.509\right)}\right]$$

F_{mc} is again 1×10^{-7}

$$1\times10^{-7} = 4.961\times10^{4}\left[\frac{D^3}{(0.1)\left(\dfrac{D}{1.856\times10^{-6}}+1.509\right)}\right]$$

$$\frac{(0.1)(1\times10^{-7})}{4.96\times10^{4}} = \frac{D^3}{(5.388\times10^{4}D)+(1.509)}$$

Let the term to the left of the = be K

$$A = 5.388 \times 10^4$$

$$B = 1.509$$

Then $$K = \frac{D^3}{AD + B}$$

$$K(AD+B) = D^3$$

Dividing both sides by AB,

$$KAD + KB = D^3$$

$$D^3 - KAD - KB = 0$$

Let $D = y$, $-KA = p$, $-KB = q$. Then: $y^3 + py + q = 0$, which is the reduced form of a cubic equation, whose real root is $\alpha + \beta$; where:

$$\alpha = \sqrt[3]{\frac{-q+\sqrt{q^2+\dfrac{4(p)^3}{27}}}{2}}$$

$$\beta = \sqrt[3]{\frac{-q - \sqrt{q^2 + \frac{4(p)^3}{27}}}{2}}$$

$$\alpha = \sqrt[3]{\frac{KB + \sqrt{K^2 B^2 + \frac{4(-KA)^3}{27}}}{2}}$$

$$KA = \frac{(0.1)(1 \times 10^{-7})}{4.961 \times 10^4}(5.388 \times 10^4) = 1.086 \times 10^{-8}$$

$$KB = \frac{(0.1)(1 \times 10^{-7})}{4.961 \times 10^4}(1.509) = 3.04 \times 10^{-13}$$

$$\alpha = \sqrt[3]{\frac{3.04 \times 10^{-13} + \sqrt{9.25 \times 10^{-26} + 4\frac{1.281 \times 10^{-24}}{27}}}{2}}$$

$$\alpha = \sqrt[3]{\frac{3.04 \times 10^{-13} + \sqrt{9.25 \times 10^{-26} + 1.897 \times 10^{-25}}}{2}}$$

$$\alpha = \sqrt[3]{\frac{3.04 \times 10^{-13} + \sqrt{2.8225 \times 10^{-25}}}{2}}$$

$$\alpha = \sqrt[3]{4.176 \times 10^{-13}}$$

$$\alpha = 7.37 \times 10^{-5}$$

$$\beta = \sqrt[3]{-1.136 \times 10^{-13}}$$

$$\beta = -4.84 \times 10^{-5}$$

$$\alpha + \beta = (7.47 \times 10^{-5}) + (-4.84 \times 10^{-5})$$

$$D = Y = 2.63 \times 10^{-5} \text{ cm}$$

This diameter is smaller than the first calculation as the latter equation corrects for the end effect and for the transitional region.

REFERENCES

1. Gutherie, A., and Wakering, R. K., *Vacuum Equipment and Techniques,* (Manhattan Project), McGraw-Hill Book Company (1949)
2. Dushman, S., *Scientific Foundations of Vacuum Technique,* Second Edition (Revised by members of the General Electric Research Staff), John Wiley & Sons (1962)
3. Poiseuille, see Lamb, H., *Hydrodynamics,* Ch. 11, Dover Publications, NY (1945)
4. Davey, J. G., "Model Calculations for Maximum Allowable Leak Rates of Hermetic Packages," *Vac. Sci. Technol.,* 12(1):423-429 (Jan./Feb., 1975)
5. Lund, M., and Berman, A. S., *J. Appl. Phys.,* 37:2489 (1966)
6. Gutherie, A., and Wakerling, R. K., *Vacuum Equipment and Techniques,* (Manhattan Project), McGraw-Hill Book Company, pp. 29–31 (1949)
7. Dushman, S., *Scientific Foundations of Vacuum Technique,* Second Edition (Revised by members of the General Electric Research Staff), John Wiley & Sons, p. 15 (1962)
8. Ibid., p. 88
9. Clausing, P., *Ann. Physik,* 12:961 (1932)
10. Dushman, S., "The Production and Measurement of High Vacua," *Gen. Elec. Rev.* (1922)
11. Dushman, S., *Scientific Foundations of Vacuum Technique,* Second Edition (Revised by members of the General Electric Research Staff), John Wiley & Sons, pp. 104–111 (1962)
12. Knudsen, M., *Ann. Physik,* 28:75 (1909); and 35:389 (1911)

3

The Flow of Gases

1.0 GENERAL FLOW CHARACTERISTICS

The flow of gases through a leak channel is governed by the following general equation:

Eq. (3-1) $$R = F\sqrt{\frac{T}{M}}(P_1 - P_2)$$

where:

R = the measured leak rate

F = the molecular conductance of the leak channel

T = the temperature in °K

M = the molecular weight of the leaking gas

P_1 = the pressure on the high pressure side of the leak

P_2 = the pressure on the low pressure side of the leak

Equation (3-1) is valid for transitional and molecular flow. The equation for viscous flow is:

Eq. (3-2) $$R_v = F_v (P_1 - P_2)$$

F_v is the viscous conduction and contains the viscosity coefficient of the gas. This coefficient is proportional to $\sqrt{T}$. An exact empirical relationship between the viscosity coefficient η and temperature was developed by W. Sutherland in 1893 and is:

$$\eta = \frac{n_0 \sqrt{T}}{1 + \frac{c}{T}}$$

where n_0 and c are constants for the specific gas

For the purposes of this book, it is sufficient to use $\sqrt{T}$ as the temperature factor, or to use the actual viscosity at a given temperature.

Viscous flow only exists when $(P_1 - P_2)$ represents a difference in the total pressures. The following examples are conditions where viscous flow could take place.

- A sealed package containing a total pressure in excess of one atm, and leaking into the normal atm
- A sealed package containing a total pressure less than one atm, and residing in an environment of one atm.
- A sealed package containing a total pressure of one atm, and being in a bomb of three atm of helium or other gas.
- A sealed package at one atm, which is being leak tested in a helium leak detector

Transitional and molecular flows are present when there is a difference in partial pressures. A partial pressure is the pressure of a single gas. The pressures in Eq. (3-1) are partial pressures when considering molecular or transitional flow. The total pressure is equal to the sum of all the partial pressures.

Eq. (3-3) $$P_{\text{total}} = p_1 + p_2 + p_3 + p_n$$

Assume a chamber having a volume of 44.8 liters. The chamber is filled with one molecular weight of helium and one molecular weight of nitrogen. Each molecular weight occupies 22.4 liters, so that the chamber now contains 50% helium and 50% nitrogen by volume. It has been shown in the derivation of Eq. (1-5), that

$$P = \frac{(v_s)^2}{3}\rho$$

$$v_s = \sqrt{(v_a)^2} = \sqrt{\left(\frac{8R_oT}{\pi M}\right)^2}$$

$$P = \frac{8R_oT}{3\pi M}\rho \text{ and } \rho = \frac{M}{V}$$

Eq. (3-4) $$P = \frac{8R_oT}{3\pi V}$$

Equation (3-4) is independent of the molecular weight of a gas and is independent of the type of gas. Only the volume is a factor. The volumes of the helium and the nitrogen are equal in the total volume of 44.8 liters. Therefore the partial pressures of the two gases are equal because their volumes are equal.

As an example of Eq. (3-2), consider the composition of dry air:

Nitrogen	= 78.084%
Oxygen	= 20.9476%
Argon	= 0.934%
Helium	= 5.24 ppm = 5.24×10^{-4}%
Neon	= 18.18 ppm = 1.818×10^{-3}%
Carbon dioxide	= 0.0314% (typical but varies)
Methane	= 0.0002% (typical but varies)
Krypton	= 0.000114%
The sum of other gases	$<$ 0.0001%

The partial pressures are:

Nitrogen	= 0.78084 atm
Oxygen	= 0.209476 atm
Argon	= 0.00934 atm
Helium	= 0.00000524atm
Neon	= 0.00001818 atm
Carbon dioxide	= 0.000314 atm
Methane	= 0.000002 atm
Total pressure	= 0.99999542 atm

2.0 MEASURED, STANDARD AND TRUE LEAK RATES

Consider two sealed packages having different internal volumes, but having the same measured leak rates. The identical measured rates indicate that the rate of helium exiting the packages are the same, but the size of the leak channels is unknown. This is because the partial pressure of helium inside the packages is unknown. The above example holds true even if the packages are the same size. Equation (3-1) has two unknown variables: F and $(P_1 - P_2)$.

A reference pressure for the pressure on the high side of the leaking gas has been established to eliminate the confusion demonstrated in the above examples. The reference pressure is one atm. Two packages having one atm of helium can now be directly compared on the basis of the their measured leak rates. This is even true for packages of different sizes. The reference pressure of one atm produces the "standard" or "true" leak rate. In this book, these terms are used interchangeably.

Equation (3-5) shows the relationship between the true and measured leak rates.

Eq. (3-5) $$L_x = \frac{R_x}{\text{atm}_x}$$

where L_x = the true leak rate of gas x

R_x = the measured leak rate of gas x atm_x

atm_x = the number of atm of gas x in the package

Examples of this equation are:

- A package filled with 0.1 atm of helium measures 1×10^{-8} atm-cc/sec. The true helium leak rate is 1×10^{-7} atm-cc/sec.
- A package filled with 1.1 atm of helium measures 1×10^{-8} atm-cc/sec. The true helium leak rate is 9.09×10^{-9} atm-cc/sec.
- A package filled with 1.0 atm of helium measures 1×10^{-8} atm-cc/sec. The true leak rate is the same as the measured value because the helium pressure inside the package is one atm.

3.0 LEAK RATES FOR DIFFERENT GASES

Leak rates for different gasses through the same leak channel are different because of the gas's different mass and size. Chapter 2 has shown that the flow of a gas is a function of

$$\sqrt{\frac{T}{M}}$$

The leak rate relationship between one gas and another, at the same temperature, is:

Eq. (3-6) $$L_1 = L_2\sqrt{\frac{M_2}{M_1}}$$

where:
L_1 = the true leak rate of gas 1
L_2 = the true leak rate of gas 2
M_1 = the molecular weight of gas 1
M_2 = the molecular weight of gas 2

for example:

$$L_{\text{nitrogen}} = L_{\text{helium}}\sqrt{\frac{M_{\text{helium}}}{M_{\text{nitrogen}}}} = L_{\text{helium}}\sqrt{\frac{4}{28}} = (0.378)L_{\text{helium}}$$

Table 3-1 shows the conversion relationships between different gasses of interest. To convert the true leak rate of a gas in the left column to a true leak rate of a gas in the horizontal row, multiply the column value by the value in the table where the two gasses intersect.

Table 3-1. Conversion of Leak Rates for Different Gases

	L_{He}	L_{N_2}	L_{O_2}	L_{Ar}	L_{H_2O}	L_{CO_2}	L_{air}*
L_{He}	–	0.378	0.354	0.316	0.471	0.301	0.373
L_{N_2}	2.646	–	0.935	0.837	1.247	0.798	0.988
L_{O_2}	2.828	1.069	–	0.894	1.333	0.853	1.056
L_{Ar}	3.162	1.195	1.118	–	1.491	0.953	1.181
L_{H_2O}	2.121	0.802	0.750	0.671	–	0.640	0.792
L_{CO_2}	3.317	1.254	1.173	1.049	1.563	–	1.238
L_{air}*	2.679	1.012	0.947	0.847	1.263	0.808	–

*The leak rate for air is an artificial rate as air does not flow as a single molecular species. The gases in air flow at different rates. It is included in this table because MIL-STD-883, Method 1014, and other documents refer to it as the equivalent standard air leak rate, with the molecular weight of air = 28.7 grams.

For example:

If $L_{He} = 2 \times 10^{-7}$ atm-cc/sec of helium,

Then $L_{O_2} = (2 \times 10^{-7})(0.354) = 7.08 \times 10^{-8}$ atm-cc/sec of oxygen

If $L_{Ar} = 3 \times 10^{-7}$ atm-cc/sec of argon,

$L_{H_2O} = (3.07 \times 10^{-7})(1.491) = 4.47 \times 10^{-7}$ atm-cc/sec of water

4.0 CHANGE OF PARTIAL PRESSURE WITH TIME

If a sealed package has a leak, the partial pressure of the leaking gas inside the package will change with time. If the partial pressure inside the package is greater than the partial pressure outside, the measured flow rate out will decrease with time as the partial pressure of the leaking gas decreases. The decrease in the partial pressure is proportional to the conductance of the leak channel and the initial partial pressure in the package. It is inversely proportional to the volume of the package. The following equation expresses this relationship:

Eq. (3-7) $$\frac{dp}{dt} = -\frac{F}{V}p$$

where: dp/dt = change of partial pressure inside the package with time

F = the conductance of the leak for a particular gas and temperature

V = the internal volume of the package

p = the partial pressure inside the package

Rewriting Eq. (3-7):

Eq. (3-8) $$\frac{dp}{p} = \frac{F}{V}dt$$

Integrating Eq. (3-8):

$$\int_{p_i}^{p_t} \frac{dp}{p} = -\frac{F}{V}\int_0^t dt$$

$$L_n p_t - L_n p_i = -(F/V)\,t$$

$$\text{or } \frac{p_t}{p_i} = e^{-\frac{Ft}{V}}$$

or

Eq. (3-9) $$p_t = p_i e^{-\frac{Ft}{V}}$$

$$R = F(p_1 - p_2) \text{ or } F = R/\Delta P$$

Substituting the value of *F* in Eq. (3-9),

$$p_t = p_i e^{-\left(\frac{Rt}{V\Delta P}\right)}$$

when $\Delta P = 1$ atm, $R = L$.

Equation (3-9) can then be written as:

Eq. (3-10) $$p_t = p_i e^{-\frac{Lt}{Vp_o}}$$

where: p_t = the partial pressure inside the package at time *t*

p_i = the initial partial pressure inside the package

L = the true leak rate

t = the time

V = internal volume of the package

P_o = 1 atm difference in partial pressure between the inside and outside of the package

A similar derivation as used to derive Eq. (3-10) yields Eq. (3-11):

Eq. (3-11) $$\Delta p_t = \Delta p_i \left(e^{-\frac{Lt}{Vp_o}} \right)$$

Equations (3-9) and (3-10) can also be written in terms of the measured leak rate. Substituting the value for *L* as indicated in Eq. (3-5):

Eq. (3-12) $$p_t = p_i e^{-\frac{Rt}{V\Delta P}}$$

Eq. (3-13)
$$\Delta p_t = \Delta p_i \left(e^{-\frac{Rt}{V\Delta P}} \right)$$

where R = the measured leak rate

p_0 = the initial partial pressure of helium

Δp_t = the partial pressure difference from inside to outside the package at time t

Δp_i = the partial pressure difference from inside to outside the package at the initial time

Δp = the partial pressure of the gas when the measured value is R

Equations (3-12) and (3-13) are not very useful as it is cumbersome to deal with measured leak rate values for gasses that are not directly measured. For example: the equivalent measured leak rate of oxygen would have to be calculated from the measured leak rate of helium. This would most likely become confusing. Converting to true leak rates is not only easier but it conveys a clearer understanding of the different leak rates. *Therefore, Eqs. (3-10) and (3-11) are the most beneficial equations to use.*

One of the uses of Eq. (3-10) is to predict the amount of helium remaining in a package as a function of time. Assume a package having an internal volume of 3 cc containing one atm of helium. The measured (also true) leak rate is 3×10^{-7} atm-cc/sec of helium. The fraction of helium remaining can be plotted versus time after sealing. This representation is Eq. (3-14):

Eq. (3-14)
$$\frac{p_t}{p_i} = e^{-\frac{Lt}{V}}$$

Such a plot is shown in Fig. 3-1. This figure indicates a helium value of 0.6 of the original value after 60 days. The one atm of helium has been reduced to a pressure of 0.6 atm.

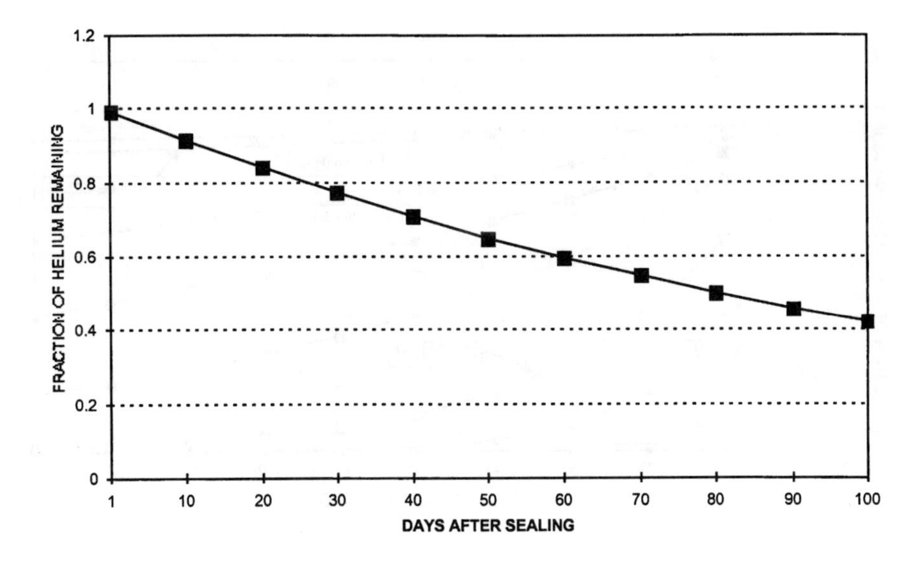

Figure 3-1. Decrease in helium with time for L = 3.0E-7 and volume = 3.0 cc.

A series of curves can be plotted for various true leak rate to volume ratios, showing the decrease of helium with time. Such data is plotted in Fig. 3-2. To illustrate the use of these curves in Fig. 3-2, consider the 5×10^{-07} plot. This curve represents the following examples:

a. A true helium leak rate of 5×10^{-07} and a volume 1 cc

b. A true helium leak rate of 2×10^{-07} and a volume of 0.4 cc

c. A true helium leak rate of 1×10^{-06} and a volume of 2 cc

d. A true helium leak rate of 5×10^{-08} and a volume of 0.1 cc

The above four examples all follow the same curve and are equivalent to each other with regard to the loss of helium with time. Note that the curve in Fig. 3-1 is the same as the 1×10^{-07} curve in Fig. 3-2.

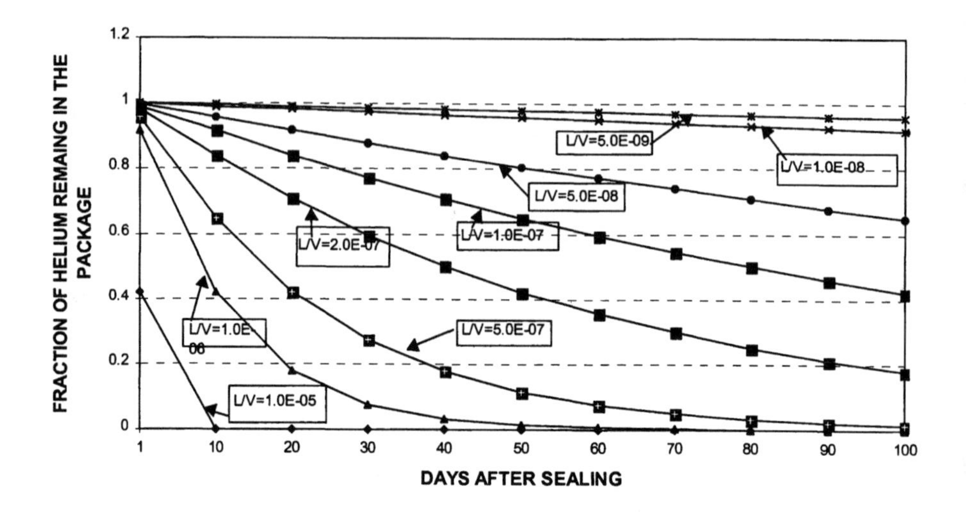

Figure 3-2. Decrease in helium in a package with time, for different true leak rate to volume ratios.

Equation (3-10), can be rearranged to solve for t:

$$L_n p_t - L_n p_i = -\frac{Lt}{V}$$

Eq. (3-15) $$t = -\left(\frac{V}{T}\right)\left[\text{Ln}\left(\frac{p_t}{p_i}\right)\right]$$

Equation (3-11) can also be rewritten:

Eq. (3-16) $$t = -\left(\frac{V}{L}\right)\left[\text{Ln}\left(\frac{\Delta p_t}{\Delta p_i}\right)\right]$$

An example of the use of Eq. (3-15) is Ex. 1:

Example 1.

$V = 3$ cc

$L = 3 \times 10^{-7}$ atm cc/sec

$p_i = 1$ atm

$p_t = 0.5$ atm

How long will it take for the partial pressure to decrease to half its initial value?

$$t = -\left(\frac{3}{3\times10^{-7}}\right)\left[\mathrm{Ln}\left(\frac{0.5}{1}\right)\right] = -1 \times 10^{-7}(-0.693) = 6{,}930{,}000 \text{ sec}$$

$$t = \frac{6{,}930{,}000}{24\times3600} = 80.2 \text{ days}$$

This value agrees with the curve in Fig. 3-1.

5.0 VISCOUS FLOW FROM SEALED PACKAGES

Viscous flow can occur when there is a difference in total pressure between the gases inside a package and the package's environment. The change in the total pressure difference, from inside to outside the package, depends upon the flow of gases in both directions. The equations describing these processes is presented in the next chapter.

Davey[1] has calculated the viscous and molecular flow for a cylindrical leak channel using Eq. (2-23). A specific application where this is important is during helium leak testing. The package is leaking into a vacuum so that there usually is one atm difference in total pressure. An example of this kind of calculation, using Eq. (2-23), is Ex. 3-2.

Example 2. Assume a cylindrical leak channel 0.1 cm in length with a diameter of 1×10^{-4} cm. The gas inside the package is 90% nitrogen and 10% helium. The average pressure is 0.5 atm (the average of the 1 atm inside and the vacuum outside).

From Table 3-2, the mean free path is 13.4×10^{-6} cm.

Table 3-2. The Mean Free Path of Gases at Selected Average Pressures*

Gas	0.5 Atm	1.0 Atm	1.5 Atm	2.0 Atm	2.5 Atm	3.0 Atm
Helium	37.0	18.5	12.3	9.25	7.40	6.17
Nitrogen	12.5	6.24	4.16	3.12	2.50	2.08
Oxygen	13.8	6.91	4.61	3.46	2.76	2.30
Water	17.6	8.81	5.87	4.41	3.52	2.94
Argon	13.1	6.56	4.37	3.28	2.62	2.19
Carbon Dioxide	8.47	4.24	2.83	2.12	1.70	1.41
80% Nitrogen-20% Oxygen	12.7	6.37	4.25	3.19	2.55	2.12
90% Nitrogen-10% Helium	13.4	6.70	4.67	3.35	2.68	2.23

*All mean free path values are in 10^{-6} cm

The total conductance, F_c is:

$$F_c = 1.249 \times 10^8 P_{\text{ave}} \left[\frac{YD^4}{\ell + \left(\frac{Y}{Z}\right) D} \right]$$

$$+ 4.961 \times 10^4 \left[\frac{D^3}{\left(\ell + \frac{4D}{3}\right)\left(\frac{D}{\text{mfp}} + 1.509\right)} \right] \quad \text{atm-cc /sec}$$

The constant 1.249 includes the viscosity of the nitrogen-helium mixture, where P_{ave} is in atm, (1 + 0)/2 = 0.5 atm during leak testing. The first term is the viscous conductance, the second term the molecular conductance.

where $$Y = 1 + \left[\frac{4.4445}{\left(\frac{D}{\text{mfp}} \right) + 1.1949} \right]$$

where $$Z = 1.6977 + \left[\frac{4.6742}{\left(\frac{D}{\text{mfp}} \right) + 2.0444} \right]$$

The mean free path (mfp) is calculated using Eq. (1-13), and values for selected average pressures are given in Table 3-2.

The viscous flow = $Q_{VC} = F_c\,(\Delta P)$

Eq. (3-17) $$Q_{vc} = 1.249 \times 10^8\, P_{\text{ave}} \left[\frac{YD^4}{\ell + \left(\frac{Y}{Z} \right) D} \right] \Delta P$$

$$\frac{D}{\text{mfp}} = \frac{1 \times 10^{-4}}{1.34 \times 10^{-5}} = 7.46$$

$$Y = 1 + \left(\frac{4.4445}{7.46 + 1.1949} \right) = 1 + 0.513 = 1.513$$

$$Z = 1.6977 + \left(\frac{4.6742}{7.46 + 2.0444} \right) = 1.6977 + 0.4918 = 2.189$$

$$\frac{Y}{Z}=\frac{1.513}{2.189}=0.691$$

$$Q_{vc}=1.249\times10^{8}(0.5)\left[\frac{(1.513)(10^{-4})^{4}}{(0.1)+(0.691\times10^{-4})}\right]\times1$$

$$Q_{vc}=1.249\times10^{8}\,(0.5)(1.513\times10^{-5})$$

$$Q_{vc}=9.45\times10^{-8}\text{ atm-cc/sec}$$

The helium content is only 10%, therefore, The viscous helium flow = $Q_{vcHE}=9.45\times10^{-9}$ atm-cc/sec. The helium molecular flow is:

$$Q_{mc}=4.961\times10^{4}\left[\frac{D^{3}}{\left(\ell+\frac{4D}{3}\right)+\left(\frac{D}{\text{mfp}}+1.509\right)}\right]\times\Delta P$$

$$Q_{mc}=4.961\times10^{4}\;\frac{1\times10^{\;12}}{0.1+\frac{4\times10^{\;4}}{3}\;(7.46+1.509)}\times0.1$$

$$Q_{mc}=4.961\times10^{4}\left[\frac{1\times10^{-13}}{(0.1001)(8.969)}\right]=4.96\times10^{4}(1.115\times10^{-13})$$

$$Q_{mc}=5.53\times10^{-9}\text{ atm-cc/sec of helium}$$

The total helium flow is $Q_{total}=9.45\times10^{-9}+5.53\times10^{-9}=1.50\times10^{-8}$ atm-cc/sec. The percent viscous flow is

$$\%V = \frac{(9.45 \times 10^{-9})(100)}{1.50 \times 10^{-8}} = 63\%$$

The measured helium leak rate is equal to the total helium flow = 1.50×10^{-8} atm-cc/sec = R_{HE}. The true helium leak, L_{HE}, is the measured rate R_{HE}/fraction of helium on the package.

$$L_{HE} = \frac{1.50 \times 10^{-8}}{0.1} = 1.50 \times 10^{-7} \text{ atm-cc/sec}$$

Figure 3-3 is a plot of percent viscous flow versus the true helium leak rate, for cylindrical leak channels 0.1 centimeter long and different diameters.

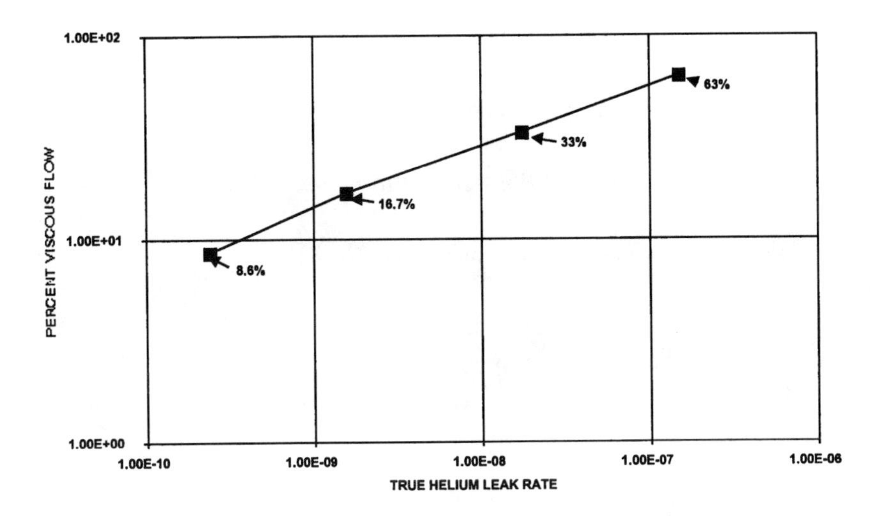

Figure 3-3. Percent viscous flow during leak testing for cylindrical leak channels 0.1 cm long.

The percent viscous flow for leak channels with a rectangular cross-section can also be calculated. The viscous flow can be calculated using Eq. (2-6):

$$F_{vr} = \frac{0.26Ga^2b^2P_{ave\mu}}{\ell} \text{ in liters per second}$$

where a and b = the sides of the rectangle in cm, $P_{ave\mu}$ is the average pressure in microns of mercury = 0.5 atm = 3.8×10^5 microns, G is a function of the a/b ratio and is given in Tables 2-1a and 2-1b.

Assume a leak channel 0.1 cm long with a cross-section of 1×10^{-5} cm by 1×10^{-3} cm. $G = 0.04$.

$$F_{vr} = \frac{(0.26)(0.04)(1\times10^{-5})^2(1\times10^{-3})^2(3.8\times10^5)}{0.1} \text{ liters/sec}$$

$$F_{vr} = 3.95 \times 10^{-12} \text{ liters/sec}$$

$$F_{vr} = 3.95 \times 10^{-9} \text{ cc/sec}$$

The package contains only 10% helium.

$$10\%\ F_{vr} = 3.95 \times 10^{-10} \text{ atm-cc/sec helium}$$

The molecular flow is calculated using Eq. (2-18):

$$F_{mr} = \frac{9.7a^2b^2}{\ell(a+b)}\sqrt{\frac{T}{M}}$$

$$\sqrt{\frac{T}{M}} = \sqrt{\frac{293}{4}} = 8.56$$

$$F_{mr} = \frac{9.7(10^{-5})^2(10^{-3})^2(8.56)}{(0.1)(10^{-5}+10^{-3})}$$

$$F_{mr} = 8.3 \times 10^{-11} \text{ liters/sec}$$

$$F_{mr} = 8.3 \times 10^{-8} \text{ cc/sec}$$

For 10% helium, $F_{mr} = 8.3 \times 10^{-9}$ atm-cc/sec.

The total helium flow is $3.95 \times 10^{-10} + 8.3 \times 10^{-9} = 8.70 \times 10^{-9}$ atm-cc/sec.

The percent viscous flow is

$$\%V = \frac{3.96 \times 10^{-10}(100)}{8.7 \times 10^{-9}} = 4.5\%.$$

The true helium leak rate = $8.7 \times 10^{-9}/0.1 = 8.7 \times 10^{-8}$ atm-cc/sec.

Figure 3-4 is a plot of the percent viscous flow versus the true helium leak rate for rectangular leak channels 0.1 centimeters deep and ratios of 10:1, 100:1 and 1000:1.

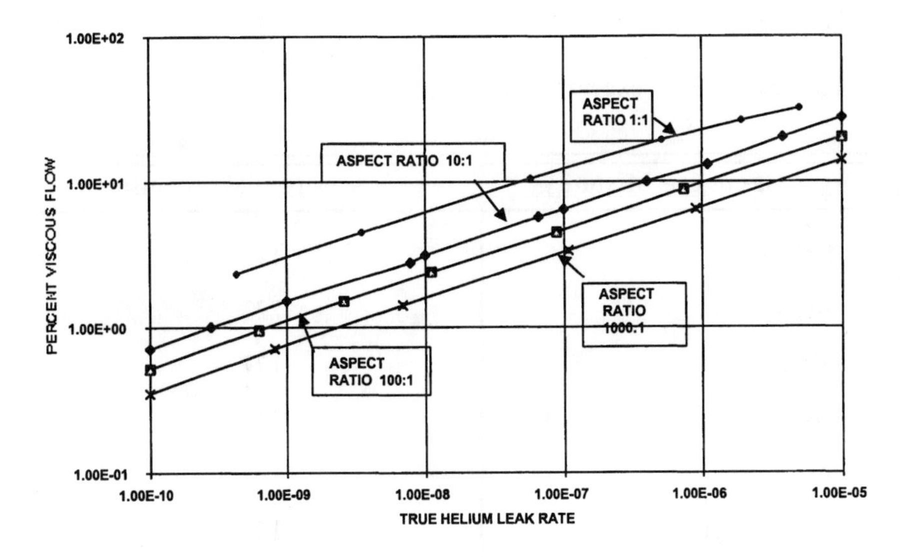

Figure 3-4. Percent viscous flow during leak testing for rectangular leak channels having four length to width ratios, all 0.1 cm long.

Figures 3-3 and 3-4 show the percent viscous flow during leak testing. Under this testing condition, the total pressure difference is 1 atm and the average pressure is 0.5 atm. Viscous conductance, and thereby viscous leak rates, are proportional to the average pressure in the leak channel, and inversely proportional to the coefficient of viscosity and the length of the leak channel. This can be seen in Eqs. (2-3)–(2-6). The average pressure is $(P_1 + P_2)/2$, where P_1 is the total pressure on the high side, and P_2 is the total pressure on the low side. The percent viscous flow as shown in Figs. 3-3 and 3-4 must be modified for a different average pressure and different length of the leak channel.

6.0 VISCOUS FLOW RATES OF DIFFERENT GASES

When there is viscous flow, all the gases flow as one gas. All individual gas species are flowing at the same rate, which is inversely proportional to the weighted average viscosity of the mixture. The viscosity of some gases are given in Ch. 2, Table 2-2 and are repeated here. All values are for 20°C except that for water vapor.

Gas or Gas Mixture	Viscosity in Micropoise
Helium	194
Argon	222
Nitrogen	177
Oxygen	202
Air	184
Water Vapor @ 100°C	125
Carbon Dioxide	148
90% Nitrogen-10% Helium	179
80% Nitrogen-20% Helium	180

The viscous flow rates relative to the viscous flow rate for helium are:

Eq. (3-18a) $L_{N_2V} = 194/177 L_{HEV} = 1.096 L_{HEV}$

Eq. (3-18b) $L_{O_2V} = 194/202\, L_{HEV} = 0.960\, L_{HEV}$

Eq. (3-18c) $L_{AIR} = 194/184\, L_{HEV} = 1.054\, L_{HEV}$

Eq. (3-18d) $L_{ARGON} = 194/222\, L_{HEV} = 0.874\, L_{HEV}$

Eq. (3-18e) $L_{90\%\, N_2\text{-}10\%\, HE} = 194/179\, L_{HEV} = 1.084\, L_{HEV}$

Eq. (3-18f) $L_{80\%\, N_2\text{-}20\%\, HE} = 194/180\, L_{HEV} = 1.078\, L_{HEV}$

The viscous flow rate of other gas mixtures can be calculated using the above viscosities.

The following example demonstrates the use of viscous flow from a package.

Example 3. A 1.0 cc package is sealed with 100% nitrogen at one atm. It is then bombed at 3 atm absolute of helium for 4 hours. Assume that 0.05 atm of helium was forced into the package. The measured leak rate, *R*, is 1×10^{-8} atm-cc/sec, and the viscous contribution is 10%. The true helium leak rate is:

$$L_{HE} = \frac{R}{0.05} = \frac{1 \times 10^{-8}}{0.05} = 2 \times 10^{-7} \text{ atm-cc/sec}$$

The leak detector does not discriminate between the molecular and viscous flow, but measures the total flow. Gases leak out of and into the package quantitatively as follows:

1. When the package is in an air environment, the leak rates for the various gases leaking out and in will be based on the total leak rate less the viscous leak rate. In this example 90% of the true helium leak rate modified for the difference in molecular weight.
2. When the package is on the moon, the leak rates of the gases out and in will be based on 100% of the true helium leak rate, just as in the leak detector.

The pressure in the package after bombing is 1.05 atm. The new percentages for nitrogen and helium are:

nitrogen = (1/1.05) × 100 = 95.24%

helium = (0.05/1.05) × 100 = 4.76%

The average viscosity of the 1.05 atm = (.9524 × 177) + (0.0476 × 194) = 177.8 micropoise. The viscous flow rate relative to the viscous helium flow = (194/177.8) L_{HE} =1.091 L_{HE}. The 10% viscous flow based on the measurement of the leak detector included the 1.091 rate factor.

For gases leaking into the air. The difference in total pressure between the inside and outside of the package is 0.05 atm There will be both molecular and viscous flow. The molecular flow rate for helium is 90% of the total rate = 0.9 × 2 × 10^{-7} = 1.8 × 10^{-7} atm-cc/sec. The molecular flow rate for nitrogen = 0.378 × 0.9 × 2 × 10^{-7} = 6.804 × 10^{-8} atm-cc/sec.

Letting $t = 1 \times 10^7$ and $V=1$, then for the amount of helium leaking out:

$$\Delta p_t = \Delta p_i \left[1 - e^{-\left(1.8\times10^{-7}\right)\left(1\times10^{7}\right)}\right]$$

$$\Delta p_t = 0.05(1 - e^{-1.8}) = 0.05(1 - 0.1653)$$

$$= 0.05(0.8307) = 0.0417 \text{ atm of helium}$$

This is the molecular contribution.

The viscous contribution is based on a difference in total pressure of 0.05 atm and an average pressure of 0.025 atm When the difference in total pressure was 1.05 atm with an average pressure of 0.525 atm, during leak testing, the viscous contribution was 10%. With only an average pressure is 0.025, the viscous leak rate is reduced to:

(0.025/0.525) × 10% = 0.476%

The viscous leak rate = 0.00476 × 2 × 10^{-7} = 9.52 × 10^{-10}

$$\Delta p_t = 0.05\left[1 - e^{-\left(9.52\times10^{-10}\right)\left(1\times10^{7}\right)}\right] = 0.05\left(1 - e^{-9.25\times10^{-3}}\right)$$

$$\Delta P_t = 0.05(1 - 0.99052) = 0.05(0.00947) = 0.00047 \text{ atm helium}$$

The total molecular helium leaking out into the air is 0.0417 + 0.0005 = 0.0422 atm. If there were no viscous contribution, then 100% of the helium leak rate, 2×10^{-7}, could be used and the helium leaking out would be 0.04323 atm.

The amount of nitrogen leaking out can be calculated in a similar manner. When gases are leaking out when on the surface of the moon. The difference in total pressure is 1.05 atm as it was during leak testing. The amount of helium leaking out is:

$$\Delta p_t = \Delta p_i \left[1 - e^{-\left(2\times10^{-7}\right)\left(1\times10^{7}\right)}\right] = 0.05\left(1 - e^{-2}\right)$$

$$\Delta p_t = 0.05\ (1 - 0.1353) = 0.05 \times 0.8647 = 0.0423 \text{ atm}$$

This includes the viscous flow, just as it did during leak testing.

7.0 PROBLEMS AND THEIR SOLUTIONS

Problem 1. An electronic assembly is sealed with a mixture of nitrogen, helium and argon. The total pressure is 1.2 atm The composition of the mixture is 80% nitrogen, 15% helium, and 5% argon. What are the partial pressures?

Solution.

The partial pressure of nitrogen = 80% of 1.2 atm = 0.96 atm

The partial pressure of helium = 15% of 1.2 atm = 0.18 atm

The partial pressure of argon = 5% of 1.2 atm = 0.06 atm

Problem 2. A package was sealed with 100% nitrogen at one atm. It was then bombed in helium at three atm, resulting in 0.05 atm of helium being forced into the package. Assuming that no nitrogen leaks out, what are the partial pressures?

Solution. The amount of nitrogen in the package has not changed so that the partial pressure of nitrogen is one atm. The partial pressure of helium is 0.05 atm. The total pressure is 1.05 atm.

Problem 3. A package is sealed with 90% nitrogen and 10% helium. The measured leak rate immediately after sealing is 2×10^{-9} atm-cc/sec of helium. Assuming there is no viscous flow, what is the true water leak rate?

Solution. First find the true helium leak rate using Eq. (3-5):

$$L_{\mathrm{HE}} = \frac{R_{\mathrm{HE}}}{\mathrm{atm}_{\mathrm{HE}}} = \frac{2 \times 10^{-9}}{0.1} = 2 \times 10^{-8} \text{ atm-cc/sec}$$

From Table 3-1, the true water leak rate $(L_{H_2O}) = 0.471 \times L_{HE}$

$$L_{H_2O} = 0.471 \times 2 \times 10^{-8} = 9.42 \times 10^{-9} \text{ atm-cc/sec of water}$$

Problem 4. A package is sealed with 0.1 atm of helium, and the total pressure inside the package is 0.1 atm. The measured leak rate immediately after sealing is 5×10^{-9} atm-cc/sec What is the true oxygen leak rate?

Solution. This is similar to Problem 3. The fact that helium is the only gas inside the package and that the total pressure is 0.1 atm, has no bearing on the problem. The true helium leak rate is:

$$L_{\mathrm{HE}} = \frac{5 \times 10^{-9}}{0.1} = 5 \times 10^{-8} \text{ atm-cc/sec}$$

From Table 3-1:

$$L_{O_2} = 0.354\ \mathrm{L_{HE}} = 0.354 \times 5 \times 10^{-8}$$

$$L_{O_2} = 1.77 \times 10^{-8} \text{ atm-cc/sec oxygen}$$

Problem 5. A package containing 10% helium and 90% nitrogen is leak tested at 20°C and measured 1×10^{-8} atm-cc/sec helium. What is the true helium leak rate when the package is at 125°C, assuming no change in the dimensions of the leak channel?

Solution. The true helium leak rate at 20°C is L_{HE}, $1 \times 10^{-8}/0.1 = 1 \times 10^{-7}$ atm-cc/sec.

Equation (3-1) shows that the leak rate is directly proportional to the square root of the absolute temperature. We can therefore write the following equation:

$$\frac{L_1}{L_2} = \frac{\sqrt{T_1}}{\sqrt{T_2}}$$

or

$$L_2 = L_1\sqrt{\frac{273+125}{273+20}} = 1\times10^{-7}\sqrt{1.358}$$

$L_2 = 1.165 \times 10^{-7}$ atm-cc/sec helium

Problem 6. A 0.5 cc package is sealed with one atm of helium. The package measures 1×10^{-7} atm-cc/sec helium immediately after sealing. What is the expected measured value 30 days later?

Solution. The measured value is directly proportional to the partial pressure of the helium inside the package. The helium pressure 30 days after sealing is expressed as Eq. (3-9):

$$p_{30} = p_i\left(e^{-\frac{Lt}{V}}\right)$$

where: p_{30} = the helium pressure 30 days after sealing

p_i = the initial helium pressure = 1 atm

L = 1×10^{-7} atm-cc/sec helium

V = 0.5 cc

t = 30 days = 30 × 24 × 3600 = 2,592,000 sec

$$\frac{Lt}{V} = \frac{1\times10^{-7}(2{,}592{,}000)}{0.5} = 0.5184$$

$$e^{-5184} = 0.595$$

$$p_{30} = 1 \text{ atm} \times 0.595 = 0.595 \text{ atm}$$

The expected measured leak rate is $0.595 \times 1 \times 10^{-7} = 5.95 \times 10^{-8}$ atm-cc/sec helium.

Problem 7. A 0.05 cc package sealed with 10% helium and 90% nitrogen measures 1×10^{-8} atm cc/sec of helium. How much helium remains in the package after 30 days? How much nitrogen?

Solution. The true helium leak rate,

$$L_{he} = \frac{1 \times 10^{-8}}{0.1} = 1 \times 10^{-7} \text{ atm-cc/sec}$$

$$t = 30 \times 24 \times 3600 = 2{,}592{,}000 \text{ sec}$$

$$p_i = 0.1 \text{ atm}$$

$$\frac{Lt}{V} = \frac{(1 \times 10^{-7})(2{,}592{,}000)}{0.05} = 5.184$$

$$p_t = 0.1(e^{-5.184}) = 0.1(0.0056)$$

$$p_t = 0.00056 \text{ atm} = 0.056\% \text{ helium}$$

The true nitrogen leak rate, $L_{N_2} = L_{HE} \times 0.378 = 3.78 \times 10^{-8}$ atm-cc/sec.

$$\frac{Lt}{V} = \frac{3.78 \times 10^{-8}(2{,}592{,}000)}{0.05} = 1.959$$

To solve this problem for nitrogen, Eq. (3-11) must be used because the nitrogen in the package is escaping into an atmosphere containing nitrogen. The initial partial pressure difference is 0.9 atm inside the package and 0.7808 atm outside the package (nitrogen in air).

$$\Delta p_t = \Delta p_i(e^{-1.959}) = (0.9 - 0.7808)(0.141) = 0.01681 \text{ atm}$$

This is the difference in nitrogen pressure from inside to outside the package at time t. The total pressure inside = the difference + the outside pressure, the total pressure inside = 0.01681 + 0.7808 = 0.7976 atm of nitrogen.

The identical answers are reached if Eqs. (3-12) and (3-13) are used instead of Eqs. (3-10) and (3-11):

$$\frac{R_o t}{V\Delta p} = \frac{(1\times10^{-8})(2{,}592{,}000)}{(0.05)(0.1)} = 5.184$$

$$p_t = 0.1(e^{-5.184}) = 0.1(0.0056)$$

$$p_t = 0.00056 \text{ atm} = 0.056\% \text{ helium}$$

$$R_{N2} = R_{he}\,(0.378) = 1 \times 10^{-8}\,(0.378) = 3.78 \times 10^{-9}$$

$$\frac{R_o t}{V\Delta p} = \frac{(3.78\times10^{-9})(2{,}592{,}000)}{(0.05)(0.1)} = 1.959$$

$$Dp_t = Dp_i(e^{-1.959}) = (0.9 - 0.7808)(0.141) = 0.01681\text{atm}$$

The total nitrogen pressure inside = 0.01681 + 0.7808 = 0.7976 atm of nitrogen.

Problem 8. An MCM with an internal volume of 4 cc is sealed with 10% helium and 90% nitrogen. The true helium leak rate acceptance limit is 5×10^{-7} atm-cc/sec. The MCM measures 4×10^{-8} immediately after sealing. Prior to delivery 60 days later, the MCM is again leak tested. What is the measured limit and the expected value at that time?

Solution. The measured leak rate limit immediately after sealing is the true leak rate limit × the number of atm of helium in the MCM (Eq. 3-5).

$$R_{he(limit)} = L_{he(limit)} \times 0.1 = 5 \times 10^{-7} \times 0.1 = 5 \times 10^{-8} \text{ atm-cc/sec helium}$$

$$t \text{ for 60 days} = 60 \times 24 \times 3600 = 5{,}184{,}000 \text{ sec}$$

The amount of helium in the MCM 60 days later is p_t. The true helium leak rate = $4 \times 10^{-8}/0.1 = 4 \times 10^{-7}$ atm-cc/sec.

$$p_t = 0.1\left[e^{-\frac{(4\times10^{-7})(5{,}184{,}000)}{4}}\right] = 0.1(e^{-0.5184}) = 0.1(0.595)$$

$$p_t = 0.06 \text{ atm helium}$$

The new measured leak rate limit is the true leak rate limit × number of atm at time *t*.

$$R_{\text{NEW}} = 5 \times 10^{-7}\ (0.06) = 3 \times 10^{-8} \text{ atm cc/sec}$$

The expected measurement value,

$$R_t = L_{\text{he}}\ (0.06) = (4 \times 10^{-7})\ (0.06)$$

$$R_t = 2.4 \times 10^{-8} \text{ atm-cc/sec helium}$$

Problem 9. An MCM with an internal volume of 5 cc is sealed with 20% helium and 80% nitrogen. The measured leak rate is 1×10^{-8} atm-cc/sec helium. Thirty days after sealing, the MCM is subjected to burn in for 168 hours at 125°C. During this time, the leak channel opens so that the true helium leak rate is 1×10^{-7} atm-cc/sec. After burn in, the leak channel returns to its original size, and the leak rate returns to its initial value. What is the expected leak rate measurement five days after removal from the burn in chamber?

Solution. The initial true helium leak rate is $(1 \times 10^{-8})/(0.2) = 5 \times 10^{-8}$ atm-cc/sec. Thirty days later, the helium pressure in the MCM is p_t.

$$t = 30 \times 24 \times 3600 = 2{,}592{,}000 \text{ sec}$$

$$p_i = 0.2 \text{ atm}$$

$$\frac{Lt}{V} = \frac{(5\times10^{-8})(2{,}592{,}000)}{5} = 0.0259$$

$$p_t = 0.2(e^{-0.0259}) = 0.2(0.974) = 0.195 \text{ atm helium}$$

This is now a new $p_{i,} = p_{i\text{NEW}} = 0.195$ atm. The new true helium leak rate during burn in = 1×10^{-7} atm-cc/sec. The new $t = 168 \times 3600 = 604{,}800$ sec.

$$\left(\frac{Lt}{V}\right)_{\text{NEW}} = \left[\frac{(1\times 10^{-7})(604{,}800)}{5}\right] = 0.0121$$

$$p_{t\text{NEW}} = 0.195(e^{-0.0121}) = (0.195)(0.988) = 0.193 \text{ atm}$$

This is the helium pressure in the MCM immediately after burn in, and becomes the final $p_{i.}$

$$p_{i\text{FINAL}} = 0.193 \text{ atm}$$

$$t_{\text{FINAL}} = 5 \text{ days} = 5 \times 24 \times 3600 = 432{,}000 \text{ sec}$$

$$L_{\text{FINAL}} = 5 \times 10^{-8} \text{ atm-cc/sec helium}$$

$$\frac{Lt}{V} = \frac{(5\times 10^{-8})(432{,}000)}{5} = 0.00432$$

$$p_{t\text{FINAL}} = p_{i\text{FINAL}}(e^{-0.00432}) = 0.193(0.996)$$

$$p_{i\text{FINAL}} = 0.192 \text{ atm}$$

The measured leak rate = the true leak rate × partial pressure of helium. The measured leak rate is $R_{\text{FINAL}} = (5 \times 10^{-8})\,(0.192) = 9.61 \times 10^{-9}$ atm-cc/sec.

Problem 10. A 5 cc package is sealed with 20% helium and 80% nitrogen. The measured leak rate immediately after sealing is 3×10^{-7} atm-cc/sec helium. Sometime later, the package is subjected to an RGA. The RGA shows 10% helium. How many days were there between the sealing and the RGA?

Solution. The true helium leak rate is: $3 \times 10^{-7}/0.2 = 1.5 \times 10^{-6}$ atm-cc/sec. Using Eq. (3-15):

$$t = \left(\frac{V}{L}\right)\left[\ln\left(\frac{p_t}{p_i}\right)\right]$$

where $V = 5$ cc
$L = 1.5 \times 10^{-6}$ atm-cc/sec
$p_t = 0.1$ atm helium
$p_i = 0.2$ atm helium

$$t = -\frac{5}{1.5\times10^{-6}}\left[\ln\left(\frac{0.1}{0.2}\right)\right] = \left(3.33\times10^{6}\right)\left(-0.69315\right) = 2.3105\times10^{6} \text{ sec}$$

$$t_{\text{DAYS}} = \frac{2.3105\times10^{6}}{24\times3600} = 346.6 \text{ days}$$

Problem 11. A package made from an aluminum casting has an internal volume of 10 cc. The walls of this package are 0.1 cm thick. Previous experiments have shown that leaks in this type of package are due to a pin hole through the wall. The package is sealed with 100% helium and measures 1×10^{-7} atm-cc/sec immediately after sealing. Considering both viscous and molecular flow, what is the helium pressure in the package after being in air for 100 days?

Solution. The measured and true helium leak rates are the same because the helium pressure in the package is one atm. The viscous contribution during leak testing for a pin hole type leak is shown in Fig. 3-3. At a true leak rate of 1×10^{-7}, the viscous contribution is 56%. The molecular contribution is (100% – 56%) = 44%. The true molecular helium leak rate is therefore $0.44 \times 10^{-7} = 4.4 \times 10^{-8}$ atm-cc/sec. When the package is leaking into the air, there is no viscous flow. The helium remaining in the package after 100 days is calculated as follows:

$$P_{100} = P_i\left(e^{-\frac{Lt}{V}}\right)$$

where: P_{100} = the helium pressure after 100 days

P_i = the helium pressure in the package = 1 atm

L = the molecular helium leak rate = 4.4×10^{-8} atm-cc/sec

t = 100 days = 100 × 24 3600 = 8,640,000 sec

V = the internal volume = 10 cc

$$P_{100} = 1.0\left[e^{-\frac{(4.4\times10^{-8})(8,640,000)}{10}}\right]$$

$$P_{100} = 1.0(e^{-0.03802}) = 1.0(0.9627)$$

$$P_{100} = 0.9627 \text{ atm of helium}$$

Problem 12.A package with an internal volume of 0.1 cc is sealed with 20% helium and 80% nitrogen. It measures 1×10^{-7} immediately after sealing. Assuming a rectangular leak channel with an aspect ratio of 10:1, how much helium is in the package after 10 days in air?

Solution. The total true helium leak rate = measured rate/atm of helium = $1 \times 10^{-7}/0.2 = 5 \times 10^{-7}$ atm-cc/sec. Figure 3-4 indicates a 10.5% viscous contribution during leak testing.

The true molecular helium leak rate:

$$= (100\% - 10.5\%)(5 \times 10^{-7})$$
$$= (0.895)(5 \times 10^{-7})$$
$$= 4.475 \times 10^{-7} \text{ atm-cc/sec helium}$$

$$P_{10} = P_i\left(e^{-\frac{Lt}{V}}\right)$$

where: P_{10} = the helium pressure after 10 days

P_i = the initial helium pressure = 0.2 atm

$L = 4.475 \times 10^{-7}$ atm-cc/sec of helium

t = 10 days = 10 × 24 × 3600 = 864,000 sec

V = 0.1 cc

$$P_{10} = 0.2\left(e^{-\frac{(4.475\times10^{-7})(864{,}000)}{0.1}}\right)$$

$$P_{10} = 0.2(e^{-3.8664})$$

$$P_{10} = 0.2(0.0209) = 0.0042 \text{ atm of helium}$$

The helium has decreased from 20% to 0.42% in only 10 days.

Problem 13. A package with an internal volume of 5 cc is sealed with 80% nitrogen, 10% helium and 10% argon. The measured leak rate immediately after sealing is 1×10^{-8}. Assume the leak channel is rectangular with an aspect ratio of 1000:1. How much argon is in the package after 100 days in air, considering that the measured value contains a viscous contribution?

Solution. The true total helium leak rate is $1 \times 10^{-8}/0.1 = 1 \times 10^{-7}$ atm-cc/sec. Figure 3-4 shows a viscous contribution of 3.3% during leak testing. The true molecular helium leak rate is $(0.967)(10^{-7}) = 9.67 \times 10^{-8}$ atm-cc/sec. The true argon molecular leak rate, using Table 3-1, is:

$$(9.67 \times 10^{-8})(0.316) = 3.06 \times 10^{-8}$$

Using Eq. (3-13):

$$\Delta P_{100} = \Delta P_i\left(e^{-\frac{Lt}{V}}\right)$$

where: ΔP_{100} = the difference in argon pressure between the inside and outside of the package after 100 days

ΔP_i = the initial difference in argon pressure = 0.1 atm – 0.00934 atm= 0.09066 atm

L = the true argon leak rate = 3.06×10^{-8} atm-cc/sec

t = 100 days = 100 × 24 × 3600 – 8,640,000 sec

V = 5 cc

$$\Delta P_{100} = 0.09066\left[e^{-\frac{(3.06\times10^{-8})(8,640,000)}{5}}\right]$$

$$\Delta P_{100} = 0.09066(e^{-0.0529})$$

$$\Delta P_{100} = 0.09066\ (0.94847) = 0.08599 \text{ atm}$$

This is the difference after 100 days. The argon pressure in the package is the difference plus the argon in the air.

$$P_{100} = 0.08599 + 0.00934 = 0.09533 \text{ atm}$$

Problem 14. A 1 cc package is sealed with 2 atm of helium. Immediately after sealing the leak rate measures 1×10^{-8} atm-cc/sec. The leak channel is cylindrical and 0.1 cm long. What is the viscous contribution when the package is leaking into the air, and what is the total amount of helium that has leaked out in 60 days?

Solution. The true helium leak rate = measured leak rate/atm of helium in the package. The true helium leak rate = $1 \times 10^{-8}/2 = 5 \times 10^{-9}$ atm-cc/sec. The average pressure during the leak test is $(2 + 0)/2 = 1$ atm. Figure 3-3 shows a 22% viscous contribution for a true helium leak rate of 5×10^{-9} atm-cc/sec. This 22% is for an average pressure of 0.5 atm. At an average pressure of 1 atm, the viscous contribution is 44% during leak testing.

When the package is leaking into the air, the average pressure is $(2 + 1)/2 = 1.5$ atm The viscous contribution is now 33%, and the molecular contribution is 67%. The molecular leak rate = $0.67 \times 5 \text{ x } 10^{-9} = 2.11 \times 10^{-9}$ atm-cc/sec. The viscous flow rate = $0.33 \times 5 \times 10^{-9} = 1.65 \times 10^{-9}$ atm/sec. This is also the viscous leak rate for helium as helium is the only gas in the package.

$$60 \text{ days} = 60 \times 24 \times 3600 = 5,184,000 \text{ sec}$$

The amount of helium leaking out of the package due to molecular flow is:

$$p_t = p_i\left[1 - e^{-\frac{(2.11\times10^{-9})(5,184,000)}{1}}\right]$$

$$p_t = 2(1 - e^{-0.010938}) = 2(1 - 0.98912)$$
$$= 2(0.01088) = 0.02176 \text{ atm helium}$$

The amount of helium that has leaked out due to viscous flow is:

$$\Delta P_t = \Delta P_i \left[1 - e^{-\frac{(1.65\times10^{-9})(5{,}184{,}000)}{1}} \right]$$

$$\Delta P_t = 1(1 - e^{-0.008554}) = (1 - 0.99148) = 0.00852 \text{ atm}$$

The total helium that leaked out into the air in 60 days = 0.0218 + 0.0085 = 0.0303 atm.

Problem 15. A 0.1 cc package is sealed with helium to a total pressure of 0.8 atm. The measured leak rate is 1×10^{-9} atm-cc/sec. The leak channel is cylindrical and 0.1 cm. long. How much helium is left in the package after 60 days, considering both viscous and molecular flow?

Solution. The true helium leak rate, $L_{HE} = 1 \times 10^{-9}/0.8 = 1.25 \times 10^{-9}$ atm-cc/sec. Figure 3-3 shows a 15% viscous contribution at a true helium leak rate of 1.25×10^{-9} atm-cc/sec. The viscous contribution during leak testing is based on an average pressure of (0.8 + 0)/2 = 0.4 atm. Figure 3-3 is based on an average pressure of 0.5 atm. Correcting for the difference in average pressure, the % viscous = 15% × 0.4/0.5 = 12%. This is the viscous contribution during leak testing. The molecular contribution is (100% – 12%) = 88% or $0.88 \times 1.25 \times 10^{-9} = 1.1 \times 10^{-9}$ atm-cc/sec. The difference in total pressure when the package is leaking into the air, is negative so that there is no viscous flow out of the package.

The helium left in the package is:

$$p_t = 0.8 \left[e^{-\frac{(1.1\times10^{-9})(5{,}184{,}000)}{0.1}} \right] = 0.8\left(e^{-0.057024}\right)$$

$$p_t = 0.8(0.94457)$$

$$p_t = 0.7557 \text{ atm of helium left in the package}$$

REFERENCES

1. Davey, J. G., "Model Calculations For Maximum Allowable Leak Rates Of Hermetic Packages," *J. Vac. Sci. Technology*, 12(1):423–429 (Jan/Feb, 1975)

4

The Flow of Gases into Sealed Packages

1.0 MOLECULAR FLOW

Most of the physics pertaining to the flow of gases into packages are similar to the physics described in the flow of gases out of packages. The leak channels and conductances are identical; however, there is one important difference. The physics governing the difference between the high partial pressure and the low partial pressure of the leaking gas is very different. When a gas is leaking out of a package, the gas is usually leaking into a gaseous environment that does not change. Helium leaking out of a package into the air does not change the helium concentration of helium in the air, it remains at zero. If nitrogen is leaking out of a package into the air, the nitrogen in the air remains at 0.78084 atm. When a gas is leaking into a package, the partial pressure of that gas on the low side does change.

One of the requirements for a gas to leak into a package is that the partial pressure outside the package be greater than the partial pressure of that gas inside the package. As this gas leaks in, the partial pressure of the gas inside increases until it equals the partial pressure outside. If the nitrogen partial pressure inside is less than 0.78084 atm, nitrogen will leak in until the inside partial pressure is 0.78084 atm. A package sealed with 90% nitrogen and 10% helium will have the following exchange of gases:

- Nitrogen leaking out
- Helium leaking out
- Oxygen leaking in
- Argon leaking in
- Neon leaking in

Carbon dioxide and methane could be leaking in either direction depending on the concentration of these gases inside the package. These two gases could be present in the package due to improper processing. Water is a unique case and will be treated in a subsequent chapter.

These gases will be flowing at different rates through the same leak channel(s) because of different partial pressure differences and different molecular weights. The primary question pertaining to a gas leaking into a package is, "how much gas leaked in, in a given time?" The basic relationship for the flow of a gas was presented in as Eq. (2-1) and is repeated here.

Eq. (4-1) $$Q = F(p_1 - p_2)$$

where: Q = the quantity of gas flowing per unit of time. The quantity of gas is expressed in volume × pressure, which is called throughput

p_1 = the pressure on the high side

p_2 = the pressure on the low side

F = the conductance of the leak channel in units of volume per sec

For molecular flow:

Eq. (4-2) $$R = F_m \sqrt{\frac{T}{M}}(p_1 - p_2)$$

where: p_1 = the partial pressure on the high side

p_2 = the partial pressure on the low side

F_m = the molecular conductance of the leak channel

R = the throughput per second

T = the temperature in degrees K

M = the molecular weight of the leaking gas

If a specific gas and its temperature are not considered, Eq. (2) becomes:

Eq. (4-3) $$R = F_m(p_1 - p_2)$$

The quantity of gas which has leaked into a package from the sealing time (t_o) to some later time (t), is the integration of the leak rate over this time period. The quantity of gas entering the package during this time is Q_{in}, where Q_{in} is:

Eq. (4-4) $$Q_{in} = \int_{t=0}^{t=t} R\Delta p_t \, dt$$

In Ch. 3 it was shown that:

$$\Delta p_t = \Delta p_i \left(e^{-\frac{Lt}{VP_o}} \right)$$

which is also equal to

$$\Delta p_i \left(e^{-\frac{Rt}{VP_R}} \right)$$

where: P_o = partial pressure of the leaking gas when it equals 1 atm

P_R = partial pressure of the leaking gas when it is other than 1 atm

Substituting Δp_t into Eq. (4-4):

Eq. (4-5) $$Q_{in} = R\Delta p_i \int_{t-0}^{t=t} e^{-\frac{Rt}{VP_R}} dt$$

Using true leak rates,

Eq. (4-6)
$$Q_{in} = L\Delta p_i \int_{t=0}^{t=t} e^{-\frac{Lt}{VP_o}} dt$$

where $P_o = 1$.

Integrating,

$$Q_{in} = L\Delta p_i \left(-\frac{V}{L}\right)\left[\left(e^{-\frac{Lt}{V}}\right)^t - \left(e^{-\frac{Lt}{V}}\right)_0\right]$$

$$Q_{in} = -V\Delta p_i \left[\left(e^{-\frac{Lt}{V}} + K\right) - (1 + K)\right]$$

$$Q_{in} = -V\Delta p_i \left[\left(e^{-\frac{Lt}{V}}\right) - (1)\right]$$

Q_{in} is the throughput, i.e. volume × pressure that has leaked in. To get the quantity of gas in terms of pressure, we divide by the fixed volume of the package.

Eq. (4-7)
$$Q_{inP} = \Delta p_i \left(1 - e^{-\frac{Lt}{V}}\right)$$

where: Q_{inP} = the quantity of gas in atm entering the package in time t
Δp_i = the initial difference in partial pressure in atm
L = the true leak rate in atm.-cc/sec
t = the time the gas is leaking into the package in sec
V = the volume of the package in cc

Q_{inP} and Δp_i must always be in the same units, such as atmospheres, percent, or parts per million (ppm).

Rewriting Eq. (4-7),

$$\frac{Q_{inP}}{\Delta p_i} = \left(1 - e^{-\frac{Lt}{V}}\right)$$

$$\left(1 - \frac{Q_{inP}}{\Delta p_i}\right) = e^{-\frac{Lt}{V}}$$

Taking the logarithm,

Eq. (4-8) $$\ln\left(1 - \frac{Q_{inP}}{\Delta p_i}\right) = -\frac{Lt}{V}$$

The minus sign indicates that the gas is leaking in. This equation is very useful, especially solving for t or L.

Eq. (4-9) $$t = -\frac{V}{L}\left[\ln\left(1 - \frac{Q_{inP}}{\Delta p_i}\right)\right]$$

Eq. (4-10) $$L = -\frac{V}{t}\left[\ln\left(1 - \frac{Q_{inP}}{\Delta p_i}\right)\right]$$

Equation (4-7) predicts the amount of gas leaking into a given package in a given time, if the true rate and the initial partial pressure are known.

Example 1.

The true leak rate of oxygen = L_{O_2} =1 × 10^{-7} atm-cc/sec

The time = t = 60 days × 24 × 3600 = 5,184,000 sec

The initial difference in the partial pressure of oxygen = 0.2095 atm

The internal volume of the package = 3 cc

$$Q_{inP} = 0.2095\left[1 - e^{-\frac{(1\times10^{-7})(5,184,000)}{3}}\right]$$

$$Q_{inP} = 0.2095\ (1 - 0.8413)$$

$Q_{inP} = 0.0332$ **atm of oxygen entering the package in time (t)**

Equation (4-9) calculates the time it takes for a given amount of gas to leak in, when the leak rate, internal package volume and initial partial pressure difference are known.

Example 2.

The internal volume = 1 cc

The true leak rate of oxygen = $L_{O_2} = 1 \times 10^{-7}$ atm-cc/sec

The initial partial pressure difference of oxygen = 0.2095 atm

The amount of oxygen that has leaked in = Q_{in} = 0.01 atm

$$t = -\frac{1}{1\times10^{-7}}\left[\ln\left(1 - \frac{0.01}{0.2095}\right)\right]$$

$$t = -1\times10^{7}\left[\ln(0.9523)\right] = -1 \times 10^{7}\ (-0.04891)$$

$$t = 489,095 \text{ sec} = 5.66 \text{ days}$$

Equation (4-10) calculates the true leak rate for a given amount of gas leaking into a given package in a given time.

Example 3.

The internal volume = 0.3 cc

The amount of oxygen which has leaked in = 0.01 atm

The initial partial pressure difference of oxygen = 0.2096 atm

The time = 90 days = 90 × 24 × 3600 =7,776,000 sec

$$L_{O_2} = -\frac{0.3}{7{,}776{,}000}\left[\ln\left(1-\frac{0.01}{0.2095}\right)\right]$$

$$L_{O_2} = -3.858 \times 10^{-8}(\ln 0.95227)$$
$$= -3.858 \times 10^{-8}(-0.04891)$$

$$L_{O_2} = 1.89 \times 10^{-9} \text{ atm-cc/sec oxygen}$$

The total quantity of gas in a package is the quantity that has leaked in, plus the amount that was there initially, minus that which has leaked out. This is expressed in the following equation:

Eq. (4-11) $$Q_{total} = Q_{initial} + Q_{inP} - Q_{out}$$

2.0 VISCOUS FLOW INTO AND OUT OF SEALED PACKAGES

All the previous equations in this chapter have been confined to molecular flow. An additional gas flow is possible when there is a difference in *total* pressure. There are four important conditions when this can take place. They are:

1. The forcing of helium into a package by subjecting the package to a helium pressure greater than that in the package (helium bombing).
2. Helium leak testing a package. The pressure outside the package is close to zero.
3. Packages that contain more than one atmosphere, leaking into the air.
4. Packages sealed under vacuum or a reduced pressure.

A similar equation to Eq. (4-4) can be written:

Eq. (4-12) $$Q_{inv} = \int_{t=0}^{t=t} R_v \Delta P_t \, dt$$

where: R_v = the viscous leak rate

ΔP_t = the difference in total pressure at time t

Q_{inv} = the quantity of gas entering the package due to viscous flow

Equation (4-12) is slightly modified when considering a specific gas:

Eq. (4-13)
$$Q_{inv} = \int_{t=0}^{t=t} R_v \Delta P_t M_f \, dt$$

where M_f is the molecular fraction for the specific gas.

For example: consider the viscous flow of helium from a mixture of 50% helium and 50% argon from a bomb at 3 atm, into a package containing 90% nitrogen and 10% helium at a total pressure of 1 atm. M_f, the molecular fraction of the gas leaking in, is 1/2 = 0.5. Then Q_{inv} for helium is:

$$Q_{inv} = \int_{t=0}^{t=t} R_v \Delta P_t (0.5) dt$$

The partial pressure of helium in the package is not an inhibiting factor for the viscous flow.

All the gases flow together like a single gas, so that:

$$\Delta P_t = \Delta P_i \left(e^{-\frac{R_v t}{V P_V}} \right)$$

Substituting for ΔP_t into Eq. (4-13):

Eq. (4-14)
$$Q_{inv} = R_v \Delta P_i \int_{t=0}^{t=t} e^{-\frac{R_v t}{V P_V}} dt$$

Adopting a true viscous leak rate, where the difference in total pressure is one atm, we have:

Eq. (4-15a) $$Q_{inv} = L_v \Delta P_i \int_{t=0}^{t=t} e^{-\frac{L_v t}{V}} dt$$

and for a specific gas,

Eq. (4-15b) $$Q_{inv} = L_v \Delta P_i (M_f) \int_{t=0}^{t=t} e^{-\frac{L_v t}{V}} dt$$

After integration:

Eq. (4-16a) $$Q_{inv} = -V \Delta P_i \left[\left(e^{-\frac{L_v t}{V}} \right) - 1 \right]$$

Eq. (4-16b) $$Q_{inv} = -V \Delta P_i (M_f) \left[\left(e^{-\frac{L_v t}{V}} \right) - 1 \right]$$

Q_{inv} is in units of volume × pressure. To get the amount of gas in units of pressure, for example atm, we divide by the volume:

Eq. (4-17a) $$Q_{invP} = \Delta P_i \left(1 - e^{-\frac{L_v t}{V}} \right)$$

Eq. (4-17b) $$Q_{invP} = \Delta P_i (M_f) \left[1 - e^{-\frac{L_v t}{V}} \right]$$

where Q_{invP} = quantity of gas in atm that has entered the package due to viscous flow.

Rearranging Eqs. (4-17a) and (4-17b),

$$\frac{Q_{invP}}{\Delta P_i} = \left(1 - e^{-\frac{L_v t}{V}}\right)$$

$$\left(1 - \frac{Q_{invP}}{\Delta P_i}\right) = e^{-\frac{L_v t}{V}}$$

Taking the logarithm,

$$\ln\left(1 - \frac{Q_{invP}}{\Delta P_i}\right) = -\frac{L_v t}{V}$$

Eq. (4-18a) $$t = -\frac{V}{L_v}\left[\ln\left(1 - \frac{Q_{invP}}{\Delta P_i}\right)\right]$$

Similarly,

Eq. (4-18b) $$t = -\frac{V}{L_v}\left[\ln\left(1 - \frac{Q_{invP}}{\Delta P_i M_f}\right)\right]$$

Solving for L_v,

Eq. (4-19a) $$L_v = -\frac{V}{t}\left[\ln\left(1 - \frac{Q_{invP}}{\Delta P_i}\right)\right]$$

Eq. (4-19b) $$L_v = -\frac{V}{t}\left[\ln\left(1 - \frac{Q_{invP}}{\Delta P_i M_f}\right)\right]$$

Some examples illustrate the use of these equations.

Example 4. A package with an internal volume of 1 cc is sealed with 100% nitrogen. The true helium viscous leak rate of the package is 1×10^{-7} atm-cc/sec. The package is bombed in 4 atm absolute of helium for 16 hours. How much helium is in the package after the bombing due to viscous flow?

Using Eq. (4-17a),

$$Q_{invP} = (4-1)\left[1 - e^{-\frac{(1\times10^{-7})(57{,}600)}{1}}\right]$$

$$Q_{invP} = (3)(1 - e^{-0.00576}) = (3)(1 - 0.99426)$$

$$Q_{invP} = 0.0172 \text{ atm of helium due to viscous flow}$$

Example 5. A 1 cc package is sealed with 90% nitrogen and 10% helium. It is then bombed in 4 absolute atm of helium for 24 hours. The true viscous leak rate (L_v) is 1×10^{-7} atm-cc/sec. The true molecular helium leak rate (L_{he}) is 5×10^{-8} atm-cc/sec of helium. How much helium is in the package after the bombing?

The viscous flow into the package is:

$$Q_{invP} = (3)\left[1 - e^{-(1\times10^{-7})(86{,}400)}\right] = (3)\left[1 - e^{-0.00868}\right] = (3)(0.0086)$$

$$Q_{invP} = 0.0258 \text{ atm}$$

The molecular flow into the package is:

$$Q_{inmP} = (4-0.1)\left[1 - e^{-(5\times10^{-8})(86{,}400)}\right]$$

$$= (3.9)(1 - e^{-0.00432}) = (3.9)(1 - 0.995689)$$

$$Q_{inmP} = 0.0168 \text{ atm}$$

The total helium in the package is:

$$Q_{total} = Q_{inv} + Q_{inmP} + Q_{he\text{-}initial}$$

$$Q_{total} = 0.0258 \text{ atm} + 0.0168 \text{ atm} + 0.1 \text{ atm}$$

$$= 0.1426 \text{ atm}$$

3.0 THE SIMULTANEOUS FLOW OF GASES IN BOTH DIRECTIONS

If a package has a leak, there will be a flow of one or more gases leaking in *and* out. The exception to this is:

- When the package is sealed with air at one atmosphere.
- When the package is sealed under vacuum
- When the package is in a vacuum environment

A package that has a leak, and sealed with 90% nitrogen and 10% helium will have the following gas transports when being in a dry air environment.

- Nitrogen leaking out
- Helium leaking out
- Oxygen leaking in
- Argon leaking in

At some equilibrium time after sealing, depending upon the size of the leak and the volume of the package, the composition of the gases inside the package will be that of air. Prior to this equilibrium time, the composition will be between the starting composition and air.

Example 6. Assume a 1 cc package sealed with 90% nitrogen and 10% helium. The measured helium leak rate is 1×10^{-8} atm-cc/sec. What is the composition of the gases inside the package after 1 year (neglecting moisture)?

$$L_{HE} = R_{HE}/0.1 = 1 \times 10^{-7} \text{ atm-cc/sec}$$

$$1 \text{ yr} = 365 \times 24 \times 3600 = 3.15 \times 10^{7} \text{ sec}$$

From Table 3-1,

$$L_{N_2} = 3.78 \times 10^{-8}$$

$$L_{O_2} = 3.54 \times 10^{-8}$$

$$L_{ar} = 3.16 \times 10^{-8}$$

The pressure difference in the package after 1 year, for the nitrogen and helium leaking out, is Δp_t.

$$\Delta p_t = \Delta p_i \left(e^{-\frac{Lt}{V}} \right)$$

For nitrogen:

$$\Delta p_t = (0.9000 - 0.78084)\left(e^{-\frac{(3.78\times10^{-8})(3.15\times10^{7})}{1}} \right)$$

$$\Delta p_t = 0.11916\left(e^{-1.192212}\right) = 0.11916(0.303549)$$

$$\Delta p_t = 0.036171 \text{ atm}$$

$$p_t = 0.036171 + 0.78084 = 0.81701 \text{ atm of nitrogen}$$

For helium:

$$\Delta p_t = (0.1)\left[e^{-(1\times10^{-7})(3.15\times10^{7})}\right]$$

$$\Delta p_t = (0.1)\left(e^{-3.15}\right)$$

$\Delta p_t =$ 0.0042681 atm of helium, and this is the amount of helium remaining inside the package as there is no helium outside

Leaking in:

$$Q_{inP} = \Delta p_i \left(1 - e^{-\frac{Lt}{V}}\right)$$

For oxygen:

$$Q_{inPO_2} = (0.209476)\left[1 - e^{-(3.54\times10^{-8})(3.15\times10^{7})}\right]$$

$$Q_{inP_{O_2}} = (0.209476)(1 - e^{-1.1165}) = (0.209476)(1 - 0.3274185)$$

$$Q_{inPO_2} = 0.1408896 \text{ atm of oxygen has leaked in}$$

For argon:

$$Q_{inP_{Ar}} = (0.00934)\left[1 - e^{-(3.16\times10^{-8})(3.154\times10^{7})}\right]$$

$$Q_{inPAR} = (0.00934)(1 - e^{0.996664}) = (0.00934)(1 - 0.3691087)$$

$$Q_{inPAr} = 0.0058925 \text{ atm of argon has leaked in}$$

Summing the atm inside the package for each gas at time t, the sum is only 0.96806 atm. The total is not one atmosphere because the helium and nitrogen are leaking out much faster than the oxygen and argon are leaking in. Eventually, when most of the helium has leaked out, the total pressure will reach one atmosphere. Figure 4-1 shows the atmosphere of the four gases as well as the total atm in the package. Figure 4-2 shows an expanded graph for the oxygen, helium and argon shown in Fig. 4-1.

The time when the total atmosphere reaches a minimum is a function of the leak rate and the volume of the package. Specifically it is this ratio which is the exact factor. Figure 4-3 is a plot of the total atmosphere in a package sealed with 90% nitrogen and 10% helium, for three different true helium leak rate to volume ratios:

$$\frac{L_{HE}}{V} = 1\times10^{-6}; \frac{L_{HE}}{V} = 1\times10^{-7}; \frac{L_{HE}}{V} = 1\times10^{-8}$$

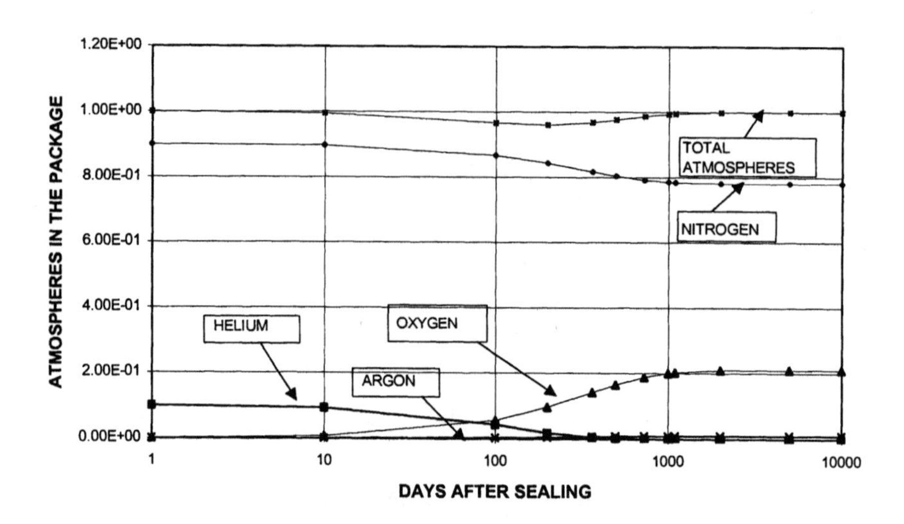

Figure 4-1. Atmospheres in a 1 cc package sealed with 90% nitrogen and 10% helium, having a true helium leak rate of 1×10^{-7}.

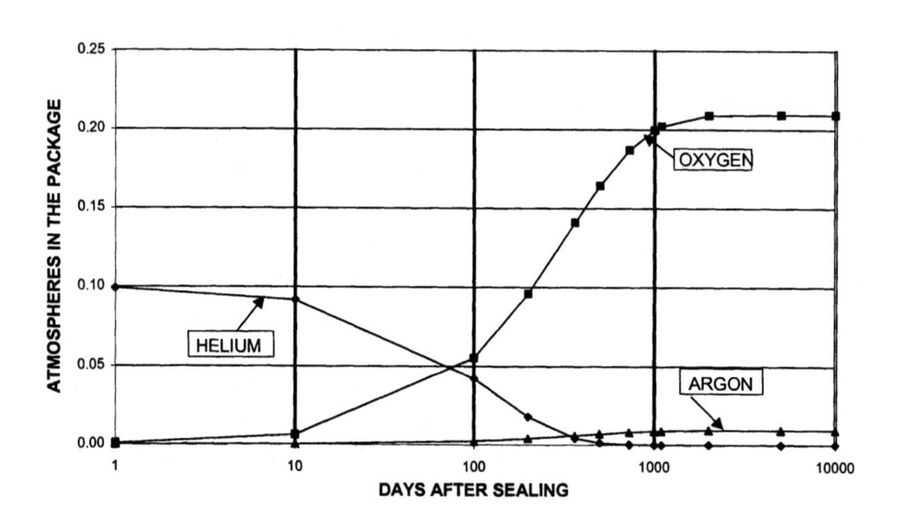

Figure 4-2. Atmospheres of oxygen, helium and argon in a 1 cc package sealed with 90% nitrogen and 10% helium, having a true helium leak rate of 1×10^{-7}.

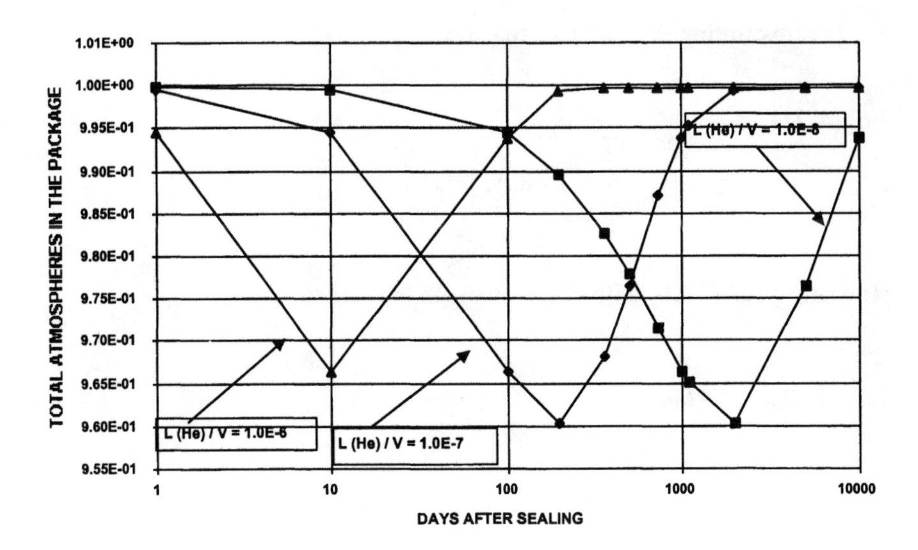

Figure 4-3. Total atmospheres in a package sealed with 90% nitrogen and 10% helium for three true helium leak rate to volume ratios.

Calculating the true helium leak rate from the measured leak rate, when the amount of helium in the package is not known, is a commonly occurring problem. The relationship between the true and measured leak rates is presented as Eq. (3-5), and is reproduced here as Eq. (4-20).

Eq. (4-20) $$L_x = \frac{R_x}{Atmospheres_x}$$

In many applications, the number of atmospheres in the above equation is the atmosphere of helium gas forced into a package when it is bombed. This is part of the process when leak testing is in accordance with Method 1014 of MIL-STD-883 which is discussed in detail in Ch. 7. However, the general solution to this problem is presented here.

If the helium in the package was due to bombing,

$$L_{\mathrm{HE}} = \frac{R_{\mathrm{HE}}}{Q_{inP}}$$

Substituting Eq. (4-7) for the atm:

Eq. (4-21)
$$L_{HE} = \frac{R_{HE}}{\Delta p_i \left(1 - e^{-\frac{Lt}{V}}\right)}$$

Rewriting Eq. (4-21) with slight changes in nomenclature:

Eq. (4-22)
$$L_{HE} = \frac{R_{HE}}{\Delta p_{HE} \left(1 - e^{-\frac{L_{HE}T}{V}}\right)}$$

where: L_{HE} = the true helium leak rate

R_{HE} = the measure helium leak rate

Δp_{HE} = the helium pressure in the bomb – the helium pressure inside the package

T = the bombing time in seconds

Calculating L_{HE} is not straight forward. One method of handling this problem has been to generate tables of L versus R for values of Δp_{HE}, V and T. A more useful method is a graphic plot of R_{HE} versus L_{HE}/V on log-log paper. This a straight line. Figure 4-4 is an example of such a plot. Table 4-1 is a partial spread sheet for such a plot.

Another problem, whose solution is not straight forward, is calculating the leak rate which would produce a certain pressure in a package at a specific time.

Example 7. A 100 cc package is sealed with 90% nitrogen and 10% helium at a total pressure of 1.20 atm. What is the maximum measured helium leak rate to guarantee a total pressure of 1.10 atm, in an air environment, a year after sealing (assume no viscous flow)?

Solution. The nitrogen and helium are both leaking out causing a decrease in the total pressure. Oxygen and argon are leaking in causing an increase in the total pressure. The net effect is a decrease in the total pressure, eventually reaching 1 atmosphere of air.

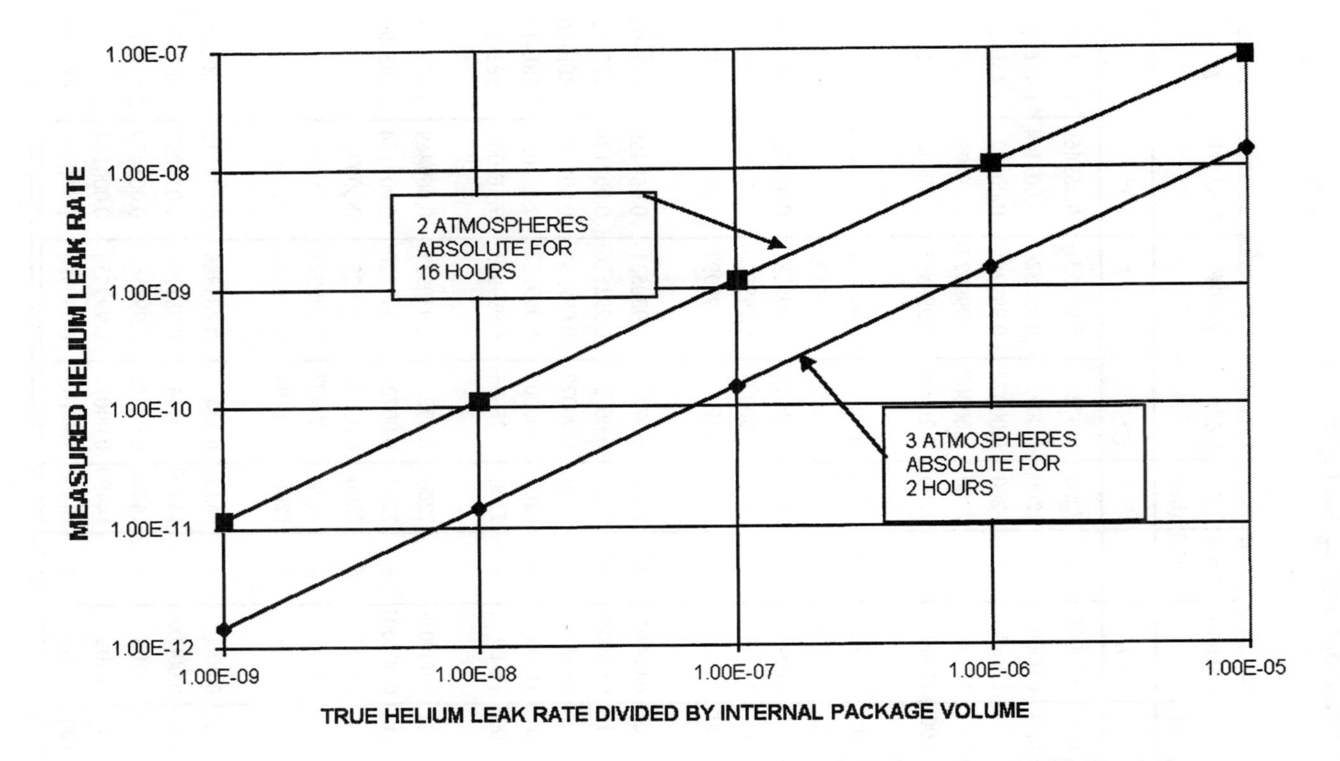

Figure 4-4. Measured helium leak rate vs true helium leak rate divided by the internal volume for two helium bombing conditions.

Table 1. Partial Spreadsheet for Figure 4-4

L (He)	*V* (cc)	*L* (He)/*V*	Delta *p*(He)	*T* (Sec)	*LT/V*	*e-LT/V*	1-*e-LT/V*	*R*
		A/B			*C*E*	EXP(-*F*)	1-*F*	*A*D*H*
1.00E-07	0.01	0.00001	2	7200	0.072	0.930531	0.069469	1.39E-08
1.00E-07	0.1	0.000001	2	7200	0.0072	0.992826	0.007174	1.43E-09
1.00E-07	1	1E-07	2	7200	0.00072	0.99928	0.00072	1.44E-10
1.00E-07	10	1E-08	2	7200	0.00007	0.999928	7.2E-05	1.44E-11
1.00E-07	100	1E-09	2	7200	7.2E-06	0.999993	7.2E-06	1.44E-12
1.00E-07	0.01	0.00001	3	7200	0.072	0.930531	0.069469	2.08E-08
1.00E-07	0.1	0.000001	3	7200	0.0072	0.992826	0.007174	2.15E-09
1.00E-07	1	1E-07	3	7200	0.00072	0.99928	0.00072	2.16E-10
1.00E-07	10	1E-08	3	7200	0.00007	0.999928	7.2E-05	2.16E-11
1.00E-07	100	1E-09	3	7200	7.2E-06	0.999993	7.2E-06	2.16E-12
1.00E-07	0.01	0.00001	4	7200	0.072	0.930531	0.069469	2.78E-08
1.00E-07	0.1	0.000001	4	7200	0.0072	0.992826	0.007174	2.87E-09
1.00E-07	1	1E-07	4	7200	0.00072	0.99928	0.00072	2.88E-10
1.00E-07	10	1E-08	4	7200	0.00007	0.999928	7.2E-05	2.88E-11
1.00E-07	100	1E-09	4	7200	7.2E-06	0.999993	7.2E-06	2.88E-12
1.00E-07	0.01	0.00001	5	7200	0.072	0.930531	0.069469	3.47E-08
1.00E-07	0.1	0.000001	5	7200	0.0072	0.992826	0.007174	3.59E-09
1.00E-07	1	1E-07	5	7200	0.00072	0.99928	0.00072	3.6E-10
1.00E-07	10	1E-08	5	7200	0.00007	0.999928	7.2E-05	3.6E-11
1.00E-07	100	1E-09	5	7200	7.2E-06	0.999993	7.2E-06	3.6E-12
1.00E-07	0.01	0.00001	2	14400	0.144	0.865888	0.134112	2.68E-08
1.00E-07	0.1	0.000001	2	14400	0.0144	0.985703	0.014297	2.86E-09
1.00E-07	1	1E-07	2	14400	0.00144	0.998561	0.001439	2.88E-10
1.00E-07	10	1E-08	2	14400	0.00014	0.999856	0.000144	2.88E-11
1.00E-07	100	1E-09	2	14400	1.44E-5	0.999986	1.44E-05	2.88E-12

Stating the leak rates of nitrogen, oxygen and argon in terms of the leak rate of helium:

$L_{N_2} = 0.378\, L_{HE}$

$L_{O2} = 0.354\, L_{HE}$

$L_{AR} = 0.316\, L_{HE}$

1 year = 3.15×10^7 sec

Initial nitrogen pressure = $0.9 \times 1.2 = 1.08$ atm

Initial helium pressure = $0.1 \times 1.2 = 0.12$ atm

Atm nitrogen = $0.78084 + \Delta p_{t(N_2)}$

$$\Delta p_{t(N_2)} = \Delta p_{i(N_2)} \left(e^{-\frac{0.378 L_{HE} t}{V}} \right) = (1.08 - 0.78084) \left(e^{-\frac{0.378 L_{HE}}{100} \times 3.15 \times 10^7} \right)$$

$$\text{Atm helium} = \Delta p_{t(HE)} = \Delta p_i \left(e^{-\frac{L_{HE} t}{V}} \right) = 0.12 \left(e^{-\frac{L_{HE}}{100} \times 3.15 \times 10^7} \right)$$

Atm oxygen = $Q_{inP(O_2)}$

$$Q_{inP(O_2)} = \Delta p_{i(O_2)} \left(1 - e^{-\frac{0.354 L_{HE} t}{V}} \right) = 0.2095 \left(1 - e^{-\frac{0.354 L_{HE}}{100} \times 3.15 \times 10^7} \right)$$

Atm argon = $Q_{inP(AR)}$

$$Q_{inP(AR)} = \Delta p_{i(AR)} \left(1 - e^{-\frac{0.316 L_{HE} t}{V}} \right) = 0.0093 \left(1 - e^{-\frac{0.316 L_{HE}}{100} \times 3.15 \times 10^7} \right)$$

The total atm in the package = ATM_{TOTAL}

$$ATM_{TOTAL} = 0.78 + 0.299\left(e^{-\frac{0.378 L_{HE}}{100}\times 3.15\times 10^{7}}\right) +$$
$$0.12\left(e^{-\frac{L_{HE}}{100}\times 3.15\times 10^{7}}\right) +$$
$$0.2095\left(1 - e^{-\frac{0.354 L_{HE}}{100}\times 3.15\times 10^{7}}\right) +$$
$$0.00934\left(1 - e^{-\frac{0.316 L_{HE}}{100}\times 3.15\times 10^{7}}\right)$$

Solving for L_{HE} when AMT_{TOTAL} is known is not straight forward. Using a spread sheet, a graphical solution can be made, and it is shown as Fig. 4-5.

Figure 4-5 shows a true helium leak rate of 3.3×10^{-6} atm-cc/sec. The measured leak rate *R*, corresponding to this true leak rate is the initial helium atmosphere in the package × the true helium leak rate. $R = 0.12 \times 3 \times 10^{-6} = 3.6 \times 10^{-7}$ atm-cc/sec.

4.0 PROBLEMS AND THEIR SOLUTIONS

Problem 1. A package with a one cc internal volume is sealed with 90% nitrogen and 10% helium. Immediately after sealing, it has a measured helium leak rate of 5×10^{-8} atm-cc/sec. How long will it take for one percent oxygen to enter the package when it is in an air environment?

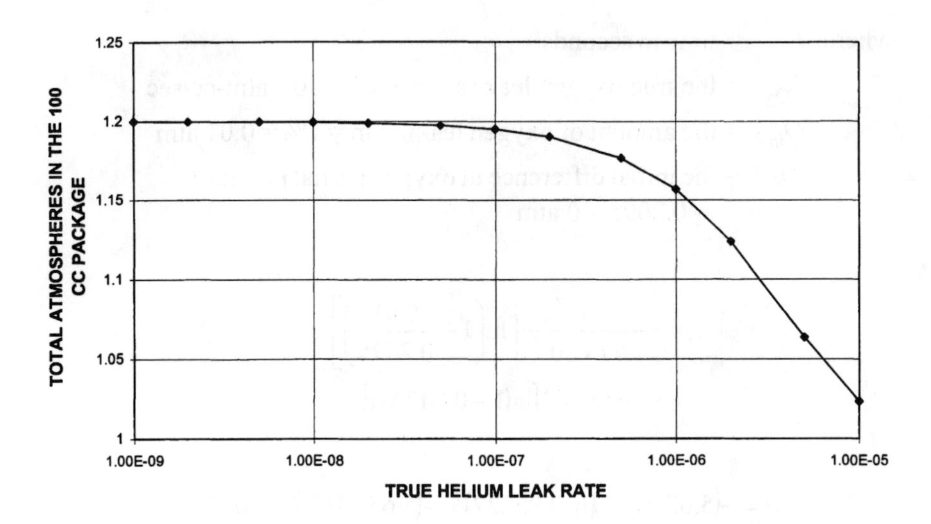

Figure 4-5. Total amospheres in a 100 cc package after one year in air. Original composition of the package is 90% nitrogen and 10% helium, at a total pressure of 1.2 atmospheres.

Solution. First, we calculate the true helium leak rate using Eq. 3-5.

$$L_{\mathrm{HE}} = \frac{R}{\mathrm{Helium(atm)}} = \frac{5 \times 10^{-8}}{0.1} = 5 \times 10^{-7} \text{ atm-cc/sec}$$

The relationship between the true helium leak rate, L_{HE}, and the true oxygen leak rate, $L_{\mathrm{O_2}}$, is given in Table 3-1, Ch. 4.

$$L_{\mathrm{O_2}} = 0.354 \times L_{\mathrm{HE}} = (0.354)(5 \times 10^{-7})$$

$$L_{\mathrm{O_2}} = 1.77 \times 10^{-7} \text{ atm-cc/sec}$$

Equation (4-9) calculates the time for a gas to leak into a package.

$$t = -\frac{V}{L_{\mathrm{O_2}}}\left[\ln\left(1 - \frac{Q_{inP}}{\Delta p_i}\right)\right]$$

where: t = time in seconds

L_{O_2} = the true oxygen leak rate = 1.77×10^{-7} atm-cc/sec

Q_{inP} = the amount of oxygen leaking in = 1% = 0.01 atm

Δp_i = the initial difference in oxygen partial pressure = 0.2095 – 0 atm

$$t = -\frac{1}{1.77 \times 10^{-7}}\left[\ln\left(1 - \frac{0.01}{0.2095}\right)\right]$$

$$= -5.65 \times 10^{6}\left[\ln(1 - 0.04773)\right]$$

$$t = -(5.65 \times 10^{6})(\ln 0.95227) = -(5.65 \times 10^{6})(-0.04891)$$

$$t = 2.7633 \times 10^{5} \text{ sec} = 76.76 \text{ hr}$$

Problem 2. To check the integrity of the sealing chamber, a hybrid package with an internal volume of 2 cc. is sealed in a 100% helium atmosphere. The measured leak rate immediately after sealing is 1×10^{-7} atm-cc/sec. Arthur Frugal, the technician who used to send out the RGA samples, kept this package in his desk waiting for additional RGA samples. The RGA is performed 30 days after sealing. What is the expected amount of nitrogen in the package?

Solution. The true nitrogen leak rate (L_{N_2}) is obtained using Table 3-1.

$$L_{N_2} = 0.378$$

$$L_{HE} = 0.378 \times 1 \times 10^{-7} = 3.78 \times 10^{-8} \text{ atm-cc/sec}$$

Equation (4-17a) can be used to calculate the quantity of nitrogen leaking into the package:

$$Q_{inP(N_2)} = \Delta p_{i(N_2)}\left[1 - e^{-\frac{L_{N_2} t}{V}}\right]$$

where: $\Delta p_{i(N_2)} = 0.78084$ atm

$L_{N_2} = 3.78 \times 10^{-8}$ atm-cc/sec

$t = 30$ days $= 30 \times 24 \times 3600 = 2{,}592{,}000$ sec

$V = 2$ cc

$$Q_{inP(N_2)} = 0{,}78084 \left[1 - e^{-\frac{(3.78\times10^{-8})(2592000)}{2}}\right]$$

$$Q_{inP(N_2)} = 0.78084\ (1 - e^{-(0.04899)})$$

$$Q_{inP(N_2)} = 0.78084\ (1\text{-}0.95219)$$

$$Q_{inP(N_2)} = 0.0373 \text{ atm} = 3.73\% \text{ nitrogen}$$

Problem 3. A 200 cc package is sealed with 90% nitrogen and 10% helium. The RGA analysis, 1000 hour after sealing, shows 0.5% oxygen. What is the true helium leak rate of the package?

Solution. The composition of the gases in the package is irrelevant to this problem as long as oxygen was not a constituent. Equation (4-10) can be used to solve this problem.

$$L_{O_2} = -\frac{V}{t}\left[\ln\left(1 - \frac{Q_{inP(O_2)}}{\Delta p_{i(O_2)}}\right)\right]$$

where L_{O_2} = the true oxygen leak rate

t = 1000 hrs = 1000 × 3600 = 3,600,000 sec

Q_{inPO_2} = 0.5% oxygen

Δp_{iO_2} = 20.95% oxygen

$$L_{O_2} = -\frac{200}{3600000}\left[\ln\left(1 - \frac{0.5}{20.95}\right)\right]$$

$$L_{O_2} = -5.556 \times 10^{-5}\left[\ln(1 - 0.023866)\right]$$

$$L_{O2} = -5.556 \times 10^{-5}(\ln 0.976134)$$
$$= -5.556 \times 10^{-5}(-0.02416)$$

$$L_{O_2} = 1.34 \times 10^{-6} \text{ atm oxygen}$$

From Table 3-1, $L_{HE} = 2.828$, $L_{O_2} = 2.828 \times 1.34 \times 110^{-6} = 3.80 \times 10^{-6}$ atm-cc/sec.

Problem 4. A 0.1 cc package is backfilled with 20% helium and 80% nitrogen. The measured helium leak rate immediately after sealing is 1×10^{-8} atm-cc/sec. How much oxygen and argon has leaked into the package after 60 days in air?

Solution. The true helium leak rate = $1 \times 10^{-8}/0.2 = 5 \times 10^{-8}$ atm-cc/sec. Using Eq. (4-7):

$$Q_{inP(O_2)} = \Delta p_{i(O_2)}\left(1 - e^{-\frac{L_{O_2}t}{V}}\right)$$

where: $Q_{inP(O_2)}$ = the atm of oxygen that has leaked in

$\Delta p_{i(O_2)}$ = the initial partial pressure difference = 0.2095 atm

L_{O2} = the true oxygen leak rate = $0.354 \times L_{HE}$

$= 0.354 \times 5 \times 10^{-8} = 1.77 \times 10^{-8}$

t = 60 days = 60 × 24 × 3600 = 5,184,000 sec

V = 0.1 cc

$$Q_{inP(O_2)} = 0.2095\left[1 - e^{-\frac{(1.77\times10^{-8})(5184000)}{0.1}}\right]$$

$$Q_{inP(O_2)} = 0.2095[1 - e^{-(0.917568)}]$$

$$Q_{inP(O_2)} = 0.2095(1 - 0.3995) = 0.2095(0.6005)$$

$$Q_{inP(O_2)} = 0.1258 \text{ atm of oxygen}$$

From Table 3-1 $L_{AR} = 0.316\ L_{HE} = 0.316 \times 5 \times 10^{-8} = 1.58 \times 10^{-8}$ atm-cc/sec.

$$\Delta\pi_{i(AR)} = 0.00934 \text{ atm}$$

$$Q_{inP(AR)} = 0.00934\left[1 - e^{-\frac{(1.58\times10^{-8})(5184000)}{0.1}}\right]$$

$$Q_{inP(AR)} = 0.00934(1 - e^{-0.81907}) = 0.00934(0.55916)$$

$$Q_{inP(AR)} = 0.00522 \text{ atm of argon}$$

Problem 5. A 1 cc package is sealed with 90% nitrogen and 10% helium. The leak rate measures 1×10^{-8} immediately after sealing. Assuming that neither oxygen nor argon is sealed in the package, what is the oxygen to argon ratio after 1000 hours in an air environment?

Solution.

$$1000 \text{ hours} = 1000 \times 3600 = 3{,}600{,}000 \text{ sec}$$

$$L_{HE} = 1 \times 10^{-8}/0.1 = 1 \times 10^{-7} \text{ atm-cc/sec}$$

$$L_{O_2} = 0.354 \times L_{HE} = 3.54 \times 10^{-8} \text{ atm-cc/sec}$$

$$L_{AR} = 0.316 \times L_{HE} = 3.16 \times 10^{-8} \text{ atm-cc/sec}$$

Calculating the amount of oxygen that has leaked in:

$$Q_{O_2} = \Delta p_i\left(1 - e^{-\frac{L_{O_2}t}{V}}\right)$$

$$Q_{O_2} = 0.2095\left[1 - e^{-\frac{(3.54\times10^{-8})(3.6\times10^{6})}{1.0}}\right]$$

$$Q_{O_2} = 0.2095(1 - e^{-0.12744})$$

$$Q_{O_2} = 0.2095(1 - 0.88035) = 0.2095(0.11965)$$

$$Q_{O_2} = 0.02507 \text{ atm}$$

Calculating the amount of argon that has leaked into the package:

$$Q_{AR} = 0.00934\left[1 - e^{-\frac{(3.16\times10^{-8})(3.6\times10^{6})}{1.0}}\right]$$

$$Q_{AR} = 0.00934\left(1 - e^{-0.11376}\right)$$

$$Q_{AR} = 0.00934(1 - 0.89247) = 0.00934(0.10753)$$

$$Q_{AR} = 0.00100 \text{ atm of argon}$$

The oxygen to argon ratio = 0.02507/0.00100 = 25.07:1

Problem 6. A 1 cc package, having a known true helium leak rate of 1×10^{-7} atm-cc/sec, is sealed with 100% nitrogen. It is bombed in three absolute atm of helium for 16 hr. Assume a rectangular leak channel with an aspect ratio of 1000:1.

- Neglecting viscous flow, how much helium is in the package after bombing?
- Neglecting viscous flow, how much helium is in the package after 30 days in air?
- Including viscous flow, how much helium is in the package after bombing?
- Including viscous flow, how much helium is in the package after 30 days in air?

Solution.

16 hr = 16 × 3600 = 57,600 sec

30 days = 30 × 24 × 3600 = 2,592,000 sec

Neglecting viscous flow: The amount of helium forced into the package during bombing is calculated using Eq. (4-7):

$$Q_{inP} = \Delta p_i \left[1 - e^{-\frac{L_{HE} t}{V P_o}} \right]$$

where Δp_i = difference in helium pressures = 3 atm, P_o =1

$$Q_{inP} = 3 \left[1 - e^{-\frac{(1\times10^{-7})(57,600)}{1\times1}} \right] = 3\left(1 - e^{-0.00576}\right)$$

Q_{inP} = 3(1 – 0.994256) = 3(0.005743)

Q_{inP} = 0.0172 atm

The helium left in the package after 30 days is calculated using Eq. (3-10):

$$p_t = p_i \left(e^{-\frac{L_{HE} t}{V P_o}} \right)$$

$$p_t = 0.0172\left[e^{-(1\times10^{-7})(2,529,000)}\right] = 0.0172\left(e^{-0.2592}\right)$$

p_t=0.0172(0.7717) = 0.0133 atm left after 30 days

Including viscous flow: The viscous flow rate into the package is found using Fig. 3-4 for the 1000:1 ratio. At a true helium leak rate of 1 × 10^{-7}, the viscous contribution is 3.3% when the average pressure was at 0.5 atm and the difference in pressure was 1 atm. This equaled 3.3 × 10^9 atm-cc/sec. When bombed at 3 absolute atm, the average pressure is (3 + 1)/2 = 2 atm. Correcting for the new average pressure, the new leak rate = (2/0.5) × 3.3 × 10^{-9} = 1.32 × 10^{-8} atm-cc/sec. After bombing for 16 hrs, we have:

$$Q_{inP} = 2\left[1 - e^{-\frac{(1.32\times10^{-8})(57{,}600)}{1x1}}\right] = 2\left(1 - e^{-7.603^{-4}}\right)$$

$$Q_{inP} = 2(1 - 0.9992340) = 2(0.00076)$$

Q_{inP} = 0.00152 atm of helium forced into the package due to viscous flow

The molecular leak rate, when there is a viscous flow, is the total leak rate less the viscous rate. The helium molecular flow rate in this case equals 1 × 10^{-7} – 3.3% = 9.67 × 10^{-8} atm-cc/sec. The amount of helium forced in due to molecular flow is:

$$Q_{inP} = 3\left[1 - e^{-\frac{(9.67\times10^{-8})(57{,}600)}{1\times1}}\right] = 3\left(1 - e^{-0.0055699}\right)$$

$$Q_{inP} = 3(1 - 0.99444556) = 3(0.00555444)$$

Q_{inP} = 0.0166 atm of helium forced in due to molecular flow

The total forced into the package when there is both molecular and viscous flow = 0.0166 + 0.00152 = 0.01812 atm.

The viscous leak rate out of the package when the average pressure was 0.5 atm was 3.3% of $L_{HE.}$ The average pressure when leaking into the air = (1.0172+1)/2 = 1.0086atm. The viscous leak rate = (1.0086/0.5) × 3.3 × 10^{-9} = 6.66 × 10^{-9} atm-cc/sec.

The viscous flow out of the package is:

$$\Delta P_t = \Delta P_i\left(1 - e^{-\frac{L_V t}{V P_o}}\right) = 0.01812\left[1 - e^{-\left(6.66\times10^{-9}\right)\left(2{,}592{,}000\right)}\right]$$

$$= 0.01812(1 - e^{-0.017256}) = 0.01812(1 - 0.98289)$$

$= 0.00031$ atm leaked out due to viscous flow

The helium flowing out of the package due to molecular flow is now different than that calculated previously. The molecular leak rate is now 93.37% of $1 \times 10^{-7} = 9.33 \times 10^{-8}$ atm-cc/sec.

$$\Delta p_t = \Delta p_i\left(1 - e^{-\frac{L_V t}{V P_o}}\right) = 0.01812\left[1 - e^{-\left(9.67\times10^{-8}\right)\left(2{,}592{,}000\right)}\right]$$

$$\Delta p_t = 0.1812(e^{-0.25065}) = 0.01812(1 - 0.77829)$$

$= 0.0040$ atm lost due to molecular flow

The total helium lost considering both viscous and molecular flow = $0.00031 + 0.0040 = 0.00432$ atm. The total remaining $= 0.01812 - 0.00432 = 0.0138$ atm of helium.

	Molecular Flow Only	**Molecular & Viscous Flow**
Helium Forced In	0.01723 atm	0.01812 atm
Helium After 30 days	0.01330 atm	0.01380 atm

Problem 7. A 5 cc package is sealed with 50% nitrogen, 20% helium, and 30% argon. The measured leak rate is 1×10^{-7} atm-cc/sec. What is the gas composition after 60 days in dry air?

Solution.

$$L_{HE} = 1 \times 10^{-7}/0.2 = 5 \times 10^{-7} \text{ atm-cc.sec}$$

$$L_{N_2} = 0.378 \times L_{HE} = 0.378 \times 5 \times 10^{-7} = 1.89 \times 10^{-7} \text{ atm-cc/sec}$$

$$L_{O_2} = 0.354 \times L_{HE} = 0.354 \times 5 \times 10^{-7} = 1.77 \times 10^{-7} \text{ atm-cc/sec}$$

$$L_{AR} = 0.316 \times L_{HE} = 0.316 \times 5 \times 10^{-7} = 1.58 \times 10^{-7} \text{ atm-cc/sec}$$

$$60 \text{ days} = 60 \times 24 \times 3600 = 5{,}184{,}000 \text{ sec}$$

Nitrogen and oxygen are leaking in. Argon and helium are leaking out. Nitrogen leaking in:

$$Q_{N_2} = \Delta p_i \left[1 - e^{-\frac{L_{N_2} t}{V}}\right] = (0.78084 - 0.5)\left[1 - e^{-\frac{(1.89\times10^{-7})(5{,}184{,}000)}{5}}\right]$$

$$Q_{N_2} = 0.28084\left(1 - e^{-0.19595}\right) = 0.28084(1 - 0.822049)$$

$$Q_{N_2} = 0.28084(0.17795) = 0.04997 \text{ atm}$$

To this is added the nitrogen already in the package. The total nitrogen in the package after 60 days = 0.04997 + 0.5 = 0.54997 atm.

Oxygen leaking in:

$$Q_{O_2} = 0.2095\left[1 - e^{-\frac{(1.77\times10^{-7})(5{,}184{,}000)}{5}}\right]$$

$$Q_{O_2} = 0.2095(1 - e^{-0.18351}) = 0.2095(1 - 0.83234)$$

$$Q_{O_2} = 0.2095(0.16766) = 0.03512 \text{ atm}$$

Helium leaking out:

$$p_{t\text{HE}} = p_i\left(e^{-\frac{L_{\text{HE}}t}{V}}\right) = 0.2\left[e^{-\frac{(5\times10^{-7})(5,184,000)}{5}}\right]$$

$$p_{t\text{HE}} = 0.2(e^{-0.5184}) = 0.2(0.59547)$$

$$p_{t\text{HE}} = 0.11909 \text{ atm}$$

Argon leaking out:

$$p_{t\text{AR}} = (0.3 - 0.00934)\left[e^{-\frac{(1.58\times10^{-7})(5,184,000)}{5}}\right]$$

$$p_{t\text{AR}} = 0.29066(e^{-0.1638144})$$

$$p_{t\text{AR}} = 0.29066\ (0.8488995)$$

$$p_{t\text{AR}} = 0.24675 \text{ atm remaining}$$

The atm in the package after 60 days is:

Argon = 0.24675
Helium = 0.11909
Nitrogen = 0.54997
Oxygen = 0.03512
Total amt= 0.95093

The percentage of these gases in the package is:

Argon = 25.948%
Helium = 12.523%
Nitrogen = 57.836%
Oxygen = 3.693%
Total % = 100.00%

Problem 8. A 0.2 cc package is backfilled with 90% nitrogen and 10% helium. The nitrogen contains an argon impurity of 500 ppm. Immediately after sealing, the leak rate measures 1×10^{-7} atm-cc/sec. What is the gas composition in the package after 60 days?

Solution.

$$L_{HE} = 1 \times 10^{-7}/0.1 = 1 \times 10^{-6} \text{ atm-cc/sec}$$

$$L_{N_2} = 0.378 \times L_{HE} = 3.78 \times 10^{-7} \text{ atm-cc/sec}$$

$$L_{O_2} = 0.354 \times L_{HE} = 3.54 \times 10^{-7} \text{ atm-cc/sec}$$

$$L_{AR} = 0.316 \times L_{HE} = 3.16 \times 10^{-7} \text{ atm-cc/sec}$$

Argon leaking in:

$$0.00934 \text{ atm} = 9340 \text{ ppm}$$

$$Q_{AR} = \Delta p_i \left(1 - e^{-\frac{L_{AR}t}{V}}\right) = (9340 - 500)_{ppm} \left[1 - e^{-\frac{(3.16\times10^{-7})(5{,}184{,}000)}{0.2}}\right]$$

$$Q_{AR} = (8840)_{ppm}\left(1 - e^{-8.19072}\right) = (8840)_{ppm}(1 - 0.000277)$$

$$Q_{AR} = (8840)_{ppm}(0.999722) = (8837)_{ppm}$$

This is the amount that has leaked in. To this is added the 500 ppm impurity for a total argon content of 9337 ppm = 0.009337 atm = 0.9337%.

Oxygen leaking in:

$$Q_{O_2} = 0.2095\left[1 - e^{-\frac{(3.54\times10^{-7})(5{,}184{,}000)}{0.2}}\right] = 0.2095\left(1 - e^{-9.17568}\right)$$

$$Q_{O_2} = 0.2095(1 - 0.0001035) = 0.2095(0.9998964)$$

$$Q_{O_2} = 0.2094 \text{ atm}$$

Nitrogen leaking out: The amount remaining of what could have leaked out is:

$$\Delta p_{N_2} = \Delta p_i \left(e^{-\frac{L_{N_2} t}{V}} \right) = (0.9 - 0.78084) \left[e^{-\frac{(3.78\times10^{-7})(5{,}184{,}000)}{0.2}} \right]$$

$$\Delta p_{N_2} = 0.11916(e^{-9.79776}) = 0.11916(0.0000557)$$

$$\Delta p_{N_2} = 0.00001 \text{ atm}$$

To this is added the atmospheric nitrogen in the air. The total nitrogen in the package after 60 days = 0.00001+ 0.78084 = 0.78085 atm.

Helium leaking out: The amount of helium remaining is:

$$p_{HE} = 0.1 \left[e^{-\frac{(1\times10^{-6})(5{,}184{,}000)}{0.2}} \right] = 0.1\left(e^{-25.92}\right) = 0.1(0)$$

$$p_{HE} = 0 \text{ atm of helium left in the package}$$

The gases in the package are:

Nitrogen = 0.78085 atm

Oxygen = 0.2094 atm

Argon = 0.009337 atm

Total = 0.999587 atm

Note: This is the composition of air.

Problem 9. A 4 cc package was sealed with 90% nitrogen and 10% helium. The measured leak rate immediately after sealing was 2×10^{-9} atm-cc/sec. A Residual Gas Analysis (RGA) was performed 1000 hr after sealing and showed 0.4% oxygen. The package was subjected to a thermal stress during screening at which time the leak channel opened. The thermal stress lasted 15 min. What was the average helium leak rate during these 15 min?

Solution.

The true helium leak rate (L_{HE})
$=2 \times 10^{-9}/0.1 = 2 \times 10^{-8}$ atm-cc/sec

The true oxygen leak rate
$= 0.354 \times L_{HE} = 7.08 \times 10^{-9}$ atm-cc/sec

$1000 \text{ hr} = 1000 \times 24 \times 3600 = 3{,}600{,}000$ sec

$15 \text{ min} = 15 \times 60 = 900$ sec

$1000 \text{ hr} - 15 \text{ min} = 3{,}599{,}100$ sec

The amount of oxygen entering the package (not counting the 15 min) is:

$$Q_{O_2} = \Delta p_i \left[1 - e^{-\frac{(7.08\times10^{-9})(3{,}599{,}100)}{4}} \right] = 0.2095\left(1 - e^{-0.006370}\right)$$

$$Q_{O_2} = 0.2095(1 - 0.9936498) = 0.2095(0.00635)$$

$$Q_{O_2} = 0.00133 \text{ atm} = 0.133\%$$

The amount of oxygen that has leaked in during the 15 min = 0.4% – 0.133% = 0.267%. The oxygen leak rate during the 15 min can be calculated using Eq. (4-10):

$$L_{15O_2} = -\frac{V}{t}\left[\ln\left(1 - \frac{Q_{inP}}{\Delta p_i}\right)\right] = -\frac{4}{900}\left[\ln\left(1 - \frac{0.267\%}{20.95\%}\right)\right]$$

$$L_{15O_2} = -0.00444 \ln(1 - 0.012746) = -0.00444 \ln(0.987255)$$

$$L_{15O_2} = -0.00444(-0.012826)$$

$L_{15O_2} = 5.69 \times 10^{-5}$ atm-cc/sec. This is the true oxygen leak rate during the 15 min

The true helium leak rate = $5.69 \times 10^{-5}/0.354 = 1.61 \times 10^{-4}$ atm-cc/sec.

Problem 10. A 20 cc package is sealed with 80% nitrogen and 20% helium to a total pressure of 1.20 atm. The measured leak rate immediately after sealing is 1×10^{-7} atm-cc/sec. What is the total pressure and composition of the gases in the package one year after sealing, neglecting viscous flow?

Solution. The atm of nitrogen in the package = 0.8 × 1.20 = 0.96. The atm of helium in the package is equal to 0.2 × 1.20 = 0.24.

$$L_{HE} = 1 \times 10^{-7}/0.24 = 4.17 \times 10^{-7} \text{ atm-cc/sec}$$

$$L_{N_2} = 0.378\, L_{HE} = 1.58 \times 10^{-7} \text{ atm-cc./sec}$$

$$L_{O_2} = 0.354\, L_{HE} = 1.48 \times 10^{-7} \text{ atm-cc/sec}$$

$$L_{AR} = 0.316\, L_{HE} = 1.32 \times 10^{-7} \text{ atm-cc/sec}$$

$$1 \text{ year} = 365 \times 24 \times 3600 = 3.15 \times 10^{7} \text{ sec}$$

Helium leaking out:

$$p_{t\text{HE}} = p_i \left[e^{-\frac{L_{\text{HE}}t}{V}} \right] = 0.24 \left[e^{-\frac{(4.17\times10^{-7})(3.15\times10^{7})}{20}} \right]$$

$$p_{t\text{HE}} = 0.24(e^{-0.6575}) = 0.24(0.5181)$$

$$p_{t\text{HE}} = 0.1243 \text{ atm remaining in the package}$$

Nitrogen leaking out:

$$\Delta p_{t\text{N}_2} = (0.96 - 0.78084) \left[e^{-\frac{(1.58\times10^{-7})(3.15\times10^{7})}{20}} \right] = 0.17916\left(e^{-0.24913}\right)$$

$$\Delta p_{t\text{N}_2} = 0.17916(0.77948) = 0.1396 \text{ atm}$$

To this is added the atm of nitrogen in the air. The total nitrogen in the package after one year = 0.1396 + 0.78084 = 0.9205 atm.

Oxygen leaking in:

$$Q_{inO_2} = \Delta p_{iO_2}\left(1 - e^{-\frac{L_{O_2}t}{V}}\right) = 0.2095\left[1 - e^{-\frac{(1.48\times10^{-7})(3.15\times10^{7})}{20}}\right]$$

$$Q_{inO_2} = 0.2095(1 - e^{-0.23336}) = 0.2095(1 - 0.7918) = 0.2095(0.2081)$$

$$Q_{inO_2} = 0.0436 \text{ atm}$$

Argon leaking in:

$$Q_{inAR} = 0.00934\left[1 - e^{-\frac{(1.32\times10^{-7})(3.15\times10^{7})}{20}}\right]$$

$$Q_{inAR} = 0.00934(1 - e^{-0.208139}) = 0.00934\ (1 - 0.812095)$$

$$Q_{inAIR} = 0.00934(0.187705)$$

$$Q_{inAR} = 0.0017 \text{ atm}$$

The quantities of the gases in the package are:

Helium = 0.1243 atm
Nitrogen = 0.9205 atm
Oxygen = 0.0436 atm
Argon = 0.0017 atm
Total = 1.0901 atm

The composition in percentages is:

Helium = 11.40
Nitrogen = 84.44
Oxygen = 4.00
Argon = 0.16
Total % = 100.00

Problem 11. This problem is the same as Problem 10, but includes viscous flow for a rectangular channel having an Aspect Ratio of 10:1.

Solution. The viscous contribution for a total true helium leak of 4.17×10^{-7} is found in Fig. 3-4, and is about 8.3%. The curve in this figure is for an average pressure of 0.5 atm. The average pressure when testing this package is (1.2 + 0)/2 = 0.6 atm. The percentage of the viscous leak rate must be increased by the ratio of the average pressures. The corrected percent viscous flow is 8.3% × (0.6/0.5) =10%.The molecular leak rates are therefore 90% of that in Problem 10.

$$L_{HE} = 0.9 \times 4.17 \times 10^{-7} = 3.75 \times 10^{-7} \text{ atm-cc/sec}$$

$$L_{N_2} = 0.9 \times 1.58 \times 10^{-7} = 1.422 \times 10^{-7} \text{ atm-cc/sec}$$

$$L_{O_2} = 0.9 \times 1.48 \times 10^{-7} = 1.332 \times 10^{-7} \text{ atm-cc/sec}$$

$$L_{AR} = 0.9 \times 1.32 \times 10^{-7} = 1.188 \times 10^{-7} \text{ atm-cc/sec}$$

Molecular helium leaking out:

$$p_{tHE} = 0.24\left[e^{-\frac{(3.753\times10^{-7})(3.15\times10^{7})}{20}}\right] = 0.24\left(e^{-0.69177}\right)$$

$$p_{tHE} = 0.24(0.55334) = 0.1328 \text{ atm. helium remaining}$$

Molecular nitrogen leaking out:

$$p_{tN_2} = (0.96 - 0.78084)\left[e^{-\frac{(1.422\times10^{-7})(3.15\times10^{7})}{20}}\right]$$

$$p_{tN_2} = 0.17916(e^{-0.223965}) = 0.17916(0.79934)$$

$$p_{tN_2} = 0.1432 \text{ atm}$$

Adding the nitrogen that is in the air, the total nitrogen remaining in the package is 0.1432 + 0.78084 = 0.92405 atm remaining.

Molecular oxygen leaking in:

$$Q_{O_2} = 0.2095\left[1 - e^{-\frac{(1.332\times10^{-7})(3.15\times10^{7})}{20}}\right]$$

$$Q_{O_2} = 0.2095(1 - e^{-0.20979}) = 0.2095(1 - 0.81075) = 0.2095(0.18924)$$

$$Q_{O_2} = 0.0396 \text{ atm of oxygen leaked in}$$

Molecular argon leaking in:

$$Q_{AR} = 0.00934\left[1 - e^{-\frac{(1.188\times10^{-7})(3.15\times10^{7})}{20}}\right]$$

$$Q_{AR} = 0.00934(1 - e^{-0.18711}) = 0.00934(1 - 0.82935) = 0.00934(0.17065)$$

$$Q_{AR} = 0.0016 \text{ atm of argon leaked in}$$

There is a viscous flow out of the package because the total pressure inside the package is greater than the total pressure in the air. The viscous leak rate is proportional to the average pressure. The average pressure when leaking into the air is (1.2 + 1)/2 = 1.1 atm. When the average pressure was 0.6 atm, the percent viscous flow was 10% of L_{HE}. The percent viscous flow is now (1.1/0.6) × 10% = 18.3%. The viscous leak rate is 0.183 × 4.17 × 10^{-7} = 7.63 × 10^{-8} atm-cc/sec.

All the gases are flowing out at the same viscous flow rate. This rate is 1.078 × 7.63 × 10^{-8} = 8.23 × 10^{-8} atm-cc/sec = L_V. (see Eq. 3-18f). The total viscous flow out is calculated as follows:

$$\Delta P_V = \Delta P_{iV}\left(1 - e^{-\frac{L_V t}{V}}\right) = 0.2\left[1 - e^{-\frac{(8.23\times10^{-8})(3.15\times10^{7})}{20}}\right]$$

$$\Delta P_V = 0.2(1 - e^{-0.129456}) = 0.2(1 - 0.87848) = 0.2(0.12152)$$

$$\Delta P_V = 0.02430 \text{ total atm leaked out due to viscous flow}$$

The amount of helium that leaked out due to viscous flow = 0.02430 × (0.24/1.2) = 0.00486 atm. The amount of nitrogen that leaked out due to viscous flow = 0.02430 × (0.96/1.2) = 0.01944 atm. The total helium that leaked out = 0.1072 (molecular) + 0.00486 (viscous) = 0.11206 atm. The helium remaining in the package = (0.24 – 0.11206) = 0.12794 atm. The total nitrogen that leaked = 0.0360 (molecular) + 0.01944 (viscous) = 0.05544 atm. The nitrogen remaining in the package = (0.96 – 0.05544) = 0.90456 atm.

The quantities of gas in the package and their percentages are shown in the table below:

Gas	Atm	Percentage
Helium	0.1279	11.912
Nitrogen	0.9046	84.257
Oxygen	0.0396	3.688
Argon	0.0016	0.149
Total	1.0737	100.000

Problem 12. A manufacturer of ceramic transmitter tubes has a contract to deliver such tubes with a maximum true helium leak rate of 5×10^{-10} atm-cc/sec. The internal volume of these tubes is 6 cc. The transmitter tubes work normally when the pressure inside the tube is less than 0.1 torr. From 0.1–1.0 torr, the tube will arc and not operate. Above 1.0 torr, the efficiency of operation is greatly reduced. Bombing 100% of the tubes to introduce helium for leak testing would increase the pressure in the tubes. Engineer Bill Bright has proposed that a sample from each lot of tubes be operating and simultaneously bombed at 5 atm of helium for a length of time. If the tubes have not arced at that time, the leak rate must be less than 5×10^{-10} atm-cc/sec of helium. What is the time?

Solution.

$$0.1 \text{ torr} = (1/7600) \text{ atm}$$

There will be no viscous flow because of this small leak rate. Using Eq. (4-9):

$$t = -\frac{V}{L}\left[\ln\left(1 - \frac{Q_{inP}}{\Delta p_i}\right)\right]$$

where: V = 6 cc

L = 5×10^{-10} atm-cc/sec of helium

Q_{inP} = (1/7600) of an atmosphere

Δp_i = 5 atm

$$t = -\frac{6}{5 \times 10^{-10}}\left[\ln\left(1 - \frac{1}{7600 \times 5}\right)\right] = -1.2 \times 10^{-10}\left[\ln(1 - 0.000026315)\right]$$

$$t = -1.2 \times 10^{-10}(\ln 0.999973684) = -1.2 \times 10^{-10}(-2.63 \times 10^{-5})$$

$$t = 315{,}000 \text{ sec} = 87.67 \text{ hr}$$

Problem 13. A cast aluminum enclosure with 0.1 cm walls has an internal volume of 10 cc. It is sealed with 80% nitrogen and 20% helium. The measured leak rate is 1×10^{-7} atm-cc/sec immediately after sealing. The geometry of the leak channel in this type of package is known to be cylindrical. This package spends 1 month in ambient air (+22°C), 12 months on the moon at an average temperature of –10°C (263°K), and then 2 months in ambient air. At the end of this time, it is again leak tested. What is the expected measured leak rate at this time and what is the gas composition in the package?

Solution. The true helium leak rate, L_{HE}, is $(1 \times 10^{-7}/0.2) = 5 \times 10^{-7}$ atm-cc/sec.

$$L_{N_2} = 0.378 \times L_{HE} = 0.378 \times 5 \times 10^{-7} = 1.89 \times 10^{-7} \text{ atm-cc/sec}$$

$$L_{O_2} = 0.354 \times L_{HE} = 0.354 \times 5 \times 10^{-7} = 1.77 \times 10^{-7} \text{ atm-cc/sec}$$

$$L_{AR} = 0.316 \times L_{HE} = 0.316 \times 5 \times 10^{-7} = 1.58 \times 10^{-7} \text{ atm-cc/sec}$$

The percent viscous flow for a cylindrical leak channel is shown in Fig. 3-3, and is 90% for $L_{HE} = 5 \times 10^{-7}$ atm-cc/sec. The molecular rate is therefore 100% – 90% = 10%.

$$L_{10HE} = 5 \times 10^{-8} \text{ atm-cc/sec}$$

$$L_{10N_2} = 1.89 \times 10^{-8} \text{ atm-cc/sec}$$

$$L_{10O_2} = 1.77 \times 10^{-8} \text{ atm-cc/sec}$$

$$L_{10AR} = 1.58 \times 10^{-8} \text{ atm-cc/sec}$$

One Month In Air. Helium is leaking out:

$$1 \text{ mo} = 30 \times 24 \times 3600 = 2{,}592{,}000 \text{ sec}$$

$$p_{1HE} = p_i\left(e^{-\frac{L_{10HE}t}{V}}\right) = 0.2\left[e^{-\frac{(5\times10^{-8})(2{,}592{,}000)}{10}}\right]$$

where: p_i = the initial atm of helium in the package

p_{1HE} = the helium in the package after 1 month in air

L_{10HE} = the 10% helium molecular leak rate

$p_{1HE} = 0.2(e^{-0.01296}) = 0.2(0.98712)$

$p_{1HE} = 0.1974$ atm of helium remaining

Nitrogen is leaking out:

$$p_{1N2} = \Delta p_{N2}\left(e^{-\frac{L_{10N2}t}{V}}\right) = (0.80 - 0.78084)\left(e^{-\frac{(1.89x10^{-8})(2{,}592{,}000)}{10}}\right)$$

$$p_{1N_2} = 0.1916(e^{-0.0048988}) = 0.1916(0.99511)$$
$$= 0.01907 \text{ atm of nitrogen}$$

To this we add the nitrogen in the air (0.78084), 0.01907 + 0.78084 = 0.79991 atm of nitrogen.

Oxygen is leaking in:

$$Q_{1O2} = \Delta p_i \left[1 - e^{-\frac{L_{1O02} t}{V}}\right] = 0.2095\left[1 - e^{-\frac{(1.77X10^{-8})(2,592,000)}{10}}\right]$$

$$Q_{1O_2} = 0.2095(1 - e^{-0.0045878}) = 0.2095(1 - 0.99542) = 0.2095\ (0.004577)$$

$$Q_{1O_2} = 0.00096 \text{ atm of oxygen leaked into the package}$$

Argon is leaking in:

$$Q_{1AR} = 0.00934\left[1 - e^{-\frac{(1.77x10^{-8})(2,592,000)}{10}}\right]$$

$$Q_{1AR} = 0.00934(1 - e^{-0.0040954})$$
$$= 0.00934(1 - 0.99591) = 0.00934(0.004087)$$

$$Q_{1AR} = 0.00004 \text{ atm of argon leaked into the package}$$

There is no viscous flow as there is no difference in total pressure. The atm of the gases and the percentages after 1 month in air are shown below:

	Atmospheres	**Percentage**
Helium	0.1974	19.78
Nitrogen	0.7999	80.12
Oxygen	0.0010	0.10
Argon	0.0000	0.00
Totals	**0.9983**	**100.00**

Twelve Months On The Moon. There will be both viscous and molecular flow from the package into the very low atmospheric pressure of the moon.

Viscous flow out of the package: the total pressure in the package after one month in air would be 0.9983 atm at room temperature (295°K). Correcting for the temperature on the moon, the pressure in the package P_0 is:

$$P_0 = (263/295)0.9983 \text{ atm} = 0.89 \text{ atm}$$

The viscous leak rate when the average pressure was 0.5 atm was 90% of $5 \times 10^{-7} = 4.5 \times 10^{-7}$ atm-cc/sec. The average pressure is now (0.89 + 0)/2 = 0.445 atm. Correcting for the average pressure, the viscous leak rate on the moon is $(0.445/0.5) \times 4.5 \times 10^{-7} = 4.00 \times 10^{-7}$ atm-cc/sec.

The total atmospheres leaving the package due to viscous flow is:

$$\Delta P_{12V} = \Delta P_i \left(1 - e^{-\frac{L_V t}{V}}\right) = 0.89\left[1 - e^{-\frac{(4.0\times10^{-7})(3.15\times10^{7})}{10}}\right]$$

where: ΔP_{12V} = total pressure leaving the package in 12 mo

ΔP_i = initial total pressure difference

L_V = the viscous leak rate

$\Delta P_{12V} = 0.89\,(1 - e^{-1.2617}) = 0.89(1 - 0.28316)$

$\Delta P_{12V} = 0.89(0.71684)$

$\Delta P_{12V} = 0.6380$ total atm due to viscous flow

The amount of the individual gas leaving the package due to viscous flow is in accordance with their percentage in the package:

Atm of helium = 19.78% × 0.6380 = 0.1262

Atm of nitrogen = 80.12% × 0.6380 = 0.5112

Atm of oxygen = 0.1% × 0.6380 = 0.0006

The amount of gas leaking out due to molecular flow during the 12 mo on the moon is in accordance with their 10% leak rate and the atmospheres of the individual gases.

Helium leaking out due to molecular flow:

$$p_{12\text{HE}m} = p_i \left(1 - e^{-\frac{L_{\text{HE}m} t}{V}}\right) = 0.89 \times 0.1974\left[1 - e^{-\frac{(5\times10^{-8})(3.15\times10^{7})}{10}}\right]$$

where: p_{12HEm} = atm of helium leaking out due to molecular flow in 12 mo

p_i = initial helium in the package on the moon

L_{HEm} = the helium molecular leak rate

$$p_{12HEm} = 0.1757(1 - e^{-0.1575}) = 0.1757(1 - 0.85428)$$

$$p_{12HEm} = 0.1757(0.1457)$$

$p_{12HEm} = 0.0256$ atm leaked out due to molecular flow

Nitrogen leaking out due to molecular flow:

$$p_{12N_2m} = 0.89 \times 0.7999\left[1 - e^{-\frac{(1.89\times10^{-8})(3.15\times10^{7})}{10}}\right]$$

$$p_{12N_2m} = 0.7119(1 - e^{-0.05953}) = 0.7119(1 - 0.94220)$$

$$p_{12N_2m} = 0.7119(0.057797)$$

$p_{12N_2m} = 0.0411$ atm leaked out due to molecular flow

Oxygen leaking out due to molecular flow:

$$p_{12O_2m} = 0.89 \times 0.0010\left[1 - e^{-\frac{(1.77\times10^{-8})(3.15\times10^{7})}{10}}\right]$$

$$p_{12O_2m} = 0.00089(1 - e^{-0.055755}) = 0.00089\,(1 - 0.94577)$$

$$p_{12O_2m} = 0.00089(0.054229)$$

$p_{12O_2m} = 0.00005$ atm of oxygen leaked out due to molecular flow

The total leaking out due to molecular flow at –10°C is equal to 0.0256 + 0.0411 + 0.00005 = 0.06675 atm.

After twelve months on the moon, the amount of each gas in the package is:

Helium = 0.1757 – (0.1262 + 0.0256) = 0.0239 atm

Nitrogen = 0.7119 – (0.5112 + 0.0411) = 0.1596 atm

Oxygen = 0.00089 – (0.0006 + 0.00005) = 0.0002 = atm

Argon = 0.0000 atm

Total = 0.1837 atm

Their percentages are:

Helium = 13.01%

Nitrogen = 86.88%

Oxygen = 0.11%

Two months in the air: Returning to 22°C, the atmospheres of the gases changes by a factor of 295°K/265°K = 1.12 = 1/0.89

Helium = 0.0239/0.89 = 0.0268

Nitrogen = 0.1596/0.89 = 0.1793

Oxygen = 0.0002/0.89 = 0.0002

For a total of 0.2063 atm in the package

There will be molecular helium leaking out, and molecular and viscous flow leaking in of nitrogen, oxygen and argon.

$$2 \text{ mo} = 60 \times 24 \times 3600 = 5{,}184{,}000 \text{ sec}$$

Molecular helium leaking out:

$$p_{2\text{HE}m} = 0.0268\left[e^{-\frac{(5.0\times10^{-8})(5{,}184{,}000)}{10}}\right]$$

$$p_{2\text{HE}m} = 0.0268(e^{-0.2592}) = 0.0268(0.9744)$$

$$p_{2\text{HE}m} = 0.0261 \text{ atm of helium remaining in the package}$$

Molecular nitrogen leaking in:

$$Q_{2N2} = \Delta p\left(1 - e^{-\frac{(1.89x10^{-8})(5,184,000)}{10}}\right) = (0.78084 - 0.1793)\left(1 - e^{-0.00980}\right)$$

$$Q_{2N_2} = 0.6015(1 - 0.99025) = 0.6015(0.00975)$$

$$Q_{2N_2} = 0.0059 \text{ atm of nitrogen leaked in}$$

The nitrogen that was already in the package (0.1793) is added for a total nitrogen in the package of 0.1852 atm.

Molecular oxygen leaking in:

$$Q_{2O_2} = (0.2095 - 0.0002)\left[1 - e^{-\frac{(1.77\times10^{-8})(5,184,000)}{10}}\right]$$

$$Q_{2O_2} = 0.2093(1 - e^{-0.00918}) = 0.2093(1 - 0.99087)$$

$$Q_{2O_2} = 0.2093\ (0.00913)$$

$$Q_{2O_2} = 0.00191 \text{ atm of oxygen leaked in}$$

To this, we add the 0.0002 that was already in the package for a total oxygen in the package of 0.0021 atm.

Molecular argon leaking in:

$$Q_{2AR} = 0.00934\left(1 - e^{-\frac{(1.58x10^{-8})(5,184,000)}{10}}\right) = 0.00934\left(1 - e^{-0.00819}\right)$$

$$Q_{2AR} = 0.00934(1 - 0.99184) = 0.00934(0.00816)$$

$$Q_{2AR} = 0.0001 \text{ atm of argon has leaked in}$$

Viscous Flow into the Package. First the new average pressure is calculated. The new average pressure = (0.2063 + 1)/2 = 0.6032 atm. Correcting for the average pressure, the viscous leak rate is: (0.6038/0.5) × 4.5 × 10^{-7} = 5.43 × 10^{-7} atm-cc/sec. The quantity of gas entering the package because of viscous flow is:

$$Q_V = \Delta P\left[1 - e^{-\frac{(5.431\times10^{-7})(5{,}184{,}000)}{10}}\right]$$

$$Q_V = (1 - 0.2063)(1 - e^{-0.28141}) = 0.7937(1 - 0.7547)$$

$$Q_V = 0.7937(0.2063)$$

Q_V = 0.1637 total atm leaked in due to viscous flow

This viscous contribution consists of nitrogen, oxygen and argon in the percentage that they are on the high pressure side, air:

viscous nitrogen = 0.78084 × 0.1637 atm = 0.1278 atm leaked in

viscous oxygen = 0.2095 × 0.1637 atm = 0.0343 atm leaked in

viscous argon = 0.00934 × 0.1637 atm = 0.0015 atm leaked in

At the end of the 2 months in air there are the following atmospheres in the package:

Helium, 0.0261 atm

Nitrogen, 0.1852 + 0.1278 = 0.3130 atm

Oxygen, 0.0021 + 0.0343 = 0.0364 atm

Argon, 0.0001 + 0.0015 = 0.0016 atm

The total atmospheres in the package at the end of the two months = 0.3771

The percentage of each gas is:

Helium	6.92%
Nitrogen	83.00%
Oxygen	9.65%
Argon	0.43%
Total	100.00%

The expected measured leak rate is the original true helium leak rate × the atmospheres of helium in the package = $5 \times 10^{-7} \times 0.0261$ atm. = 1.31×10^{-8} atm-cc/sec.

Problem 14. A temperature controlled miniature crystal oscillator (TCMXO) consists of a microcircuit and crystal, each in their enclosure, which are in an outer TCMXO enclosure. The internal volume of the outer enclosure, less the two parts, is 13 cc. The TCMXO is pinched off at a pressure of 1×10^{-9} Torr. The crystal and microcircuit enclosures, each have an internal volume of 1.0 cc and a maximum true helium leak rate of 5×10^{-10} atm-cc/sec. The pressure within these two enclosures is 1.0 Torr, the residual gas being air. The TCMXO contains a getter that traps all the gases except argon and helium. There is sufficient getter material to trap all the gases being out gassed on the inside of the outer enclosure, and all the nitrogen and oxygen that leaks in during a years time. The $\pm 1 \times 10^{-8}$ temperature stability of the TCMXO depends upon the pressure in the TCMXO. This stability degrades when the pressure rises to 5×10^{-4} Torr. What is the maximum true helium leak rate of the TCMXO enclosure so that the pressure will remain below 5×10^{-4} Torr after 1 year?

Solution. First the amount of argon leaking out of the crystal and microcircuit enclosures is calculated. The argon leaking out of one of the enclosures is:

$$Q_{AR1} = \Delta p_{AR}\left(1 - e^{-\frac{L_{AR}t}{V}}\right)$$

where Q_{AR1} = the argon leaking out of the crystal and microcircuit enclosure into the TCMXO enclosure in one year.

The difference in total pressure between the inside of the crystal and microcircuit enclosures and the TCMXO enclosure = 1.0 torr − 1×10^{-9} torr = 1.0

$$\text{torr} = \left(\frac{1.0 \text{ torr}}{\frac{760 \text{ torr}}{\text{atm}}}\right) = 0.00136 \text{ total atm in the crystal enclosure}$$

The atm of argon in the microcircuit and the crystal enclosure

$= 0.00934$ atm of argon per total atm × the total atm

$= 0.00934 \times 0.00136$
$= 1.23 \times 10^{-5}$ atm of argon

L_{AR} = the true argon leak rate $= 0.316 \times 5 \times 10^{-10}$
$= 1.58 \times 10^{-10}$ atm-cc/sec

$t = 1 \text{ yr} = 3.15 \times 10^{7}$ sec

V = the volume of the crystal enclosure = 1 cc

$$Q_{AR1} = 1.23 \times 10^{-5}\left[1 - e^{-\frac{(1.58\times10^{-10})(3.15\times10^{7})}{1}}\right]$$

$$Q_{AR} = 1.23 \times 10^{-5}(1 - e^{-0.004977}) = 1.23 \times 10^{-5}\,(1 - 0.995036)$$

$$Q_{AR} = 1.23 \times 10^{-5}(0.00496) = 6.10 \times 10^{-8} \text{ atm of argon}$$

Converting to torr, 6.10 atm × 760 torr/atm. $= 4.64 \times 10^{-5}$ torr.

An equal amount of argon is leaking out of the microcircuit enclosure, so the total argon leaking into the TCMXO enclosure $= 2 \times 4.64 \times 10^{-5} = 9.28 \times 10^{-5}$ torr.

The volume of the TCMXO enclosure is 13 cc, therefore the pressure rise in this enclosure is $9.28 \times 10^{-5}/13 = 7.14 \times 10^{-6}$ torr. This is much less than the 5×10^{-4} torr limit and can be neglected.

Next, the leak rate for the TCMXO enclosure to allow 5×10^{-4} torr to leak into the TCMXO enclosure in one year is calculated:

$$5 \times 10^{-4} \text{ torr} = (5 \times 10^{-4} \text{ torr})/(760 \text{ torr/atm}) = 6.579 \times 10^{-7} \text{ atm}$$

$$L_{AR} = -\frac{V}{t}\left[\ln\left(1 - \frac{Q_{AR}}{\Delta p_{AR}}\right)\right]$$

where: L_{AR} = true argon leak rate

V = 13 cc

t = 3.15×10^{7} sec

Q_{AR} = the atm of argon equal to 5×10^{-4} torr = 6.579×10^{-7} atm

Δp_{AR} = the argon concentration in the air = 0.00934 atm

$$L_{AR} = -\frac{13}{3.15 \times 10^{7}}\left[\ln\left(1 - \frac{6.579 \times 10^{-7}}{0.00934}\right)\right]$$

$$L_{AR} = -4.13 \times 10^{-7}[\mathrm{Ln}(1 - 0.000070438)]$$
$$= -4.13 \times 10^{-7}[\mathrm{Ln}(0.9990296)]$$

$$L_{AR} = -4.13 \times 10^{-7}(0.00007044) = 2.91 \times 10^{-11} \text{ atm-cc/sec}$$

This is the maximum allowed argon leak rate so that the pressure in the outer enclosure is less than 5×10^{-4} torr. The equivalent true helium leak rate is:

$$L_{HE} = (2.91 \times 10^{-11})/(0.316) = 9.21 \times 10^{-11} \text{ atm-cc/sec}$$

Helium leaking into the TCMXO from the air can be neglected because of its low concentration of 5.24 ppm. The Δp multiplier is 5.24×10^{-6}.

Problem 15. A 100 cc package contains 50% nitrogen, 49.4% helium and 0.6% argon at a total pressure of 1.56 atm. What is the measured leak rate limit to guarantee a pressure of at least 1.20 atm 5 years later, neglecting viscous flow?

Solution. The atm of the gases in the package immediately after sealing are:

Nitrogen, 50% of 1.56 atm = 0.78 atm

Argon, 0.6% of 1.56 atm = 0.00936 atm

Helium, 49.4% of 1.56 atm=0.77064 atm

The nitrogen and argon partial pressure in the package are almost equal to the partial pressures outside (0.78084 and 0.00934), so that will be no molecular nitrogen or argon flowing.

The helium remaining in the package is:

$$p_{HE} = p_i\left(e^{-\frac{L_{HE}t}{V}}\right)$$

where: p_{HE} = the helium remaining after 5 years

p_i = the initial helium pressure = 0.77063 atm

L_{HE} = the true helium leak rate, presently unknown

$t = 5 \text{ yr} = 5 \times (3.15 \times 10^7 \text{ sec/year}) = 1.575 \times 10^8 \text{ sec}$

$V = 100$ cc

The oxygen leaking in is:

$$Q_{O_2} = \Delta p_i\left(1 - e^{-\frac{L_{O_2}t}{V}}\right)$$

where: Q_{O_2} = the amount of oxygen leaking into the package in 5 yr

Δp_i = the initial oxygen pressure difference

= 0.2095 – 0 = 0.2095 atm

L_{O2} = the true oxygen leak rate, presently unknown

$t = 5 \text{ yr} = 1.575 \times 10^8 \text{ sec}$

$V = 100$ cc

The most straight forward approach to this problem is graphical (see Fig. 4-6). The total pressure in the package in 5 years versus the true helium leak rate can be plotted. The remaining helium and the oxygen which has leaked in can also be plotted on the same graph. The use of a spread sheet hastens this solution. The plot shows a true helium leak rate of 5×10^{-7} atm-cc/sec at 1.20 atm. The measured helium leak rate, R, will be the atm of helium in the package × the true helium leak rate, L_{HE}.

At the start of the 5 years,

$$R = 0.7706 \times 5 \times 10^{-7} = 3.85 \times 10^{-7} \text{ atm-cc/sec}$$

At the end of the 5 years,

$$R = 0.35 \times 5 \times 10^{-7} = 1.75 \times 10^{-7} \text{ atm-cc/sec}$$

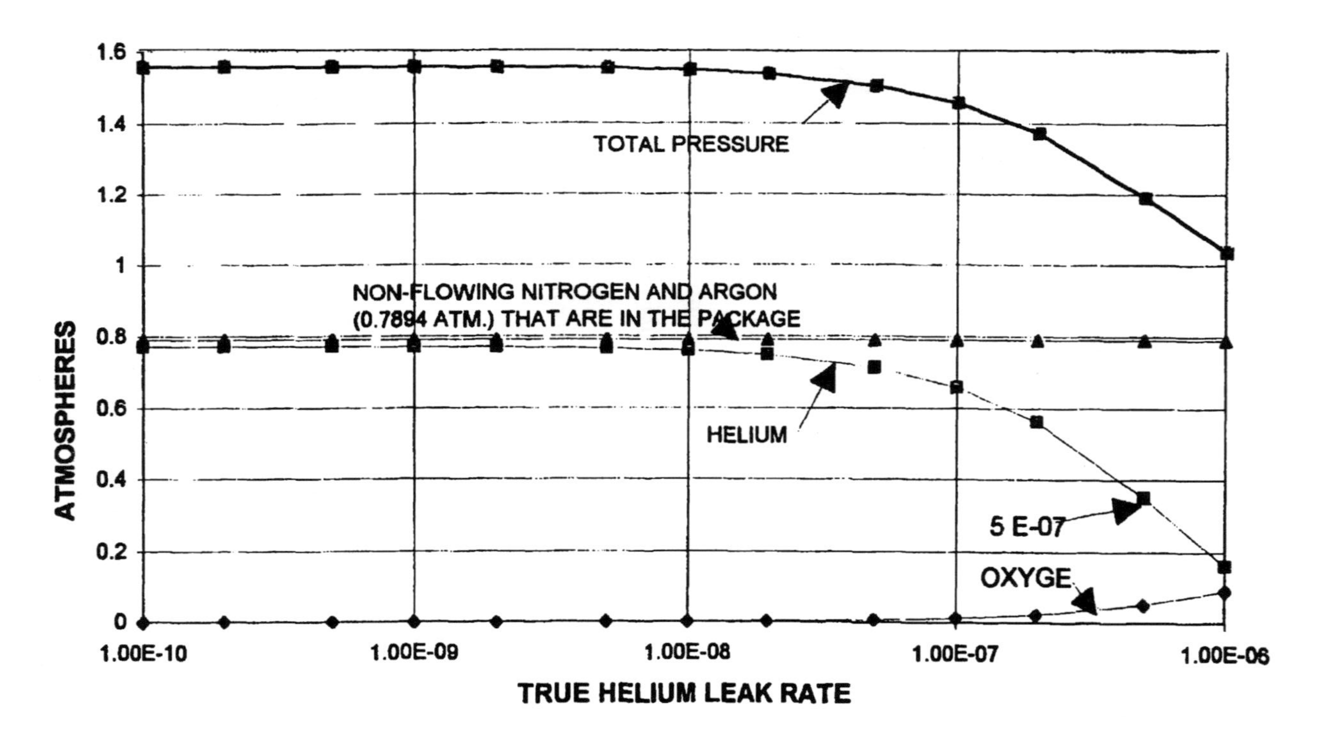

Figure 4-6. Atmospheres versus true helium leak rate for Problem 15. After five years in air, a 100 cc package, at 1.56 atm with 50% nitrogen, 49.4% helium, 0.6% argon.

5

Water in Sealed Packages

1.0 WATER RELATED CORROSION AND CIRCUIT FAILURES

Water in microcircuits, hybrids, multichip modules (MCM) and relays is a contributing factor in the failure of these devices. Water alone will not cause such a failure nor cause corrosion inside the package. Perkins, Licari, and Buckelew have shown, in a carefully planned experiment, that water alone will not cause electrical failures nor corrosion.[1] Packages containing various electrical functions were seeded with water concentrations of 1,000, 6,000, and 10,000 ppm, and subjected to various environments, including 1000 hour life tests. Similar evidence has been seen in operating circuits after 1000 hours operating life tests at an ambient of 125°C. The hybrids were successfully electrically tested at –55°C, +125°C, and +25°C at the end of the life tests. The residual gas analyses (RGA) shown in Table 1 are examples of such evidence. Not only do these packages have large amounts of water but also large amounts of ammonia, methanol, hydrogen, and MEK (methyl ethyl ketone). The samples are hybrids from three different manufacturers, representing three processing technologies and four circuit types.

Table 5-1. RGA of Hybrids after Successful 1000 Hour Life Tests at 125°C, that Contain Large Amounts of Water

Gas	Units	Sample #1	Sample #2	Sample #3	Sample #4	Sample #5
Nitrogen	%	80.3	78.1	92.2	68.8	89.03
Oxygen	PPM	ND	ND	ND	ND	120
Argon	PPM	231	ND	558	ND	110
CO_2	PPM	42,000	3103	2,760	1,896	ND
Water	PPM	6,509	10,600	16,500	6,626	6,900
Hydrogen	PPM	234	104,000	ND	683	6,140
Helium	%	12.5	10.1	5.77	25.2	5.03
Fluor.C.	PPM	ND	ND	ND	ND	ND
Ammonia	%	2.22	ND	ND	ND	2.108
Hydro.C.	PPM	374	709	ND	236	290
Methanol	PPM	1012	ND	ND	13,900	4,490
MEK	%	ND	ND	ND	2.90	ND

Water in devices is a justifiable concern because it is the vehicle for contaminates that cause corrosion and ultimately failure. Numerous reports in the literature have reported such corrosion and failures.[2]–[17] Lowry described the detailed mechanisms for such corrosions.[18] United States Government Report RADC-TR-87-210, by D. Kane and H. Domingos also describes the corrosion mechanisms as well as the experimental procedure to measure the phenomenon.[19]

The contaminants that cause the corrosion are either negative or positive ions, using water as the carrier. The corrosion is often enhanced by electrical current in the circuit. Negative ion corrosion is acidic and can occur with the following ions:

- Cl^- from hydrochloric acid (HCl)
- F^- from hydrofluoric acid (HF)
- NO_3^- from nitric acid (HNO_3)
- SO_4^{2-} from sulfuric acid (H_2SO_4)
- PO_4^{3-} from phosphoric acid (H_3PO_4)

Positive ion corrosion is basic and can occur with the following ions:

- Na^+ from sodium hydroxide (NaOH)
- K^+ from potassium hydroxide (KOH)

It is also possible to have non-aqueous corrosion due to the negative OH^- radical from an alcohol, or from the negative Cl^- or F^- ions dissociated from halogenated organic solvents. These type of corrosions are infrequent and are only relevant to unusual processing procedures.

Two examples of corrosive chemical reactions are:

Eq. (5-1) $3\ Al + 3\ Cl^- + 6\ H_2O = AlCl_3 + 2\ Al(OH)_3 + 3\ H_2 + 3\ e^-$

Eq. (5-2) $2\ Al + Na^+ + 7\ H_2O = 2\ Al(OH)_3 + NaOH + 3\ H_2 + H^+$

Corrosion not only requires a contaminant and an electrolytic vehicle, but is also pH sensitive. Little if any corrosion takes place when the pH is between 4.5 and 7.5. The pH is related to the disassociation of water. The dissociation of water into H^+ and OH^- ions is represented by equilibrium Eq. (5-3).

Eq. (5-3) $$\frac{[H^+][OH^-]}{H_2O} = K$$

where: $[H^+]$ = the concentration of hydrogen ions

$[OH^-]$ = the concentration of hydroxyl ions

K = the equilibrium constant

The activity of the H_2O is constant when the aqueous solution is dilute. Therefore Eq. (3) reduces to:

Eq. (5-4) $$[H^+][OH^-] = K_w$$

where K_W is the ionization constant for water and $= 1 \times 10^{-14}$ at 24°C. K_W is dependent on temperature. The value of K_W as a function of degrees Celsius is:

$$K_W = 14.926 - 0.0420\,t + 0.00016\,t^2$$

$$K_W = 14.926 \text{ at } 0°C, \text{ and } 12.326 \text{ at } 100 \text{ at } 0°C$$

Pure water is electronically neutral so that:

Eq. (5-5) $\quad [H^+]=[OH^-]=1 \times 10^{-7}$

The pH is defined as the negative log of the hydrogen ion concentration or:

Eq. (5-6) $\quad \mathrm{pH} = -\mathrm{Log}\left[H^+\right] = -\mathrm{Log}\left(1\times10^{-7}\right) = 7.0$

Similarly,

Eq. (5-7) $\quad \mathrm{pOH} = -\mathrm{Log}\left[OH^-\right] = -\mathrm{Log}\left(1\times10^{-7}\right) = 7.0$

Considering a sodium contamination similar to Eq. (5-2), an estimate can be made for the amount of sodium needed to cause corrosion in a microelectronic package. Assume a 1 cc package with 5,000 ppm of water, containing semiconductor devices having aluminum pads. The available water is:

$$1 \text{ cc} \times 5{,}000 \times 10^{-6} \times 1 \times 10^{-3} \text{ liters/cc} = 5 \times 10^{-6} \text{ liters}$$

The ionization for the sodium corrosion is:

$$\left[Na^+\right]\left[OH\ \right] = K_W = 1\times10^{\ 14}$$

Assuming that the corrosion will not take place until the pOH increases to 8.0, we can write:

$$\left[Na^+\right] = \frac{1\times10^{\ 14}}{1\times10^{\ 8}} = 1\times10^{\ 6} \text{ moles per liter}$$

[Equilibrium and ionization constants are expressed in moles (molecular weights) and liters.]

The amount of impurity that could be available for corrosion is the concentration of the contaminant × the available water. The moles of sodium in the 1 cc package having 5×10^{-6} liters of water is:

$$(1 \times 10^{-6} \text{ moles/liter}) \times (5 \times 10^{-6} \text{ liters}) = 5 \times 10^{-12} \text{ moles}$$

All the moles will not be available to attack an aluminum pad. If we assume that 0.1% is available, then the number of moles is $0.001 \times 5 \times 10^{-12} = 5 \times 10^{-15}$ moles. One mole of sodium weighs 23 g, so that the number of grams available to corrode a pad is:

$$(23 \text{ g/mole}) \times (5 \times 10^{-15} \text{ moles}) = 1.15 \times 10^{-13} \text{ g}$$

This is an extremely small amount, detectable only with sophisticated analytical equipment.

Experiments that have tried to correlate moisture with failure rates have been non-conclusive, unless the contaminant levels were also controlled. Stroehle,[20] and Paulson and Kirk,[12] showed variations in failure rates for the same environment and water in the package. This variation was due to the variation of the contaminant (phosphorous) in the samples resulting in variations of the semiconductor processing. Graves and Gurany[21] reported large variations of fluoride impurities on aluminum bonding pads resulting from variations in plasma processing.

The above calculations show that it is important to keep the inside of microelectronic packages very clean or without water. The hybrids in Table 5-1 were very clean. All hybrids passed the Particle Impact Noise Detection (PIND) test during screening prior to the 1000 hour life test.

The consensus, among scientists and engineers, is that the amount of liquid water necessary to promote corrosion is three monolayers. Converting the three monolayers of liquid to water vapor yields different ppm for different size packages. This is demonstrated in the following two example calculations.

Example 1. Consider a package with inside dimensions of 0.2 cm by 0.2 cm by 0.2 cm. The volume = 0.008 cc. The inside surface area = $2(0.2 \text{ cm} \times 0.2 \text{ cm}) + 4(0.2 \text{ cm} \times 0.2 \text{ cm}) = 0.24 \text{ cm}^2$. Three monolayers of water is 1.2×10^{-7} cm thick. Multiplying this thickness by the surface area yields the volume of liquid water.

$$1.2 \times 10^{-7} \text{ cm} \times 0.24 \text{ cm}^2 = 2.88 \times 10^{-8} \text{ cc}$$

One cc of water weighs 1 g so the weight of the three monolayers is 2.88×10^{-8} g. One mole of water weighs 18 g and has a gaseous volume of 22.4 liters. The volume of water vapor equivalent to the three monolayers is:

$$\frac{2.88 \times 10^{-8}\,\mathrm{g} \times 22.4\,\mathrm{l}}{18\,\mathrm{g}} = 3.584 \times 10^{-8}\,\mathrm{l} = 3.584 \times 10^{-5}\,\mathrm{cc}$$

Dividing this by the volume of the package will give the ppm in the package.

$$\mathrm{ppm} = \frac{\text{Volume of Water} \times 10^{6}}{\text{Volume of Package}} = \frac{3.584 \times 10^{-5} \times 10^{6}}{0.008} = 4480\,\mathrm{ppm}$$

Example 2. Now consider a larger package with inside dimensions of 10 cm × 10 cm × 0.4 cm. The volume is 40 cc. The surface area is 2(10 cm × 10 cm) + 4(0.4 cm × 10 cm) = 216 cm^2. The volume of liquid water = 1.2×10^{-7} cm × 216 cm^2 = 2.59×10^{-5} cc = 2.59×10^{-5} g.

The volume of water vapor =

$$\frac{2.59 \times 10^{-5} \times 22.4}{18} = 3.223 \times 10^{-5}\,\mathrm{l} = 0.03223\,\mathrm{cc}$$

$$\mathrm{ppm} = \frac{0.03223 \times 10^{6}}{40} = 806$$

Similar calculations for different size packages can be made, and when the ppm is plotted against the surface area to volume ratio, we get a straight line. Such a plot is shown in Fig. 5-1. Figure 5-2 is the same information plotted on a semilog graph showing the ppm of water vapor versus the internal volume of the package. The volume of the × axis is logarithmic. There are two curves in this figure, for different internal package heights.

The above calculations and Figs. 5-1 and 5-2 show that the amount of water vapor to produce three monolayers of liquid water is different for different size packages.

The low ppm of water for the large packages could be misleading. Table 5-2 compares the package volume, the ppm of water, and the absolute amount of water in the package corresponding to three monolayers of water. The volume of water in the large packages is much greater than the volume of water in the small packages. If these packages had the same leak rate, it would take much longer for the larger package to leak in an amount equivalent to 3 monolayers than a smaller package.

Water inside a package can have three origins.

1. Water vapor leaking into the package from outside the package.
2. Water outgassing from the inside walls or from material or components inside the package.
3. Oxygen and hydrogen inside the package reacting to form water.

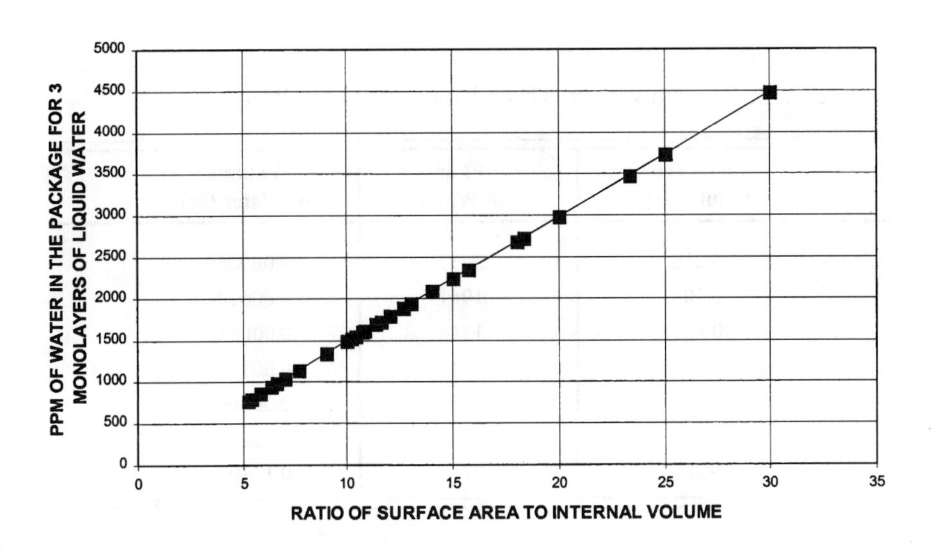

Figure 5-1. PPM of water vapor in a package for three monolayers of liquid water as a function of the internal surface area to internal volume ratio.

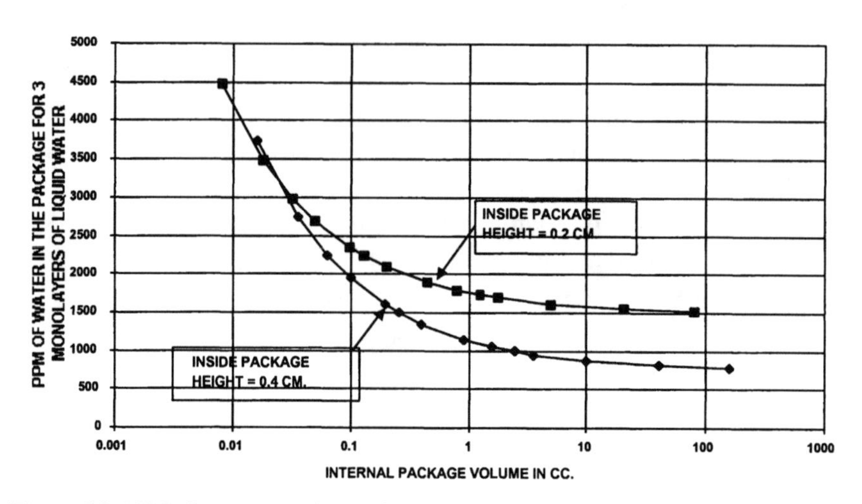

Figure 5-2. PPM of water vapor in a package equal to three monolayers of liquid water, for packages with different volumes.

Table 5-2. Comparison of PPM of Water and the Volume of Water for Several Size Packages when Three Monolayers are Present

Package Volume (cc)	PPM of Water	Volume of Water (cc)
0.016	3732	0.000059
0.10	1941	0.000194
0.40	1344	0.000537
1.6	1045	0.001672
10	866	0.008658
40	806	0.032244
160	776	0.12401

2.0 WATER LEAKING INTO A SEALED PACKAGE FROM THE OUTSIDE ENVIRONMENT

Water leaking into a sealed package is calculated in the same manner as was done for other gases in Ch. 4.

Eq. (5-8)

$$Q_{H_2O} = \Delta pi_{H_2O}\left(1 - e^{\frac{L_{H_2O}t}{V}}\right)$$

Eq. (5-9)

$$t = -\frac{V}{L_{H_2O}}\left[\ln\left(1 - \frac{Q_{H_2O}}{\Delta pi_{H_2O}}\right)\right]$$

Eq. (5-10)

$$L_{H_2O} = -\frac{V}{t}\left[\ln\left(1 - \frac{Q_{H_2O}}{\Delta pi_{H_2O}}\right)\right]$$

where: Q_{H_2O} = the water that has leaked in

V = the available internal volume of the package (volume of the parts inside the package should be subtracted) in cc

t = the time in seconds

L_{H_2O} = the true water leak rate
= 0.471 L_{HE} (see Table 3-1) in atm-cc/sec

Δpi_{H_2O} = the initial difference in the water partial pressure on the outside less the partial pressure on the inside of the package

The vapor pressure of the water on the inside of the package could be from outgassed materials in the package or from the water in the sealing chamber when the package was sealed.

The vapor pressure on the outside of the package is directly related to the relative humidity (R.H.) of the environment. Figure 5-3 shows the atmospheres of water in the air for 100% R.H. as a function of temperature. Figure 5-3 and Table 5-2 are based on data from the *Handbook of Chemistry and Physics*.[22] The atmospheres of water in the air for selected temperatures and R.H. is shown in Table 5-3.

The average relative humidity at 25°C is 65%, corresponding to 0.02 atm. This is the average relative humidity in the United States, and is a convenient value for comparing calculations. Figure 5-4 shows how long it will take to accumulate three monolayers of water in packages of different sizes and for different true water leak rates. The partial pressure of water outside the package is assumed to be 0.02 atm. Figure 5-5 is a log-log presentation of the same information as a single curve. In this figure, the time in seconds × the true water leak rate is plotted versus the internal package volume.

A "picture" of the water ingress with time, can be visualized if we solve Eq. (5-8) for different times, as is done in Ex. 5-3.

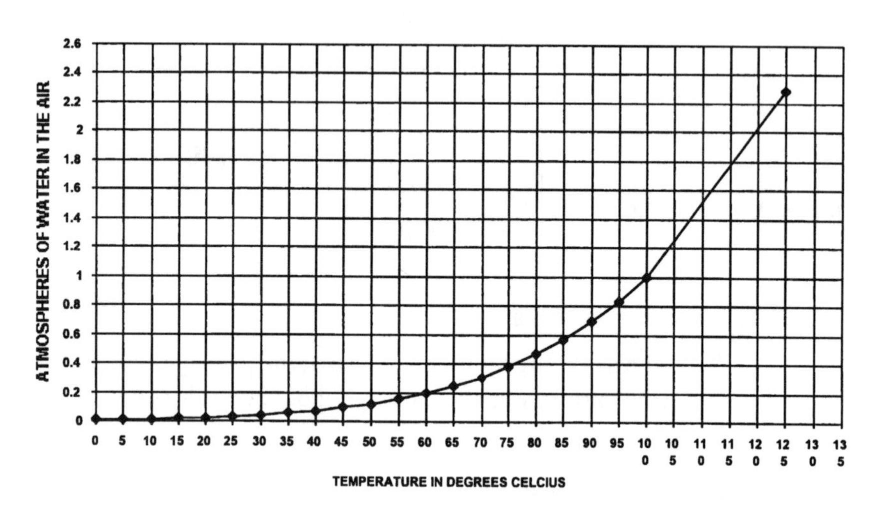

Figure 5-3. 100% relative humidity as a function of temperature.

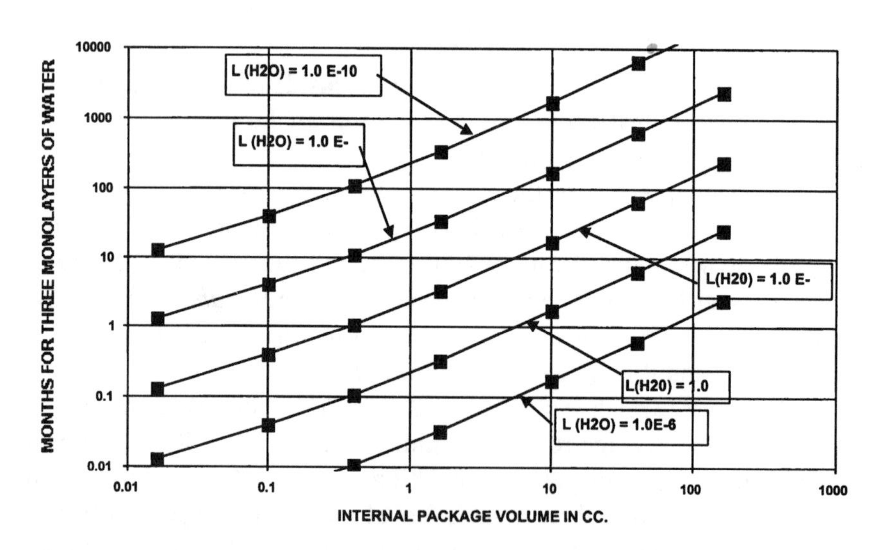

Figure 5-4. Log-log plot of time versus internal package volume, to leak in three monolayers of water for different true water leak rates.

Table 5-3. Atmospheres of Water Vapor for Selected Temperatures and Relative Humidity

% R.H.	0°C	20°C	25°C	30°C	85°C	100°C	125°C
25	0.0015	0.0058	0.0078	0.0105	0.1426	0.2506	0.5724
30	0.0018	0.0069	0.0094	0.0126	0.1711	0.3000	0.6868
35	0.0021	0.0081	0.0109	0.0147	0.1997	0.3500	0.8031
40	0.0024	0.0092	0.0125	0.0168	0.2282	0.4000	0.9158
45	0.0027	0.0104	0.0141	0.0189	0.2567	0.4500	1.0303
50	0.0030	0.0116	0.0156	0.0209	0.2853	0.5000	1.1448
55	0.0033	0.0127	0.0172	0.0230	0.3138	0.5500	1.2592
60	0.0036	0.0139	0.0188	0.0251	0.3423	0.6000	1.3739
65	0.0039	0.0150	0.0203	0.0272	0.3708	0.6500	1.4882
70	0.0042	0.0162	0.0219	0.0293	0.3994	0.7000	1.6027
75	0.0045	0.0173	0.0235	0.0314	0.4279	0.7500	1.7171
80	0.0048	0.0185	0.0250	0.0335	0.4564	0.8000	1.8316
85	0.0051	0.0196	0.0266	0.0356	0.4849	0.8500	1.9461
90	0.0054	0.0208	0.0282	0.0377	0.5135	0.9000	2.0606
95	0.0057	0.0219	0.0297	0.0398	0.5420	0.9500	2.1750
100	0.0060	0.0231	0.0313	0.0419	0.5705	1.0000	2.2895

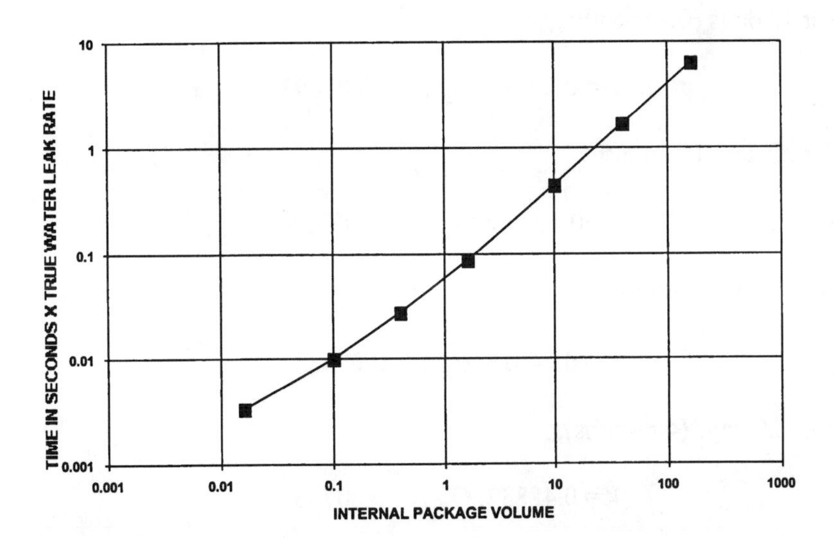

Figure 5-5. Log-log plot of time in seconds × true water leak rate, versus the internal volume of packages, to accumulate three monolayers when the environment is 65% R.H. at 25°C.

Example 3. Assume a package with a 1 cc internal volume having a true helium leak rate of 1×10^{-7} atm-cc/sec($L_{H_2O} = 4.71 \times 10^{-8}$ atm-cc/sec). There is no water in the package immediately after sealing, and the atmospheres of water outside the package is 0.02. Using Eq. (5-8):

$$Q_{H_2O} = 0.02\left(1 - e^{-\frac{4.71\times10^{-8}t}{1}}\right)$$

Solving the exponent Lt/V for various values of t. For 3 days,

$$t = 3 \times 24 \times 3600 = 259{,}200 \text{ sec}$$

$$Lt/V = 0.012208;\ e^{-0.012208} = 0.98787$$

$$Q_{H_2O} = 0.000243 \text{ atm of water}$$

For 6 days (0.2 months),

$$Lt/V = 0.024416;\ Q_{H_2O} = 0.00048$$

For 12 days (0.4 months),

$$Lt/V = 0.048832;\ Q_{H_2O} = 0.000932$$

For 30 days (1 month),

$$Lt/V = 0.12208;\ Q_{H_2O} = 0.002298$$

For 60 days (2 months),

$$Lt/V = 0.24416;\ Q_{H_2O} = 0.00433$$

For 120 days (4 months),

$$Lt/V = 0.48832;\ Q_{H_2O} = 0.00773$$

For 240 days (8 months),

$$Lt/V = 0.97664;\ Q_{H_2O} = 0.01247$$

For 300 days (10 months),

$$Lt/V = 1.2208;\ Q_{H_2O} = 0.01410$$

For 480 days (16 months),

$$Lt/V = 1.95328;\ Q_{H_2O} = 0.01716$$

For 600 days (20 months),

$$Lt/V = 2.4416;\ Q_{H_2O} = 0.01826$$

For 960 days (32 months),

$$Lt/V = 3.90656;\ Q_{H_2O} = 0.01956$$

For 1920 days (64 months),

$$Lt/V = 7.81312;\ Q_{H_2O} = 0.01999$$

This data is plotted in Fig. 5-6 and shows the shape of the water ingress. This curve will move to the left as the internal volume of the package gets smaller, and to the right as it gets larger. The curve will also move to the left as the leak rate gets larger, and to the right as the leak rate gets smaller.

Figure 5-7 shows the effect of package volume on the time to reach a given water content. The true water leak rate is constant and equals 1×10^{-6} atm-cc/sec. The partial pressure of water external to the package is 0.02 atm. Curves for four different internal volumes are presented. The figure shows that a package with a volume of 0.1 cc reaches a water content a little greater than 0.018 atm in 0.1 months. For a package with a volume of 100 cc, it takes 100 months.

Figure 5-8 shows the effect of the leak rate on a 1.0 cc package. The curves are for an external partial pressure of water equal to 0.02 atm. When the true water leak rate is 1.0×10^{-6}, the water in the package after one month is slightly greater than 0.018 atm. At the leak rate equal to 1.0×10^{-8}, it takes 100 months to reach the same amount of water. The water ingress is much less for even smaller leak rates.

Figure 9 is a similar plot as Fig. 8, but for a smaller size package equal to 0.01 cc.

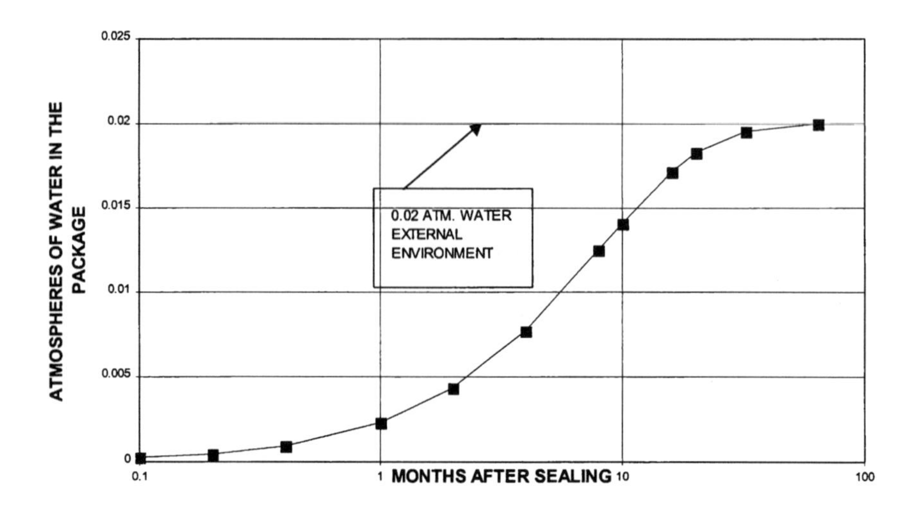

Figure 5-6. Semilog plot of water leaking into a package having a 1.0 cc internal volume. Water true leak rate = 4.71×10^{-8}.

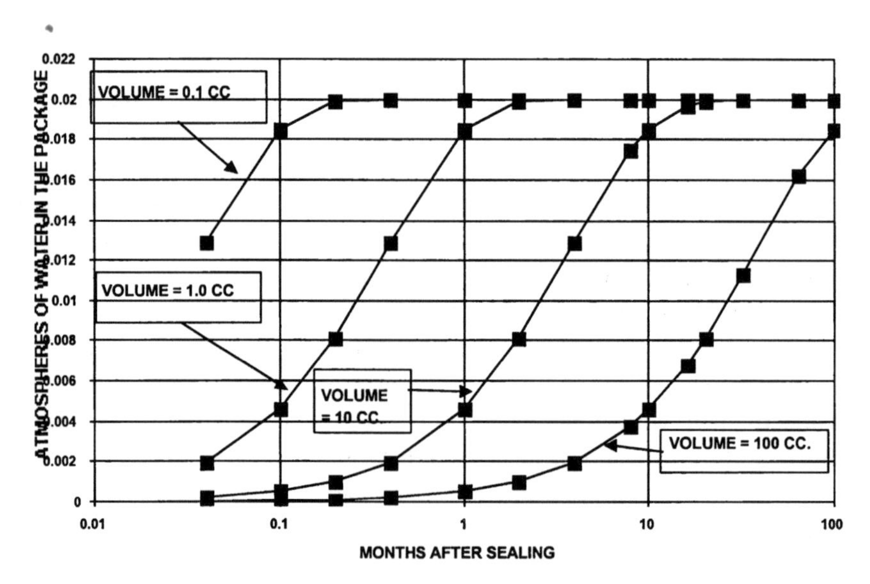

Figure 5-7. Atmospheres of water leaking into a package as a function of time and a true water leak rate of 1.0×10^{6}, for different internal volumes.

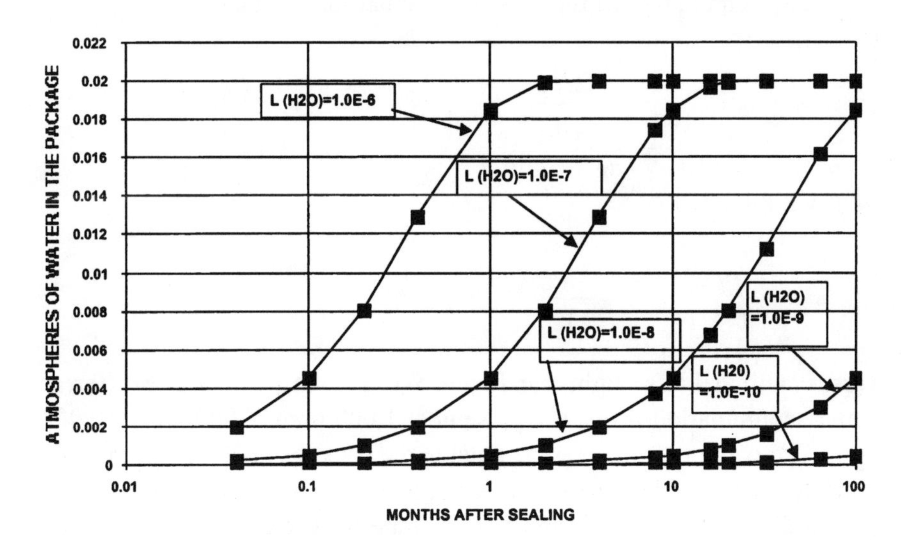

Figure 5-8. Water ingress with time for a 1.0 cc internal volume, for five different true water leak rates.

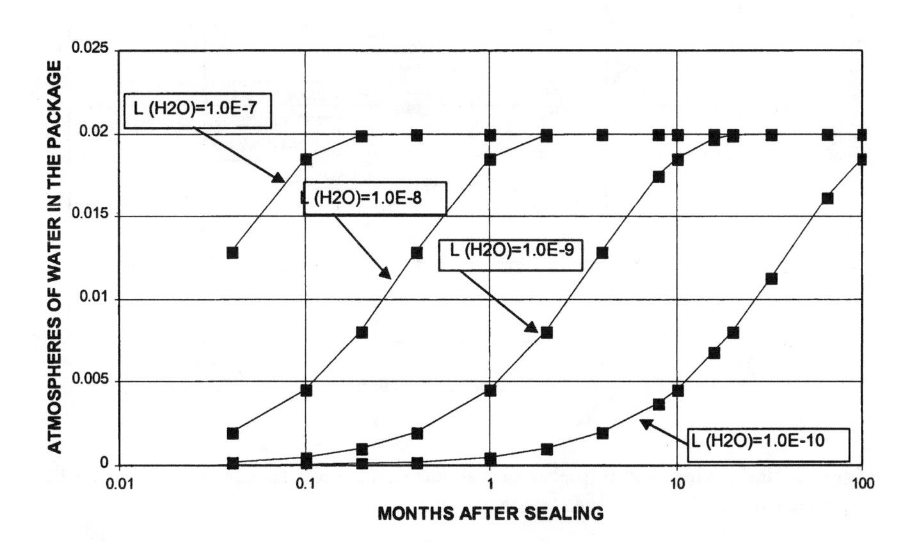

Figure 5-9. Water ingress with time for a 0.01 cc internal volume, for four different true water leak rates.

An examination of Eq. (5-9) shows that for a fixed

$$\frac{Q_{H_2O}}{\Delta pi_{H_2O}}$$

ratio, a log-log plot of

$$\frac{L_{H_2O}}{V}$$

versus t will yield a straight line. Figure 5-10 is such a plot. Three different water accumulation levels are presented: 1,000 ppm, 5,000 ppm, 10,000 ppm. The water partial pressure outside the package is 20,000 ppm (0.02 atm). It is assumed that there is no water inside the package from any other source. Examples 5-4, 5-5, and 5-6 illustrate the use of Fig. 5-10, and are indicated in Fig. 5-10.

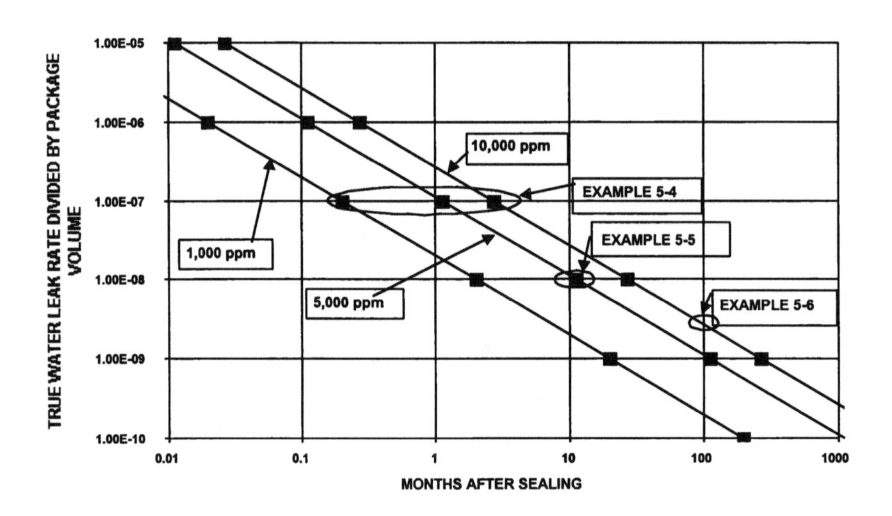

Figure 5-10. Log-log plot of true water leak rate divided by package volume, versus the time to reach specific water levels in the package.

Example 4. A package having a volume of 0.1 cc has a true water leak rate of 1×10^{-8} atm-cc/sec ($L_{H_2O}/V = 1 \times 10^{-8}/0.1 = 1 \times 10^{-7}$). The three points in Fig. 5-10 corresponding to this value are shown as "EXAMPLE 4."

Example 5. How long will it take for a 0.05 cc package with a true water leak rate of 5×10^{-10} atm-cc/sec to accumulate 5,000 ppm of water?

Solution. $L_{H_2O}/V = 5 \times 10^{-10}/0.05 = 1 \times 10^{-8}$, and from the curve, the time is 10 months.

Example 6. How long will it take for a 10 cc package to accumulate 10,000 ppm of water, if the true water leak rate is 1×10^{-8}?

Solution. $L_{H_2O}/V = 1 \times 10^{-8}/10 = 1 \times 10^{-9}$. The time from Fig. 10 equals 100 months.

3.0 WATER OUTGASSING INSIDE THE PACKAGE

The preceding equations and figures in this chapter assumed no water being generated inside the package. This is almost never true. There is some water in the sealing chamber, usually from 20 to 100 ppm. A larger amount of water comes from material outgassing. The extent of the outgassing depends on the kind of material, their preparation and their pre-seal treatment.

A vacuum bake chamber directly attached to the sealing chamber is a must for the outgassing to be kept to a reasonable value. The two chambers should be connected so that no atmosphere, other than that of the sealing chamber, can contact the unsealed parts once they enter the vacuum bake chamber. This is the assumption for the remainder of this book.

Czanderna et al. studied the weight gain and weight loss of various epoxies, for different processing and curing conditions.[23] Roberts studied the outgassing characteristics of organic materials in sealed packages relative to their pre-seal history, especially with regard to the relative humidity.[24] Fancher and Horner,[25] and Koudounoris and Wargo,[26] studied moisture levels in hybrids by the use of moisture sensor chips which were built into the hybrids. These investigations included water outgassing from within as well as water ingress. The use of such sensors allowed the monitoring of moisture levels during the screening and operational use of hybrids and MCMs.

Outgassing of water vapor is of two kinds: water adsorbed on the surface, and water that is throughout the material. The former is removed fairly easily in a short time. The water that is in the material takes a long time (days) to remove.

Water from properly cured epoxies can be removed by a 24 hr vacuum bake at 150°C. If the epoxy is insufficiently cured, a longer bake or higher temperature is required.

Encapsulated parts inside the package most always contain water throughout the material. Even a week at 125°C in air followed by a 24 hr vacuum bake at 150°C will not remove enough water to keep a 1.0 cc package below 2000 ppm. A 72 hr bake at 150°C is necessary to get below 2000 ppm.

Water from the plating inside the package can be a very serious problem for small packages. Water is trapped in the plating during the plating process. The amount of water being outgassed is directly proportional to the inside surface area of the plating (usually the same as the inside surface area of the package). Small packages have a much higher surface to volume ratio than larger packages. This has been shown in Examples 5-1 and 5-2. A package with a volume of 0.008 cc has a surface area to volume ratio of 0.24:0.008 = 30:1. A 40 cc package has a surface area to volume ratio of 216:40 = 5.4. For equivalent plating, the smaller package will have 30/5.4 = 5.56 times the percentage of water in the package than the larger package for the same outgassing process.

The amount of water trapped in gold plating varies greatly with the plating operation. A practical method of qualifying and screening plating lots consists of sealing a sample of empty packages and then baking them for 168 hours at 125°C. The samples are then subjected to RGA. Tables 5-4 and 5-5 show the RGA for several lots of headers after such a screening treatment.

The empty packages in Table 5-4 were vacuum baked for 24 hr at 150°C, and then sealed in an atmosphere of 90% nitrogen, 10% helium, and less than 50 ppm of water. The sealed units were then baked for 168 hours at 125°C. At the end of the bake, they were leak tested and sent to a laboratory for RGA. All packages had true helium leak rates of less than 2 $\times 10^{-8}$ atm-cc/sec. The package headers were plated with 50–100 microinches of nickel, followed with 100–120 microinches of gold.

Table 5-4. PPM of Water in Empty Sealed Packages after a 168 hour Bake at 125°C

Header Manufacturer	Package Volume cc	Plater*	Sample #1	Sample #2	Sample #3	Sample #4	Sample #5
A	0.0408	X	1347	1384	1429	1645	1796
A	0.0408	X	1197	1588	1821	1893	1045
A	0.0408	X	2504	2422	3664	3858	–
A	0.0408	A	846	671	344	491	567
B	0.0408	B	719	787	972	821	566
B	0.0408	B	865	764	606	100	382
B	0.0408	B	398	578	636	616	520
B	0.0408	B	631	943	780	831	420
B	0.0408	B	287	445	323	409	443
B	0.0408	B	235	403	230	681	753
C	0.0408	?	6933	1713	3830	3345	4168
C	0.0408	?	625	870	1729	898	2147
C	0.0408	?	1067	826	2083	2589	1304
A	0.0860	X	2504	2422	3664	3858	–
A	0.0860	X	974	1352	1336	1472	1403
A	0.0860	A	846	671	344	491	567
B	0.0860	B	409	261	730	725	372
B	0.0860	B	224	561	293	423	464
B	0.0860	B	316	482	419	386	489
C	0.0860	?	1800	2100	4000	1900	–
C	0.0860	?	615	1465	1154	1161	3316
C	0.0860	?	1128	1033	2263	1256	1188

*Plater "A" is manufacturer "A"
Plater "X" is a plating subcontractor to "A"
Plater "B" is manufacturer "B"
Plater "?" is an unknown plater, but not "C"

The packages with an internal volume of 0.0408 cc, had a surface area to volume ratio of 18.57, and had a internal height of 0.22 cm. The packages with an internal volume of 0.0860 cc, had a surface area to volume ratio of 16.93, and had an internal height of 0.22 cm.

When manufacturers "A" and "B" plate the headers themselves, the plating has a low water content. Plater "X" and the unknown platers cannot control the water in the plating.

These packages were used in the manufacture of hybrids that had maximum water limits of 2000 ppm after 168 hours of burn-in at 125°C, and 5000 ppm after a subsequent 1000 hours operation at 125°C. These requirements were met by setting the water limit of the five samples from each plating lot at 1000 ppm. Table 5-5 shows some examples of these hybrids after the 1000 hour life test.

Table 5-5. RGA of Microwave Hybrids after 1168 Hours of Burn-in at 125°C

	Units	Volume = 0.036 cc*	Volume = 0.070 cc*
Nitrogen	%	84.0	82.7
Oxygen	PPM	ND	ND
Argon	PPM	ND	ND
CO_2	%	1.28	3.44
Moisture	PPM	1384	953
Hydrogen	PPM	1.68	6723
Helium	%	12.6	11.8
Fluorocarbons	PPM	ND	ND
Ammonia	%	3159	1.26

*Volume of the empty package less the volume of the parts inside

Figures 5-4–5-10 assumed no water initially within the package and no water being generated within the package. When initial water is present, the total water is:

$$Q_{H_2OTOT} = Q_{H_2OINI} + Q_{H_2O}$$

where: Q_{H_2OTOT} = the total water in the package

Q_{H_2OINI} = the initial water in the package

Q_{H_2O} = the water that has leaked into the package

Incorporating Eq. (5-8):

Eq. (5-11) $$Q_{H_2OTOT} = Q_{H_2OINI} + \left[\Delta pi \left(1 - e^{-\frac{L_{H_2O}t}{V}} \right) \right]$$

where: V = the available internal volume of the package (volume of the parts inside the package should be subtracted) in cc

t = the time in seconds

L_{H_2O} = the true water leak rate = 0.471 L_{HE} (see Table 3-1) in atm-cc/sec

Dpi = the initial difference in the water partial pressure on the outside less the partial pressure on the inside of the package

The amount of water initially in the package is very small and approximately equal to the water in the sealing chamber. This value is usually between 20–100 ppm. Most of the outgassing occurs during burn-in or high temperature operation. The outgassing is logarithmic with time for a given temperature. This is for two reasons: (1) the water available for outgassing is decreasing, and (2) as the outgassing progresses, the water vapor pressure within the package increases, thereby slowing the rate of outgassing.

Typical curves of water in a package with time, when the initial and outgassing water are included, are shown in Fig. 5-11. The smallest value of atmospheres for these curves is 0.001, corresponding to the 1000 ppm of water due to outgassing and including the 100 ppm from the sealing operation.

Similar curves to that shown in Fig. 5-10 can be plotted, while incorporating a fixed amount of water from outgassing and sealing. Figure 5-12 is a plot of such curves assuming the package contains 1000 ppm total from outgassing and from the sealer, after 168 hr of burn-in.

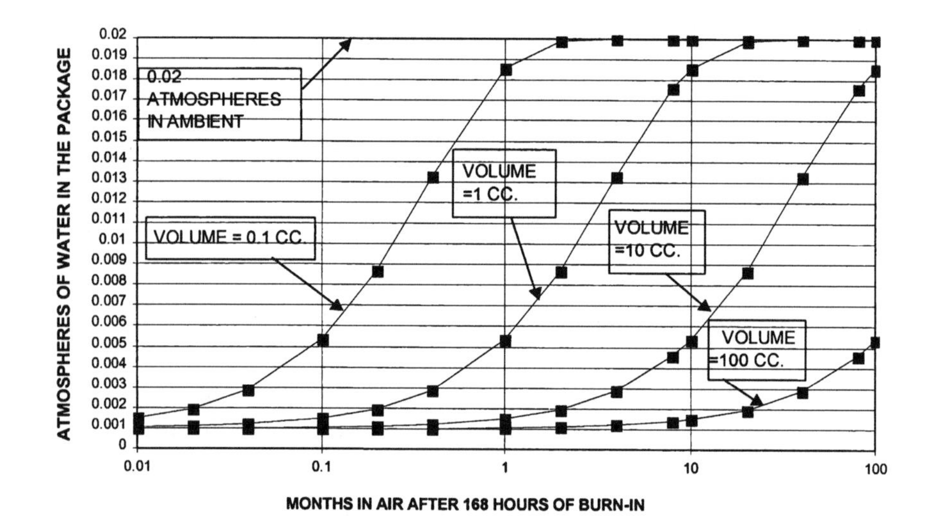

Figure 5-11. Atmospheres of water in different size packages as a function of time. True water leak rate = 1×10^{-7}; initial water = 100 PPM; 900 PPM outgassed after 168 hr at 125°C.

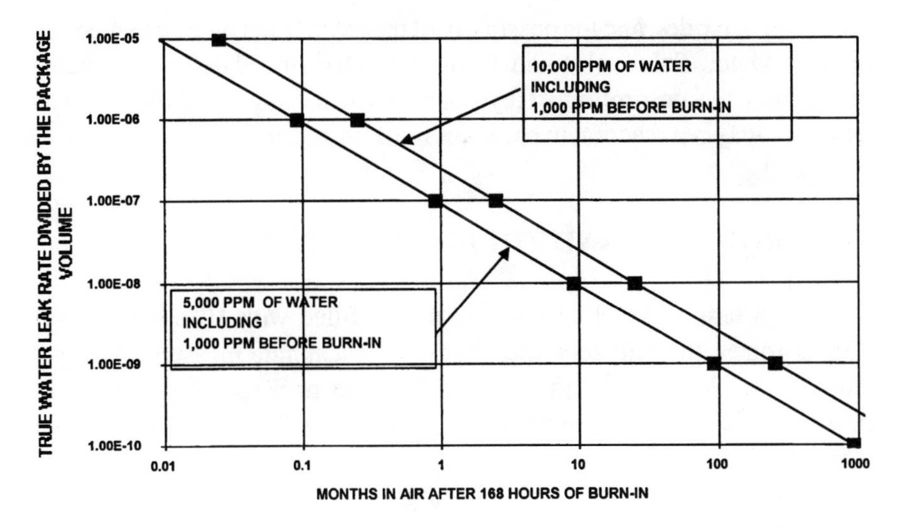

Figure 5-12. Log-log plot of true water leak rate divided by package volume, versus time to reach specific water levels; starting after 168 hr of burn-in.

4.0 WATER AS A RESULT OF A CHEMICAL REACTION WITHIN THE PACKAGE

The direct reaction between gaseous hydrogen and oxygen is theoretically possible as indicated in the following chemical reaction.

$$2\ H_2 + O_2 = 2\ H_2O$$

The free energy of this equation is highly negative (–54 K calories), so that the reaction is possible.[27] The rate of this reaction is extremely slow at ordinary temperatures and pressures, so that "no detectable amount of water would be formed in years."[28] However, a suitable catalyst could cause the reaction to take place.

Schuessler and Feliciano-Welpe published a paper on the effects of hydrogen on the reliability of electronic devices.[29] Their investigation reported on the outgassing of hydrogen from metallic packages, the reduction of oxides, and the possibility of the oxygen and hydrogen reacting to form water. Schuessler and Gonya reported on related experiments concerning the desorption of hydrogen from packaging alloys.[30] The hydrogen-oxygen reaction in packages remains questionable.

5.0 PROBLEMS AND THEIR SOLUTIONS

Problem 1. A 0.1 cc package is backfilled with 10% helium and 90% nitrogen. Immediately after sealing, the leak rate measures 5×10^{-9} atm-cc/sec. How much water leaks into the package after 1000 hr? Assume 65% relative humidity at 25°C (0.0203 atm), and no water from other sources.

Solution. The true helium leak rate,

$$L_{HE} = R/\text{atm of helium} = 5 \times 10^{-9}/0.1 = 5 \times 10^{-8} \text{ atm-cc/sec}$$

The true water leak rate = $0.471 \times L_{HE}$ (see Table 3-1) = $0.471 \times 5 \times 10^{-8}$ = 2.357×10^{-8} atm-cc/sec.

$$1000 \text{ hr} = 1000 \times 3600 = 3{,}600{,}000 \text{ sec}$$

Using Eq. (5-8):

$$Q_{H_2O} = \Delta pi_{H_2O}\left(1 - e^{-\frac{L_{H_2O}t}{V}}\right)$$

where: Δpi_{H_2O} = 0.0203 atm

L_{H_2O} = 2.357×10^{-8} atm-cc/sec

t = 3,600.000 sec

V = 0.1 cc

$$Q_{H_2O} = 0.0203\left(1 - e^{-\frac{2.357\times10^{-8}\times3{,}600{,}000}{0.1}}\right)$$

$$Q_{H_2O} = 0.0203\ (1 - e^{-0.8485}) = 0.0203\ (1 - 0.4280)$$

$$Q_{H_2O} = 0.0203\ (0.57196)$$

$$Q_{H_2O} = 0.0116 \text{ atm of water} = 1.16\% = 11{,}600 \text{ ppm}$$

Problem 2. A package with an internal volume of 2.5 cc is sealed with 10% helium and 90% nitrogen. Immediately after sealing the leak rate measures 1×10^{-8} atm-cc/sec. How long will it take for 5000 ppm of water to enter the package if it is in an ambient of 65% relative humidity at 20°C? Assume no water from any other source.

Solution.

$$L_{HE} = R/0.1 = 1 \times 10^{-8}/0.1 = 1 \times 10^{-7} \text{ atm-cc/sec}$$

From Table 3-1,

$$L_{H_2O} = 0.471\ L_{HE} = 0.471 \times 1 \times 10^{-7} = 4.71 \times 10^{-8} \text{ atm-cc/sec}$$

From Table 5-3,

65% R.H. at 20°C = 0.0150 atm of water

Using Eq. (5-9):

$$t = -\frac{V}{L_{H_2O}}\left[\ln\left(1 - \frac{Q_{H_2O}}{\Delta pi_{H_2O}}\right)\right]$$

where: $V = 2.5$ cc

$L_{H_2O} = 4.71 \times 10^{-8}$ atm-cc/sec

$Q_{H_2O} = 5{,}000$ ppm of water

$\Delta pi_{H_2O} = 0.015$ atm $= 15{,}000$ ppm

$$t = -\frac{2.5}{4.71 \times 10^{-8}}\left[\ln\left(1 - \frac{5{,}000}{15{,}000}\right)\right]$$

$$t = -5.308 \times 10^7(\text{Ln } 0.6667) = 5.308 \times 10^7(-0.4055)$$

$$t = 2.152 \times 10^7 \text{ sec} = 5{,}978.7 \text{ hr} = 249.1 \text{ days}$$

Problem 3. A 0.04 cc package is sealed with 20% helium and 80% nitrogen. The measured leak rate is 1×10^{-8} atm-cc/sec. The package is placed in a sealed desiccator which contains water. The desiccator is put in an oven which is at 60°C for 10 days. At the end of that time, the package is sent for a RGA. What is the expected water in the package assuming all the water has leaked in?

Solution.

$$L_{HE} = 1 \times 10^{-8}/0.2 = 5 \times 10^{-8} \text{ atm-cc/sec}$$

$$10 \text{ days} = 10 \times 24 \times 3600 = 864{,}000 \text{ sec}$$

$$L_{H_2O} = 0.471 \times L_{HE} = 0.471 \times 5 \times 10^{-8} = 2.35 \times 10^{-8} \text{ atm-cc/sec}$$

The desiccator contains saturated water vapor at 60°C. From Fig. 5-3, the atmospheres of water vapor in the desiccator is 0.20. Using Eq. (5-8):

$$Q_{H_2O} = \Delta p i_{H_2O}\left(1 - e^{-\frac{L_{H_2O}t}{V}}\right)$$

$$Q_{H_2O} = 0.2\left[1 - e^{-\frac{(2.35\times10^{-8})(864{,}000)}{0.04}}\right]$$

$$Q_{H_2O} = 0.2(1 - e^{-0.5076}) = 0.2(1 - 0.60194) = 0.2(0.3981)$$

$$Q_{H_2O} = 0.0796 \text{ atm} = 7.96\% \text{ water}$$

Problem 4. An Army program required a new MCM package design. The package has an internal dimension of 5 cm × 5 cm × 0.5 cm. The true water leak rate limit is 1×10^{-7} atm-cc/sec. The Army tests several packages for six months at its Panama Proving Grounds where the temperature is 85°F and the relative humidity is 85%. The MCM manufacturer tests samples at 85°C and at a relative humidity of 85% for 1000 hours. If the packages are at their leak rate limit, how much water will be in the packages at the end of the tests?

Solution.

Volume of the package = 12.5 cc

Panama Proving Grounds

$$85°\mathrm{F} = 29.4°\mathrm{C}$$

$$6 \text{ mo} = 6 \times 30 \times 24 \times 3600 = 1.555 \times 10^{7} \text{ sec}$$

Using Eq. (5-8):

$$Q_{\mathrm{H_2O}V} = \Delta pi_{\mathrm{H_2O}}\left(1 - e^{-\frac{L_{\mathrm{H_2O}}t}{V}}\right)$$

where: $\Delta pi_{\mathrm{H_2O}}$ = 0.0345 atm (linear extrapolation from Table 5-3)

$L_{\mathrm{H_2O}} = 1 \times 10^{-7}$ atm-cc/sec

$t = 1.555 \times 10^{7}$ sec

$V = 12.5$ cc

$$Q_{\mathrm{H_2O}} = 0.0345\left[1 - e^{-\frac{(1\times10^{-7})(1.555\times10^{7})}{12.5}}\right]$$

$$Q_{\mathrm{H_2O}} = 0.0345(1 - e^{-0.1244}) = 0.0345(1 - 0.88303)$$

$$Q_{\mathrm{H_2O}} = 0.0345(0.11697) = 0.004036 \text{ atm} = 4{,}036 \text{ ppm}$$

MCM Manufacturer:

$$t = 1000 \text{ hours} = 1000 \times 3600 = 3{,}600{,}000 \text{ sec}$$

$$\Delta pi_{H_2O} = 0.4849 \text{ (see Table 5-3)}$$

$$Q_{H_2O} = 0.4849\left[1 - e^{-\frac{(1\times10^{-7})(3,600,000)}{12.5}}\right]$$

$$Q_{H_2O} = 0.4849(1 - e^{-0.0288}) = 0.4849(1\text{-}0.9716)$$

$$Q_{H_2O} = 0.4849(0.0284) = 0.01377 \text{ atm} = 13,770 \text{ ppm}$$

Problem 5. A MCM having an internal volume of 2.5 cc is sealed with 10% helium and 90% nitrogen. The measured leak rate immediately after sealing is 5×10^{-9} atm-cc/sec. The MCM is placed on the roof of a building for one year, where the average water vapor pressure is 0.02 atm. Assuming no water outgassing within the package, What is the composition of the gas in the package after the year?

Solution.

$$L_{HE} = 5 \times 10^{-9}/0.1 = 5 \times 10^{-8} \text{ atm-cc/sec}$$

$$L_{O_2} = 0.354\, L_{HE} = 0.354 \times 5 \times 10^{-8} = 1.77 \times 10^{-8} \text{ atm-cc/sec}$$

$$L_{N_2} = 0.378\, L_{HE} = 0.378 \times 5 \times 10^{-8} = 1.89 \times 10^{-8} \text{ atm-cc/sec}$$

$$L_{H_2O} = 0.471 \times L_{HE} = 0.471 \times 5\ 10^{-8} = 2.36 \times 10^{-8} \text{ atm-cc/sec}$$

$$L_{AR} = 0.316 \times \mathrm{L}_{HE} = 0.316 \times 5 \times 10^{-8} = 1.58 \times 10^{-8} \text{ atm-cc/sec}$$

$$1 \text{ year} = 365 \times 24 \times 3600 = 3.154 \times 10^{7} \text{ sec}$$

Water, oxygen, and argon are leaking in. Nitrogen and helium are leaking out. Water leaking in:

$$Q_{H_2O} = 0.02\left[1 - e^{-\frac{(2.36\times10^{-8})(3.154\times10^{7})}{2.5}}\right]$$

$$Q_{H_2O} = 0.02(1 - e^{-0.29774}) = 0.02(1 - 0.7425)$$

$$Q_{H_2O} = 0.02(0.2575) = 0.00515 \text{ atm}$$

Oxygen leaking in:

$$Q_{O_2} = 0.2095\left[1 - e^{-\frac{(1.77\times10^{-8})(3.154\times10^{7})}{2.5}}\right]$$

$$Q_{O_2} = 0.2095(1 - e^{-0.2233}) = 0.2095(1 - 0.79987) = 0.2095(0.20013)$$

$$Q_{O_2} = 0.04193 \text{ atm}$$

Argon leaking in:

$$Q_{AR} = 0.00934\left[1 - e^{-\frac{(1.58\times10^{-8})(3.154\times10^{7})}{2.5}}\right]$$

$$Q_{AR} = 0.00934(1 - e^{-0.19933})$$
$$= 0.00934(1 - 0.819277) = 0.00934(0.1807228)$$

$$Q_{AR} = 0.00169 \text{ atm}$$

Nitrogen remaining:

$$\Delta p_t = \Delta p_i\left(e^{-\frac{L_{N_2}t}{V}}\right)$$

where: Δp_i = (N_2 partial pressure in the package) – (N_2 partial pressure in the air)

= $0.9 - 0.78084 = 0.11916$ atm

$$\Delta p_{tN2} = 0.11916\left(e^{-\frac{(1.89X10^{-8})(3.154X10^{7})}{2.5}}\right)$$

$$\Delta p_{tN_2} = 0.11916(e^{-0.23844}) = 0.11916(0.78785)$$
$$\Delta p_{tN_2} = 0.09388 \text{ atm}$$

To this must be added the nitrogen partial pressure in the air:

Total nitrogen = 0.09388 + 0.78084 = 0.87472 atm

Helium remaining:

$$\Delta p_{tHE} = 0.1\left[e^{-\frac{(5\times10^{-8})(3.154\times10^{7})}{2.5}} \right]$$

$$\Delta p_{tHE} = 0.1(e^{-6308}) = 0.1(0.532166)$$

$$\Delta p_{tAHE} = 0.05322 \text{ atm}$$

The composition at the end of the year is:

Gas	Atmospheres	Percentage
water	0.00515	0.527
oxygen	0.04193	4.293
argon	0.00169	0.173
nitrogen	0.87472	89.558
helium	0.05322	5.449
	Total = 0.97671	100.00

Problem 6. A MCM package has internal dimension of 2 cm × 3 cm × 0.15 cm. If the true water leak rate is 1×10^{-8} atm-cc/sec, how long will it take to accumulate 3 monolayers of liquid water? Assume a $\Delta p = 0.02$ atm.

Solution. The internal surface area is equal to the top + bottom + 4 sides = SA

$$\text{top} + \text{bottom} = 2(2 \text{ cm} \times 3 \text{ cm}) = 12 \text{ cm}^2$$

$$\text{sides} = 2(0.15 \text{ cm} \times 2 \text{ cm}) + 2\,(0.15 \text{ cm} \times 3 \text{ cm})$$
$$= 0.6 + 0.9 = = 1.5 \text{ cm}^2$$

$$SA = 13.5 \text{ cm}^2$$

3 monolayers is 1.2×10^{-7} cm thick

The volume of liquid water = thickness of the 3 monolayers × surface area. Therefore the volume of the liquid water is 1.2×10^{-7} cm × 13.5 cm^2 = 1.62×10^{-6} cc. This is also the weight of three monolayers as the density specific gravity of water is one.

$$\text{volume of the package} = 2 \text{ cm} \times 3 \text{ cm} \times 0.15 \text{ cm} = 0.9 \text{ cc}$$

One molecular weight of water vapor weighs 18 grams and occupies 22.4 liters. The volume of water vapor in the package equals the weight of water in the package × volume of 1 mole of vapor divided by the weight of 1 mole.

$$\begin{aligned}\text{volume of } H_2O \text{ vapor} &= (1.62 \times 10^{-6}\text{ g})(22.4\text{ l})/18\text{ g}\\ &= 2.016 \times 10^{-6}\text{ l}\\ &= 2.016 \times 10^{-3}\text{ cc}\end{aligned}$$

PPM of water vapor in the package = volume of water vapor in the package × 10^6, divided by the volume of the package.

$$\text{ppm} = 2.016 \times 10^{-3}\text{ cc} \times 10^6/0.9\text{ cc} = 2{,}240$$

Using Eq. (5-9):

$$t = -\frac{V}{L_{H_2O}}\left[\ln\left(1 - \frac{Q_{H_2O}}{\Delta pi}\right)\right]$$

$$t = -\frac{0.9}{10^{-8}}\left[\ln\left(1 - \frac{2{,}240}{20{,}000}\right)\right]$$

$$t = -0.9 \times 10^8[\ln(1 - 0.112)] = -0.9 \times 10^8(\ln 0.888)$$

$$t = -0.9 \times 10^8(-0.11878)$$

$$t = 1.069 \times 10^7 \text{ sec} = 123.7 \text{ days}$$

Problem 7. A MCM with an internal volume of 4.0 cc is backfilled with 10% helium and 90% nitrogen. After 1,000 hr in air containing 2.0% water, it is sent out for RGA. The RGA shows 1.00% water and 1,000 ppm of oxygen. How much water has outgassed, assuming no chemical reactions?

Solution. Use the 1,000 ppm of oxygen to calculate the true oxygen leak rate, and then the water leak rate. Calculate Q_{H_2O} and subtract Q_{H_2O} from 1% water.

$$L_{O_2} = -\frac{V}{t}\left[\ln\left(1 - \frac{Q_{O_2}}{\Delta p_{O_2}}\right)\right]$$

where:

L_{O_2} = true oxygen leak rate

V = 4.0 cc

t = 1,000 hr = 1,000 × 3,600 = 3,600,000 sec

Q_{O_2} = 1,000 ppm oxygen = 0.1%

Δp_{O_2} = 20.95%

$$L_{O_2} = -\frac{4.0}{3{,}600{,}000}\left[\ln\left(1 - \frac{0.1}{20.95}\right)\right] = -1.111 \times 10^{-6}[\ln(1 - 0.004773)]$$

$$L_{O_2} = -1.111 \times 10^{-6}(\text{Ln } 0.995226) = -1.111 \times 10^{-6}(-0.004784)$$

$$L_{O_2} = 5.315 \times 10^{-9} \text{ atm-cc/sec}$$

$$L_{H_2O} = 1.333\, L_{O_2} = 1.333 \times 5.315 \times 10^{-9} = 7.09 \times 10^{-9} \text{ atm-cc/sec}$$

$$Q_{H_2O} = \Delta p_{H_2O}\left(1 - e^{-\frac{L_{H_2O}t}{V}}\right) = 0.02\left[1 - e^{-\frac{(7.09\times10^{-9})(3{,}600{,}000)}{4.0}}\right]$$

$$Q_{H_2O} = 0.02(1 - e^{-0.006381}) = 0.02(1 - 0.99364)$$

$$Q_{H_2O} = 0.02(0.00636) = 0.000127 \text{ atm} = 0.0127\%$$

Water outgassed = total water – water leaked in

Water outgassed = 1.0% – 0.0127% = 0.9873%

Problem 8. A MCM with an internal volume of 2.0 cc is sealed with 100% nitrogen. It is bombed for 16 hours with 30 psi (gage) of helium. The measured leak rate is 4×10^{-9} atm-cc/sec helium. How long will it take for 1.0% water to enter the package when it is in an environment of 65% relative humidity at 25°C? Assume no viscous flow.

Solution. L_{HE} can not be calculated because the amount of helium in the package is unknown. The value of R can be calculated for different values of L_{HE}, using:

$$R = L_{HE} \Delta p_{HE} \left[1 - e^{-\frac{L_{HE} T}{V}} \right]$$

where: R = the measured helium leak rate

L_{HE} = the true helium leak rate

$R/L_{HE} = Q$ = the atmospheres of helium in the package

Δp_{HE} = the difference in helium partial pressures
= 45 psia = 3 atm (the difference in total pressure is only 30 psia = 2 atm)

T = 16 hr = 16 × 3600 = 57,600 sec

V = 2.0 cc

T/V = 28,800

For: $L_{HE} = 1 \times 10^{-7}$; $LT/V = 28{,}800 \times 10^{-7}$

$$R = 1 \times 10^{-7}(3)(1 - e^{-0.00283}) = 3 \times 10^{-7}(1 - 0.997124)$$

$$R = 3 \times 10^{-7}(0.002876) = 8.63 \times 10^{-10} \text{ atm-cc/sec}$$

For: $L_{HE} = 1 \times 10^{-6}$; $LT/V = 0.0288$

$$R = 3 \times 10^{-6}(1 - e^{-0.0288}) = 3 \times 10^{-6}(1 - 0.97161)$$

$$R = 3 \times 10^{-6}(0.02839) = 8.52 \times 10^{-8} \text{ atm-cc/sec}$$

For: $L_{HE} = 1 \times 10^{-5}$; $LT/V = 0.288$

$$R = 3 \times 10^{-5}(1 - e^{-0.288}) = 3 \times 10^{-5}(1 - 0.7498)$$

$$R = 3 \times 10^{-5}(0.2502) = 7.51 \times 10^{-6} \text{ atm-cc/sec}$$

If we plot R versus L_{HE} on log-log paper, a straight line will result. The curve indicates an approximate value of 2–4 $\times 10^{-7}$ for an R value of 4×10^{-9}.

Making additional calculations of R for values of L_{HE} = 2, 3, and 4 $\times 10^{-7}$ will give a more accurate value.

For: $L_{HE} = 2 \times 10^{-7}$; $LT/V = 2 \times 10^{-7} \times 57{,}600/2 = 0.00576$

$$R = 2 \times 10^{-7}(3)(1 - e^{-0.00576}) = 6 \times 10^{-7}(1 - 0.994257)$$

$$R = 6 \times 10^{-7}(0.005743) = 3.47 \times 10^{-9} \text{ atm-cc/sec}$$

For: $L_{HE} = 3 \times 10^{-7}$; $LT/V = 0.00863$

$$R = 3 \times 10^{-7}(3)(1 - e^{-0.00863}) = 9 \times 10^{-7}(1 - 0.991397)$$

$$R = 9 \times 10^{-7}(0.008603) = 7.74 \times 10^{-9} \text{ atm-cc/sec}$$

Using a linear interpolation:

$$2.1 \times 10^{-7} = (3.47 + 0.43) \times 10^{-9} = 3.9 \times 10^{-9} \text{ atm-cc/sec}$$

$$2.2 \times 10^{-7} = (3.47 + 0.86) \times 10^{-9} = 4.33 \times 10^{-9} \text{ atm-cc/sec}$$

The value for L_{HE} is a little larger than 2.1×10^{-7}, about 2.12×10^{-7} atm-cc/sec.

$$L_{H_2O} = 0.471\, L_{HE} = 0.471 \times 2.12 \times 10^{-7} = 1.00 \times 10^{-7} \text{ atm-cc/sec}$$

The time for 1% water ingress can now be calculated.

$$t = -\frac{V}{L_{H_2O}}\left[\ln\left(1 - \frac{Q_{H_2O}}{\Delta p_{H_2O}}\right)\right]$$

where: Q_{H_2O} = 1%

Δp_{H_2O} = 2.03% (see Table 5-3)

$$t = -\frac{2.0}{1\times10^{-7}}\left[\ln\left(1 - \frac{1.0}{2.03}\right)\right] = -2\times10^{-7}\left[\ln(1 - 0.49261)\right]$$

$$t = -2 \times 10^{-7}(\ln 0.50739) = -2 \times 10^{-7}(-0.678475)$$

$$t = 1.35695 \times 10^{7}\ \text{sec} = 3769.3\ \text{hr} = 157\ \text{days}$$

Problem 9. Two identical hybrids, each with a volume of 0.1 cc are in an air environment having 0.02 atm of water. The hybrids were sealed with 100% nitrogen containing 100 ppm of water. One hybrid is tested for residual gases after seven days, the RGA showing 1000 ppm of water. The second hybrid is analyzed 30 days after sealing. What is the expected water in the 30 day RGA?

Solution. First, the true water leak rate of the first hybrid is calculated. Next, the amount of water that leaked into the second hybrid will be calculated, using the water leak rate which was previously calculated.

$$L_{H_2O} = -\frac{V}{t}\left[\ln\left(1 - \frac{Q_{H_2O}}{\Delta pi_{H_2O}}\right)\right]$$

where: V = 0.1 cc

t = 7 days = 7 × 24 × 3600 = 604,800 sec

Q_{H_2O} = (1000 – 100) = 900 ppm

Δpi_{H_2O} = 0.02 atm – 100 ppm

= 20,000 ppm – 100 ppm = 19,900 ppm

$$L_{H_2O} = -\frac{0.1}{604{,}800}\left[\ln\left(1 - \frac{900}{19{,}900}\right)\right]$$

$$L_{H_2O} = -1.65 \times 10^{-7}[\ln(1 - 0.04523)]$$

$$L_{H_2O} = -1.65 \times 10^{-7}(\ln 0.954774)$$

$$L_{H_2O} = -1.65 \times 10^{-7}(-0.04628) = 7.636 \times 10^{-9} \text{ atm-cc/sec}$$

Q_{H_2O} for the second hybrid:

$$Q_{H_2O} = \Delta p_{H_2O}\left[1 - e^{-\frac{(L_{H_2O}t)}{V}}\right]$$

where: $L_{H_2O} = 7.636 \times 10^{-9}$ atm-cc/sec

$\Delta p_{H_2O} = 20{,}000 - 100 = 19{,}900$ ppm

$V = 0.1$ cc

$t = 30 \text{ days} = 30 \times 24 \times 3600 = 2{,}592{,}000$ sec

$$Q_{H_2O} = 19{,}900\left[1 - e^{-\frac{(7.636\times10^{-9})(2{,}592{,}000)}{0.1}}\right]$$

$$Q_{H_2O} = 19{,}900(1 - e^{-0.197925}) = 19{,}900(1 - 0.82043)$$

$$Q_{H_2O} = 19{,}900(0.17957)$$

$$Q_{H_2O} = 3573 \text{ ppm}$$

Problem 10. A commercial MCM manufacturer is trying to enter the hi-rel market. NASA will consider this manufacturer, if a 1.0 cc package will have less than 5,000 ppm of water after 320 hours at 100°C in a relative humidity of 50%. The company modifies their sealer so that they sealed the 1.0 cc sample with 2 atm of helium and 100 ppm of water. The measured leak rate just after being sealed was 2×10^{-7} atm-cc/sec. What is the expected water after the 320 hr assuming 200 ppm is generated inside the package?

Solution.

$$L_{HE} = R/\text{atm of helium in the package}$$
$$= 2 \times 10^{-7}/2 = 1 \times 10^{-7} \text{ atm-cc/sec}$$

$$L_{H_2O} = 0.471\, L_{HE} = 0.471 \times 10^{-7} = 4.71 \times 10^{-8} \text{ atm-cc/sec}$$

The atmospheres of water at 100°C and 50% R.H. = 0.50 (see Table 5-3)

$$Q_{H_2O} = \Delta pi_{H_2O}\left(1 - e^{-\frac{L_{H_2O}t}{V}}\right)$$

where: Q_{H_2O} = the amount of water leaking in

t = 320 hours = 320 × 3600 = 1,152,000 sec

V = 1.0 cc

Δpi_{H_2O} = water partial pressure outside – water partial pressure inside

The package is sealed with 100 ppm. and 200 ppm is generated inside. Initially there is 100 ppm and at the end of the 320 hours there is 300ppm. Assuming the average ppm inside is 200,

$$\Delta pi_{H_2O} = 0.50 \text{ atm} - 200 \text{ ppm} = 500{,}000 \text{ ppm} - 200 \text{ ppm}$$
$$= 499{,}800 \text{ ppm}$$

$$Q_{H_2O} = 499{,}800\left[1 - e^{-\frac{(4.71\times10^{-8})(1{,}152{,}000)}{1.0}}\right]$$

$$Q_{H_2O} = 499{,}800(1 - e^{-0.054259})$$
$$= 499{,}800(1 - 0.94719)$$
$$= 499{,}800(0.05281)$$

$$Q_{H_2O} = 26{,}396 \text{ ppm} = 2.6396\%$$

Problem 11. The amount of water entering the package in Problem 10 is far in excess of the NASA requirement. What are the true and measured helium leak rates required to meet the NASA requirement?

Solution. Calculate the water leak rate for the requirement.

$$L_{H_2O} = -\frac{V}{t}\left[\ln\left(1 - \frac{Q_{H_2O}}{\Delta p}\right)\right]$$

where: V = 1.0 cc

t = 1,152,000 sec

Q_{H_2O} = 5,000 ppm – 100 ppm – 200 ppm = 4,700 ppm

Δp = 499,800 (again an average)

$$L_{H_2O} = -\frac{1.0}{1{,}152{,}000}\left[\ln\left(1 - \frac{4{,}700}{499{,}800}\right)\right]$$

$$L_{H_2O} = -8.68 \times 10^{-7}[\ln(1 - 0.009404)]$$

$$L_{H_2O} = -8.68 \times 10^{-7}(\ln 0.990596)$$

$$L_{H_2O} = -8.68 \times 10^{-7}(-0.00945)$$
$$= 8.20 \times 10^{-9} \text{ atm-cc/sec}$$

$$L_{HE} = 2.12 \times L_{H2O} = 2.12 \times 8.20 \times 10^{-9}$$

$$L_{HE} = 1.74 \times 10^{-8} \text{ atm-cc/sec}$$

$$R = L_{HE} \times \text{atm He in the package}$$

$$R = 1.74 \times 10^{-8} \times 2 = 3.48 \times 10^{-8} \text{ atm-cc/sec}$$

Problem 12. An MCM has an internal volume of 2.5 cc, and was sealed with 10% helium, 0.01% water and the balance nitrogen, to a total pressure of one atm. The measured leak rate of the MCM immediately after sealing is 5×10^{-9} atm-cc/sec. The MCM is kept in ambient air having a water vapor pressure of 0.02 atm for 60 days. At the end of the 60 days, the MCM is placed in an electronic enclosure. The electronic enclosure having an internal volume of 4,000 cc is pressurized to a total absolute pressure of three atm and sealed. The composition of gas in the enclosure is 10% helium, 0.1% water, and the balance nitrogen. The measured leak rate of the enclosure immediately after sealing is 1×10^{-5} atm-cc/sec. The Army sends the enclosure to its Panama Proving Grounds where the temperature is 85°F and the relative humidity is 85%. After one year, the MCM is removed and sent for a RGA. What is the composition of the gases in the MCM at that time? Assume no viscous flow.

Solution. For the enclosure:

$$L_{HE} = R/\text{atm of Helium in the enclosure} = 1 \times 10^{-5}/0.3$$

$$L_{HE} = 3.333 \times 10^{-5} \text{ atm-cc/sec}$$

$$L_{H_2O} = 0.471 \times L_{HE} = 0.471 \times 3.333 \times 10^{-5}$$

$$L_{H_2O} = 1.570 \times 10^{-5} \text{ atm-cc/sec}$$

$$L_{O_2} = 0.354 \times L_{HE} = 0.354 \times 3.333 \times 10^{-5}$$

$$L_{O_2} = 1.179 \times 10^{-5} \text{ atm-cc/sec}$$

$$L_{N_2} = 0.378 \times L_{HE} = 0.378 \times 3.333 \times 10^{-5}$$

$$L_{N_2} = 1.259 \times 10^{-5} \text{ atm-cc/sec}$$

$$L_{AR} = 0.316 \times L_{HE} = 0.316 \times 3.333 \times 10^{-5}$$

$$L_{AR} = 1.05 \times 10^{-5} \text{ atm-cc/sec}$$

$$1 \text{ yr} = 365 \times 24 \times 3600 = 3.154 \times 10^{7} \text{ sec}$$

For the MCM:

L_{HE} = R/atm helium in the MCM = $5 \times 10^{-9}/0.1$

$L_{HE} = 5 \times 10^{-8}$ atm-cc/sec

$L_{H_2O} = 0.471 \times L_{HE} = 0.471 \times 5 \times 10^{-8}$

$L_{H_2O} = 2.355 \times 10^{-8}$ atm-cc/sec

$L_{O_2} = 0.354\, L_{HE} = 0.354 \times 5 \times 10^{-8}$

$L_{O_2} = 1.77 \times 10^{-8}$ atm-cc/sec

$L_{N_2} = 0.378\, L_{HE} = 0.378 \times 5 \times 10^{-8}$

$L_{N_2} = 1.89 \times 10^{-8}$ atm-cc/sec

$L_{AR} = 0.316\, L_{HE} = 0.316 \times 5 \times 10^{-8}$

$L_{AR} = 1.58 \times 10^{-8}$ atm-cc/sec

60 days = $60 \times 24 \times 3600$ = 5,184,000 sec

First, the composition of the gases in the MCM after the 60 days is calculated. Second, the composition of the enclosure after one year is calculated. Third, the average value of the composition of the enclosure over the year is used as the environment surrounding the MCM.

Nitrogen leaking out of the MCM in 60 days:

$$\Delta p_{N_2-60} = \Delta pi_{N_2}\left(e^{-\frac{L_{N_2}t}{V}}\right)$$

where: Δpi_{N_2} = initial N_2 partial pressure difference

$\Delta pi_{N_2} = (0.89990 - 0.78084) = 0.11906$ atm

$L_{N_2} = 1.89 \times 10^{-8}$ atm-cc/sec

t = 5,184,000 sec

V = 2.5 cc

$$\Delta p_{N_2-60} = 0.11906\left[e^{-\frac{(1.89\times10^{-8})(5{,}184{,}000)}{2.5}}\right]$$

$$\Delta p_{N_2\text{-}60} = 0.11906(e^{-0.039191}) = 0.11906(0.961567)$$

$$\Delta p_{N_2\text{-}60} = 0.11448 \text{ atm at the end of the 60 days}$$

Total nitrogen in the MCM at the end of 60 days is the $\Delta p_{N_2\text{-}60}$ plus the nitrogen outside the MCM = 0.11448 atm + 0.78084 atm = 0.89532 atm.

Helium leaking out of the MCM the first 60 days:

$$p_{HE-60} = p_i\left(e^{-\frac{L_{HE}t}{V}}\right)$$

where: $p_{HE\text{-}60}$ = helium in the MCM after 60 days

p_I = initial helium in the MCM = 0.1 atm

L_{HE} = the true helium leak rate

L_{HE} = 5 × 10^{-8} atm-cc/sec

t = 60 days = 5,184,000 sec

V = 2.5 cc

$$p_{HE-60} = 0.1\left[e^{-\frac{(5\times10^{-8})(5{,}184{,}000)}{2.5}}\right]$$

$$p_{HE\text{-}60} = 0.1(e^{-0.10368}) = 0.1(0.9015)$$

$$p_{HE\text{-}60} = 0.09015 \text{ atm helium in the MCM after 60 days}$$

Oxygen leaking into the MCM the first 60 days:

$$Q_{O_2-60} = \Delta p_{O_2}\left(1 - e^{-\frac{L_{O_2}t}{V}}\right)$$

where: $Q_{O_2\text{-}60}$ = the oxygen that leaked into the MCM in the first 60 days
Δp_{O_2} = the difference in oxygen partial pressure
Δp_{O_2} = 0.2095 atm
L_{O_2} = 1.77×10^{-8} atm-cc/sec
t = 5,184,000 sec

$$Q_{O_2-60} = 0.2095\left[1 - e^{-\frac{(1.77\times10^{-8})(5{,}184{,}000)}{2.5}}\right]$$

$$Q_{O_2\text{-}60} = 0.2095(1 - e^{-0.0367027}) = 0.2095\ (1 - 0.963963)$$

$$Q_{O_2\text{-}60} = 0.2095(0.036037)$$

$Q_{O_2\text{-}60}$ = 0.007550 atm of oxygen after the first 60 days

Argon leaking into the MCM the first 60 days:

$$Q_{AR-60} = \Delta p_{AR}\left(1 - e^{-\frac{L_{AR}t}{V}}\right)$$

where: $Q_{AR\text{-}60}$ = the argon that has leaked into the MCM the first 60 days
Δp_{AR} = the initial argon partial pressure difference
Δp_{AR} = 0.00934 atm
L_{AR} = 1.58×10^{-8} atm-cc/sec
t = 5,184,000 sec
V = 2.5 cc

$$Q_{AR-60} = 0.00934\left[1 - e^{-\frac{(1.58\times10^{-8})(5{,}184{,}000)}{2.5}}\right]$$

$$Q_{AR\text{-}60} = 0.00934(1 - e^{-0.0327629}) = 0.00934(1 - 0.967768)$$

$$Q_{AR\text{-}60} = 0.00934(0.032232)$$

$Q_{AR\text{-}60} = 0.000301$ atm of argon leaked into the MCM the first 60 days

Water leaking into the MCM the first 60 days:

$$Q_{H_2O-60} = \Delta p_{H_2O}\left(1 - e^{-\frac{L_{H_2O}t}{V}}\right)$$

where: $Q_{H_2O\text{-}60}$ = the amount of water that leaked into the MCM the first 60 days

Δp_{H_2O} = the water partial pressure difference

$\Delta p_{H_2O} = (0.02 - 0.0001) = 0.0199$ atm

$L_{H_2O} = 2.355 \times 10^{-8}$ atm-cc/sec

t = 5,184,000 sec

V = 2.5 cc

$$Q_{H_2O-60} = 0.0199\left[1 - e^{-\frac{(2.355\times10^{-8})(5{,}184{,}000)}{2.5}}\right]$$

$$Q_{H_2O\text{-}60} = 0.0199(1 - e^{-0.0488333}) = 0.0199\,(1 - 0.9523399)$$

$$Q_{H_2O\text{-}60} = 0.0199(0.04766)$$

$Q_{H_2O\text{-}60} = 0.00095$ atm of water leaked into the MCM to first 60 days

The total water in the MCM after the 60 days is what has leaked in plus what was there initially = 0.00095 + 0.0001 = 0.00105 atm.

The partial pressures of the gases in the MCM after the first 60 days are given in the table below:

Gas	Atmospheres
nitrogen	0.89532
oxygen	0.007550
helium	0.090150
argon	0.000301
water	0.00105
	Total = 0.994371

Nitrogen leaking out of the enclosure into the air:

$$\Delta p_{N_2} = \Delta pi_{N_2}\left(e^{-\frac{L_{N_2}t}{V}}\right)$$

where: Δp_{N_2} = the partial pressure difference of nitrogen at the end of the year

Δpi_{N_2} = the initial partial pressure difference of nitrogen = (2.697 – 0.78084) = 1.91616 atm

L_{N_2} = the true nitrogen leak rate = 1.26×10^{-5} atm-cc/sec

t = 3.154×10^{7} sec

V = 4,000 cc

$$\Delta p_{N_2} = 1.91616\left[e^{-\frac{(1.26\times10^{-5})(3.154\times10^{7})}{4,000}}\right]$$

$$\Delta p_{N_2} = 1.91616(e^{-0.09252}) = 1.91616(0.9055)$$

$$\Delta p_{N_2} = 1.7357 \text{ atm}$$

The total nitrogen in the enclosure at the end of the year equals the partial pressure difference plus the partial pressure of the nitrogen outside the enclosure = 1.7357 atm + 0.78084 atm = 2.51581 atm. During the year, the nitrogen partial pressure changes from 2.697 atm to 2.51581 atm The average value during the year is (2.697 + 2.51581)/2 = 2.6064 atm.

Oxygen leaking into the enclosure:

$$Q_{O_2} = \Delta p_{O_2}\left(1 - e^{-\frac{L_{O_2} t}{V}}\right)$$

where: Q_{O_2} = the oxygen that has leaked into the enclosure during the year

Δp_{O_2} = the initial partial pressure difference

Δp_{O_2} = 0.2095 atm

L_{O_2} = the true leak rate of oxygen = 1.179×10^{-5} atm-cc/sec

t = 3.154×10^{7} sec

V = 4,000 cc

$$Q_{O_2} = 0.2095\left[1 - e^{-\frac{(1.179\times10^{-5})(3.154\times10^{7})}{4,000}}\right]$$

$$Q_{O_2} = 0.2095(1 - e^{-0.092964}) = 0.2095(1 - 0.911226)$$

$$Q_{O_2} = 0.2095(0.088774)$$

Q_{O_2} = 0.01860 atm of oxygen in the enclosure at the end of the year

The average during the year = (0.01860 + 0)/2 = 0.00930 atm.

Helium leaking out of the enclosure:

$$p_{HE} = p_i\left(e^{-\frac{L_{HE} t}{V}}\right)$$

where: p_{HE} = the partial pressure of helium remaining in the enclosure at the end of the year

p_i = the initial partial pressure of helium in the enclosure = 0.3 atm

$L_{HE} = 3.333 \times 10^{-5}$ atm-cc/sec

$t = 3.154 \times 10^7$ sec

$V = 4{,}000$ cc

$$p_{HE} = 0.3\left[e^{-\frac{(3.333\times10^{-5})(3.154\times10^{7})}{4{,}000}}\right]$$

$$p_{HE} = 0.3(e^{-0.2625701}) = 0.3(0.76902)$$

$$p_{HE} = 0.2307 \text{ atm remaining at the end of the year}$$

The average during the year = (0.3 + 0.2307)/2 = 0.2654 atm.

Argon leaking into the enclosure:

$$Q_{AR} = pi_{AR}\left(1 - e^{-\frac{L_{AR}t}{V}}\right)$$

where: Q_{AR} = the argon that has leaked into the enclosure during the year

pi_{AR} = the initial partial pressure difference

pi_{AR} = 0.00934 atm

L_{AR} = the true argon leak rate = 1.05×10^{-5} atm-cc/sec

$t = 3.154 \times 10^7$ sec

$V = 4{,}000$ cc

$$Q_{AR} = 0.00934\left[1 - e^{-\frac{(1.05\times10^{-5})(3.154\times10^{7})}{4{,}000}}\right]$$

$$Q_{AR} = 0.00934\,(1 - e^{-0.082792}) = 0.00934\,(1 - 0.920542)$$

$$Q_{AR} = 0.00934(0.079458)$$

$Q_{AR} = 0.000742$ atm of argon in the enclosure after the year

The average during the year = (0.000742 + 0)/2 = 0.000371 atm.

Water leaking into the enclosure:

$$Q_{H_2O} = \Delta pi\left(1 - e^{-\frac{L_{H_2O}t}{V}}\right)$$

where: Q_{H_2O} = the atmospheres of water that has leaked in during the year

Δpi = the initial partial pressure difference of water = 0.0345 atm(see Table 5-3) – 0.003 atm

$\Delta pi = 0.0315$ atm

$L_{H_2O} = 1.57 \times 10^{-5}$ atm-cc/sec

$t = 3.154 \times 10^{7}$ sec

$V = 4{,}000$ cc

$$Q_{H_2O} = 0.0315\left[1 - e^{-\frac{(1.57\times10^{-5})(3.154\times10^{7})}{4{,}000}}\right]$$

$$Q_{H_2O} = 0.0315(1 - e^{-0.123794})$$
$$= 0.0315\,(1 - 0.88356) = 0.0315\,(0.116438)$$

$Q_{H_2O} = 0.0037$ atm leaked in during the year

The total water after the year = 0.0037 amt. + 0.003 atm = 0.0067 atm. The average during the year = (0.0067 + 0.003)/2 = 0.00485 atm.

The partial pressure of the gases in the enclosure at the end of the year, and their average values during the year are:

Gas	At End of Year	Average During Year
nitrogen	2.51581 atm	2.6064 atm
oxygen	0.01860 atm	0.00930 atm
helium	0.2307 atm	0.2654 atm
argon	0.000742 atm	0.000371 atm
water	0.0067 atm	0.00485 atm
Total	**2.772552 atm**	**2.886321 atm**

Nitrogen leaking into the MCM from the enclosure:

$$Q_{N_2} = \Delta p_{N_2}\left(1 - e^{-\frac{L_{N_2} t}{V}}\right)$$

where: Q_{N_2} = the nitrogen that has leaked into the MCM during the year

Δp_{N_2} = the partial pressure difference using the average for the enclosure less the amount in the MCM at the beginning of the year

Δp_{N_2} = 2.6064 – 0.89532 = 1.71108 atm

L_{N_2} = 1.89 × 10^{-8} atm-cc/sec

t = 3.154 × 10^{7} sec

V = 2.5 cc

$$Q_{N_2} = 1.71108\left[1 - e^{-\frac{(1.89\times10^{-8})(3.154\times10^{7})}{2.5}}\right]$$

$$Q_{N_2} = 1.71108(1 - e^{-0.23848})$$

$$= 1.71108(1 - 0.78785)$$

$$= 1.71108\,(0.212159)$$

Q_{N_2} = 0.363021 atm leaked in during the year

The total nitrogen in the MCM = 0.363021 atm + 0.89532 = 1.258341 atm.

Oxygen leaking into the MCM from the enclosure:

$$Q_{O_2} = \Delta pi_{O_2}\left(1 - e^{-\frac{L_{O_2}t}{V}}\right)$$

where:

Q_{O_2} = the oxygen that has leaked into the MCM during the year

Δpi_{O_2} = the partial pressure difference using the average for the enclosure less the amount in the MCM at the beginning of the year

Δpi_{O_2} = 0.00930 – 0.00755 = 0.00175 atm

L_{O_2} = 1.77×10^{-8} atm-cc/sec

t = 3.154×10^{7} sec

V = 2.5 cc

$$Q_{O_2} = 0.00175\left[1 - e^{-\frac{(1.77\times10^{-8})(3.154\times10^{7})}{2.5}}\right]$$

$$Q_{O_2} = 0.00175(1 - e^{-0.223303})$$

$$= 0.00175(1 - 0.79987) = 0.00175\,(0.20013)$$

Q_{O_2} = 0.0003502 atm leaked into the MCM

The total oxygen in the MCM = 0.0003502 + 0.00755 = 0.0079002 atm.

Helium leaking into the MCM from the enclosure:

$$Q_{\mathrm{HE}} = \Delta pi_{\mathrm{HE}}\left(1 - e^{-\frac{L_{\mathrm{HE}}t}{V}}\right)$$

where: Q_{HE} = the helium that has leaked into the MCM during the year

Δpi_{HE} = the partial pressure difference using the average for the enclosure less the amount in the MCM at the beginning of the year

$\Delta pi_{\mathrm{HE}} = 0.2654 - 0.09015 = 0.17525$ atm

$L_{\mathrm{HE}} = 5 \times 10^{-8}$ atm-cc/sec

$t = 3.154\text{x } 10^{7}$ sec

$V = 2.5$ cc

$$Q_{\mathrm{HE}} = 0.17525\left[1 - e^{-\frac{(5\times10^{-8})(3.154\times10^{7})}{2.5}}\right]$$

$$\begin{aligned} Q_{\mathrm{HE}} &= 0.17525(1 - e^{-0.6308}) \\ &= 0.17525(1 - 0.53217) \\ &= 0.17525(0.46783) \end{aligned}$$

$Q_{\mathrm{HE}} = 0.08198721$ atm leaked into the MCM

The total helium in the MCM = 0.08198721 + 0.09015 = 0.172137 atm.

Argon leaking into the MCM from the enclosure:

$$Q_{\mathrm{AR}} = \Delta pi_{\mathrm{AR}}\left(1 - e^{-\frac{L_{\mathrm{AR}}t}{V}}\right) \text{ the enclosure}$$

where: Q_{AR} = the argon that has leaked into the MCM during the year

Δpi_{AR} = the partial pressure difference using the average for the enclosure less the amount in the MCM at the beginning of the year

$\Delta pi_{AR} = 0.000371 - 0.000301 = 0.00007$ atm

$L_{AR} = 1.58 \times 10^{-8}$ atm-cc/sec

$t = 3.154 \times 10^{7}$ sec

$V = 2.5$ cc

$$Q_{AR} = 0.00007\left[1 - e^{-\frac{(1.58\times10^{-8})(3.154\times10^{7})}{2.5}}\right]$$

$$\begin{aligned} Q_{AR} &= 0.00007(1 - e^{-0.1993328}) \\ &= 0.00007\,(1 - 0.819277) \\ &= 0.00007\,(0.180723) \end{aligned}$$

$Q_{AR} = 0.00001265$ atm leaked into the MCM from the enclosure

The total argon in the MCM = 0.00001265 + 0.000301 = 0.00031365 atm.

Water leaking into the MCM from the enclosure:

$$Q_{H_2O} = \Delta pi_{H_2O}\left(1 - e^{-\frac{L_{H_2O}t}{V}}\right)$$

where: Q_{H_2O} = the water that has leaked into the MCM during the year

Δpi_{H_2O} = the partial pressure difference using the average for the enclosure less the amount in the MCM at the beginning of the year

$\Delta pi_{H_2O} = 0.00485 - 0.00105 = 0.00380$ atm

$t = 3.154 \times 10^{7}$ sec

$V = 2.5$ cc

$$Q_{H_2O} = 0.0038\left[1 - e^{-\frac{(2.355\times10^{-8})(3.154\times10^{7})}{2.5}}\right]$$

$$Q_{H_2O} = 0.0038\,(1 - e^{-0.297107}) = 0.0038\,(1\text{-}0.74296) = 0.0038\,(0.2570)$$

$$Q_{H_2O} = 0.0009766 \text{ atm leaked into the MCM}$$

Total water in the MCM = 0.0009766 + 0.00105 = 0.002066 atm.

The MCM at the end of the year contains the following gas composition:

Gas	Atmospheres	Percentage
nitrogen	1.258341	87.338826
oxygen	0.007900	0.548336
helium	0.172137	11.947670
argon	0.000314	0.021700
water	0.002066	0.143397
Total	**1.440758**	**99.999999**

Problem 13. The navy has started a new development program for a shipboard electronic system. The system contains several MCMs and microwave hybrids of 4 package sizes. The navy wants each package to have less than 5,000 ppm of water after one year in an environment of 80% relative humidity at 20°C. The packages are sealed with 10% helium and 90% nitrogen. Assuming no water being generated inside the packages, what is the limit for the measured helium leak rate for each package size? The internal dimensions in cm and volume of the four packages are:

Package 1: 0.7 × 0.4 × 0.2, volume = 0.56 cc

Package 2: 1.0 × 1.0 × 0.2, volume = 0.20 cc

Package 3: 3.0 × 3.0 × 0.2, volume = 1.8 cc

Package 4: 5.0 × 5.0 × 0.4, volume = 10.0 cc

Solution. First calculate the true water leak rate (L_{H2O}) for each size package. Next calculate the true helium leak rate (L_{HE}) for each size package. Then calculate the measured leak rate (R) for each size package.

Using Eq. (5-10):

$$L_{H_2O} = -\frac{V}{t}\left[\ln\left(1 - \frac{Q_{H_2O}}{\Delta p_{H_2O}}\right)\right]$$

where: L_{H_2O} = the true water leak rate

V = the internal volume of the package

t = 1 year = 3.154×10^7 sec

Δp_{H_2O} = the difference in partial pressure between the outside of the package and the inside of the package. There is no water on the inside so the difference is the 80% R.H. at 20°C = 0.0185 atm

Δp_{H_2O} = 18,500 ppm (see Table 5-3)

Q_{H_2O} = the 5,000 ppm limit

The expression $\frac{1}{t}\left[\ln\left(1 - \frac{Q_{H_2O}}{\Delta p_{H_2O}}\right)\right]$ is the same for every package size and equals

$$\frac{1}{3.154 \times 10^7}\left[\ln\left(1 - \frac{5{,}000}{18{,}500}\right)\right] = \frac{1}{3.154 \times 10^7}\left[\ln(0.72973)\right]$$

$$= \frac{-0.31508}{3.154 \times 10^7} = -9.99 \times 10^{-9}$$

L_{H_2O} now equals $-9.99 \times 10^{-9} \times -(V)$

For the package with volume = 0.0504 cc, $L_{H_2O} = -9.99 \times 10^{-9} \times -(0.0504)$ = 5.04×10^{-10} atm-cc/sec, $L_{HE} = 2.12\ L_{H2O} = 2.12 \times 5.04 \times 10^{-10} = 1.07 \times 10^{-9}$ atm-cc/sec, $R = 0.1 \times L_{HE} = 1.07 \times 10^{-10}$ atm-cc/sec and is too small to measure routinely.

For the package with volume = 0.18 cc, $L_{H2O} = -9.99 \times 10^{-9} \times -(0.18) = 1.80 \times 10^{-9}$ atm-cc/sec, $L_{HE} = 2.12 \times 1.80 \times 10^{-9} = 3.82 \times 10^{-9}$ atm-cc/sec, $R = 0.1 \times 3.82 \times 10^{-9} = 3.82 \times 10^{-10}$ atm-cc/sec and this is too small to measure routinely.

For the package with volume = 1.62 cc, $L_{H2O} = -9.99 \times 10^{-9} \times -(1.62) = 1.62 \times 10^{-8}$ atm-cc/sec, $L_{HE} = 2.12 \times 1.62 \times 10^{-8} = 3.43 \times 10^{-8}$ atm-cc/sec, $R = 0.1 \times 3.43 \times 10^{-8} = 3.43 \times 10^{-9}$ atm-cc/sec.

For the package with volume = 10.0 cc, $L_{H2O} = -9.99 \times 10^{-9} \times -(10.0) = 9.99 \times 10^{-8}$ atm-cc/sec, $L_{HE} = 2.12 \times 9.99 \times 10^{-8} = 2.12 \times 10^{-7}$ atm-cc/sec, $R = 0.1 \times 2.12 \times 10^{-7} = 2.12 \times 10^{-8}$ atm-cc/sec.

Problem 14. The navy has changed its mind about the water limit in Problem 13. Instead of the 5,000 ppm limit in each package, it has set the limit to three monolayers of liquid water. What is the measured helium leak rate limit for each package size so that no more than three monolayers of water is present?

Solution. First calculate the volume of liquid water in three monolayers. This equals the internal surface area × the thickness of three monolayers (1.2×10^{-7} cm). The volume in cc equals the weight in grams of the three monolayers. Now calculate the volume of water vapor, multiply it by 10^6 and then divide by the package volume to get the ppm of water in the package corresponding to three monolayers. An approximate value for the ppm of water for three monolayers can be obtained from Fig. 5-1. Next calculate the water leak rate, then the helium leak rate, and finally the measured leak rate.

For the package with volume = 0.0504 cc, the internal surface area (*SA*) is:

$$\text{Top} + \text{bottom} = 2(0.7 \times 0.4) = 0.56\ \text{cm}^2$$

$$\text{Sides} = 2(0.7 \times 0.2) + 2(0.4 \times 0.2) = 0.28 + 0.16 = 0.44\ \text{cm}^2$$

$$\text{Total } SA = 1.00\ \text{cm}^2$$

$$SA/V = 1.00/0.0504 = 19.84\ \text{(from Fig. 5-1, ppm = 3,000)}$$

$$\text{Volume of liquid water} = 1.00\ \text{cm}^2 \times 1.2 \times 10^{-7}\ \text{cm} = 1.2 \times 10^{-7}\ \text{cc}$$

$$\text{Weight} = 1.2 \times 10^{-7}\ \text{g}$$

The volume of H_2O vapor equals:

$$\frac{1.2\times10^{-7}\,\text{g}\times22{,}400\text{cc}}{18\text{g}}=1.493\times10^{-4}\,\text{cc}$$

(18 g of water occupies 22,400 cc)

$$\text{ppm}=\frac{1.4933\times10^{-4}\,\text{cc}\times10^{6}\,\text{ppm}}{0.0504\text{cc}}$$

$$\text{ppm}=2963$$

The water leak rate is:

$$L_{H_2O}=-\frac{V}{t}\left[\ln\left(1-\frac{Q_{H_2O}}{\Delta pi_{H_2O}}\right)\right]$$

where: L_{H_2O} = the true water leak rate
V = 0.0504 cc
t = 3.154×10^7 sec
Q_{H_2O} = the water that has leaked into the package = 2963 ppm
Δpi_{H_2O} = the difference in the water partial pressure = 18,500 ppm

$$L_{H_2O}=-\frac{0.0504}{3.154\times10^{7}}\left[\ln\left(1-\frac{2963}{18{,}500}\right)\right]$$

$$L_{H_2O}=-1.60\times10^{-9}[\ln(1-0.14216)]=-1.60\times10^{-9}(\ln 0.98578)$$

$$L_{H_2O}=-1.60\times10^{-9}(-0.014318)$$

$$L_{H_2O}=2.29\times10^{-11}\ \text{atm-cc/sec}$$

$$L_{HE} = 2.12\, L_{H_2O} = 2.12 \times 2.29 \times 10^{-11}$$

$$= 4.86 \times 10^{-11} \text{ atm-cc/sec}$$

$$R = 0.1\, L_{HE} = 0.1 \times 4.86 \times 10^{-11}$$

$$R = 4.86 \times 10^{-12} \text{ atm-cc/sec}$$

This is too small to measure.

For the package with volume = 0.18 cc, the internal surface area (SA) is:

$$\text{Top + bottom} = 2(1 \times 1) = 2 \text{ cm}^2$$

$$\text{Sides} = 4(1 \times 0.2) = 0.8 \text{ cm}^2$$

$$\text{Total } SA = 2.8 \text{ cm}^2$$

$$SA/V = 2.8/0.18 = 15.56 \text{ (from Fig. 5-1, ppm =2,100)}$$

$$\text{Volume of liquid water} = 2.8 \text{ cm}^2 \times 1.2 \times 10^{-7} \text{ cm} = 3.36 \times 10^{-7} \text{ cc}$$

$$\text{Weight} = 3.36 \times 10^{-7} \text{ g}$$

$$\text{Volume of water vapor} = \frac{3.36 \times 10^{-7} \text{ g} \times 22{,}400 \text{ cc}}{18 \text{ g}} = 4.181 \times 10^{-4} \text{ cc}$$

$$\text{ppm} = \frac{4.181 \times 10^{-4} \text{ cc} \times 10^{6}}{0.18 \text{ cc}}$$

$$\text{ppm} = 2323$$

The water leak rate is:

$$L_{H_2O} = -\frac{0.18}{3.154 \times 10^{7}}\left[\ln\left(1 - \frac{2{,}323}{18{,}500}\right)\right]$$

$$L_{H_2O} = -5.71 \times 10^{-9}[\ln(1 - 0.125568)] = -5.71 \times 10^{-9}(\ln 0.874432)$$

$$L_{H_2O} = -5.71 \times 10^{-9}(-0.13418)$$

$$L_{H_2O} = 7.66 \times 10^{-10} \text{ atm-cc/sec}$$

$$L_{HE} = 2.12 \times 7.66 \times 10^{-10} = 1.62 \times 10^{-9} \text{ atm-cc/sec}$$

$$R = 0.1 \times 1.62 \times 10^{-9} = 1.62 \times 10^{-10} \text{ atm-cc/sec}$$

This is too small to measure routinely.

For the package with volume = 1.62 cc the internal surface area is:

$$\text{Top + bottom} = 2(3 \times 3) = 18 \text{ cm}^2$$

$$\text{Sides} = 4(3 \times 0.2) = 2.4 \text{ cm}^2$$

$$\text{Total } SA = 20.4 \text{ cm}^2$$

$$SA/V = 20.4/1.62 = 12.59 \text{ (from Fig. 5-1, ppm = 1700)}$$

The volume of liquid water = 20.4 cm^2 × 1.2 × 10^{-7} cm = 2.444 × 10^{-6} cc

$$\text{The weight} = 2.444 \times 10^{-6} \text{ g}$$

The volume of water vapor equals:

$$\frac{2.4448 \times 10^{-6} \text{ g} \times 22{,}400 \text{ cc}}{18 \text{ g}} = 3.0464 \times 10^{-3} \text{ cc}$$

$$\text{ppm} = \frac{3.0466 \times 10^{-3} \text{ cc} \times 10^{6}}{1.62 \text{ cc}}$$

$$\text{ppm} = 1{,}880$$

The water leak rate is:

$$L_{H_2O} = -\frac{1.62}{3.154 \times 10^7}\left[\ln\left(1 - \frac{1{,}880}{18{,}500}\right)\right]$$

$$L_{H_2O} = -5.14 \times 10^{-8}[\ln (1\text{-}0.10162)] = -5.14 \times 10^{-8}(\ln 0.898378)$$

$$L_{H_2O} = -5.14 \times 10^{-8}(-0.107164)$$

$$L_{H_2O} = 5.51 \times 10^{-9} \text{ atm-cc/sec}$$

$$L_{HE} = 2.12 \times 5.51 \times 10^{-9} = 1.17 \times 10^{-8} \text{ atm-cc/sec}$$

$$R = 0.1 \times 1.77 \times 10^{-8} = 1.17 \times 10^{-9} \text{ atm-cc/sec}$$

For the package with volume = 10 cc, the internal surface area (*SA*) is:

Top + bottom = 2(5 × 5) = 50 cm^2

Sides = 4(5 × 0.4) = 8 cm^2

Total *SA* = 58 cm^2

SA/V = 58/9 = 6.44 (from Fig. 5-1, ppm = 800)

The volume of liquid water = 58 cm^2 × 1.2 × 10^{-7} cm = 6.96 × 10^{-6} cc.

The weight = 6.96 × 10^{-6} g

The volume of water vapor equals:

$$\frac{6.96\text{g} \times 22{,}400 \text{ cc}}{18 \text{ g}} = 8.66 \times 10^{-3} \text{ cc}$$

$$ppm = \frac{8.66 \times 10^{-3}\ cc \times 10^{6}}{10\ cc}$$

$$ppm = 866$$

The water leak rate is:

$$L_{H_2O} = -\frac{10}{3.154 \times 10^{7}}\left[\ln\left(1 - \frac{866}{18{,}500}\right)\right]$$

$$L_{H_2O} = -3.171 \times 10^{-7}[\ln(1 - 0.046811)] = -3.171 \times 10^{-7}[\ln(0.953189)]$$

$$L_{H_2O} = -3.171(-0.04794)$$

$$L_{H_2O} = 1.52 \times 10^{-8}\ \text{atm-cc/sec}$$

$$L_{HE} = 2.12 \times 1.52 \times 10^{-8} = 3.22 \times 10^{-8}\ \text{atm-cc/sec}$$

$$R = 0.1 \times 3.22 \times 10^{-8} = 3.22 \times 10^{-9}\ \text{atm-cc/sec}$$

Summarizing:

Package Volume	*R* with 0.1 atm He
0.0504 cc	4.86×10^{-12} atm-cc/sec
0.18 cc	1.62×10^{-10} atm-cc/sec
1.62 cc	1.17×10^{-9} atm-cc/sec
10 cc	3.22×10^{-9} atm-cc/sec

The smallest package can not be measured to that limit. Backfilling with 100% helium at 1 atm would only raise the measured limit to 4.86×10^{-11}, still not measurable with a helium leak detector. The 0.18 cc volume is measurable when backfilled with 100% helium. The two largest packages are measurable with the 10% backfill.

REFERENCES

1. Perkins, K. L., Licari, J. J., and Buckelew, R. L., "Investigation of Moisture Effects on Selected Microelectronic Devices," *Proc. 1978 ISHM Symposium*
2. Crawford, W. M., and Weigand, B. L., "Contamination of Relay Internal Ambients," *Fifteenth Annual Relay Conference,* Stillwater, OK (1956)
3. Lee, S. M., Licari, J. J., and Valles, A., *Physics of Failure in Electronics,* 4:464–492 (1966)
4. Eisenberg, P. H., Brandewie, G. V., and Meyer, R. A., "Effect of Ambient Gases and Vapors at Low Temperature on Solid State Devices," *Seventh New York Conference on Electronic Reliability, IEEE,* NY (May 20, 1966)
5. Rowe, W. M., and Eisenberg, P. H., "Factors Affecting the Reliability of Wet Tantalum Capacitors," *Reliability Physics Synposium,* Los Angeles, CA (1967)
6. Cannon, D. L., and Trapp, O. D., "The Effect of Cleanliness of Integrated Circuit Reliability," *Sixth Annual Reliability Physics Symposium Proceedings,* IEEE, NY (1968)
7. Young, M. R. P., and Peterman, D. A., "Reliability Engineering, Microelectronics and Reliability," 7:93–103 (1968)
8. Thomas, R. W., *Government Microcircuits Digest Applications Conference,* San Diego, CA, pp. 31–36 (1972)
9. Paulson, W., "Further Studies on the Reliability of Thin Film Nickel-Chromium Resisters," *Eleventh Annual Proceedings on Reliability Physics* (1973)
10. Thomas, R. W., and Meyer, D. E., "Moisture in IC Packages," *Solid State Technology* (1974)
11. Koelmans, H., *Twelfth Annual Proceedings Reliability Physic*, pp.168–171 (1974)
12. Paulson, W., and Kirk, L. W., *Twelfth Annual Proceedings Reliability Physics*, pp. 172–179 (1974)
13. Schumpka, A., and Piety, R. R., "Migrated-Gold Resistive Shorts in Microcircuits," *Proceedings Reliability Physics Symposium,* Las Vegas, NV(Apr., 1975)
14. Thomas, R. W., "Moisture Myths and Microcircuits," *Proceedings 26th Electronic Components Conference,* San Francisco, CA (Apr., 1976)
15. Comizzoli, R. B., *RCA Review,* 37:483 (1976)

16. Der Marderosian, A., and Murphy, C., "Humidity Threshold Variations for Dendritic Growth on Hybrid Substates," *Proceedings International Reliability Physics Symposium,* Las Vegas, NV (Apr., 1977)
17. Somerville, D. T., "The Role of Hybrid Construction Techniques on Sealed Moisture Levels," *Proceedings International Reliability Physics Symposium,* Las Vegas, NV (Apr., 1977)
18. Lowry, R. K., "Microcircuit Corrosion and Moisture Control," *Microcontamination* (May, 1985)
19. Kane, D., and Domingos, H., "Nondestructive Moisture Measurements in Microelectronics," *Final Report, RADC-TR-87-210* (Nov., 1987)
20. Stroehle, D., "On the Penetration of Gases and Water Vapor into Packages with Cavities and Maximum Allowable Leak Rates," *Proceedings International Reliability Physics Symposium,* Las Vegas, NV (Apr., 1977)
21. Graves, J. F., and Gurany, W., "Reliability Effects of Fluoride Contamination of Aluminum Bonding Pads on Semiconductor Chips," *Solid State Technology* (Oct., 1983)
22. *Handbook of Chemistry and Physics*, The Chemical Rubber Company
23. Czanderna, A. W., Vasofsky, R., and Czanderna, K. K., "Mass Changes of Adhesives During Curing, Exposure to Water Vapor, Evacuation and Outgassing," *ISHM Proceedings* (1977)
24. Roberts, S. C., "Predicting Post-seal Atmospheric Moisture Content in Hybrid Packages," *ISHM Proceedings* (1991)
25. Fancher, D. R., and Horner, R. G., "Hybrid Package Hermeticity and Moisture Level Monitoring," *ISHM Proceedings* (1978)
26. Koudounaris, A., and Wargo, E. E., "Detection and Control of Moisture in Space Hybrids," *ISHM Proceedings (*1981)
27. Glasstone, S., *Text Book of Physical Chemistry,* (D. Van Norstand Company*)*
28. Glasstone, S., *Thermodynamics for Chemists*, (D. Van Nostrand Company) p. 286 (1947)
29. Schuessler, P., and Feliciano-Welpe, D. D., "The Effects of Hydrogen on Device Reliability," *Hybrid Circuit Technology* (Jan., 1991)
30. Schuessler, P., and Gonya, S. G., "Hydrogen Desorption from Base and Processed Packaging Alloys," *RL/NIST Workshow on Moisture Measurement and Control for Microelectronics,* NISTIR 5241 (1993)

6

Understanding Helium Fine Leak Testing in Accordance with Method 1014, MIL-STD-883

1.0 PURPOSE OF THE TEST

The purpose of the test is to measure the hermeticity of sealed packages. Method 1014 includes leak rate limits for different ranges of internal package volumes. This chapter will explain the relationship between the fixed and flexible methods of testing.

2.0 BASIS OF THE TEST

If a package has a leak and it is placed in a helium pressurized vessel (bomb), some of the pressurized gas will enter the package through the leak channel. The absolute amount of helium entering the package depends upon the size of the leak channel, the time under pressure, and the pressure difference between the helium in the bomb and the helium pressure in the package.

After removal from the bomb, the package is connected to a helium leak detector. Helium now escapes from the package and is detected. The absolute amount of helium escaping depends upon the size of the leak channel and the helium pressure within the package. The helium pressure in the package depends upon the absolute amount of helium *and the internal volume of the package.*

The mathematical relationship representing this physical phenomenon is known as the "Howell-Mann Equation" which D. A. Howel and C. A. Mann proposed in 1965.[1] A form of this equation is given in Method 1014 and is reproduced here as Eq. (6-1).[2]

Eq. (6-1)
$$R_1 = \frac{LP_E}{P_0}\left(\frac{M_A}{M}\right)^{\frac{1}{2}}\left\{1-e^{-\left[\frac{Lt_1}{VP_0}\left(\frac{M_A}{M}\right)^{\frac{1}{2}}\right]}\right\}e^{-\left[\frac{Lt_2}{VP_0}\left(\frac{M_A}{M}\right)^{\frac{1}{2}}\right]}$$

where: R_1 = the measured leak rate of tracer gas (He) through the leak in atm-cc/sec of He

L = the equivalent standard leak rate in atm-cc/sec of air

P_E = the pressure of exposure in atmospheres absolute of helium

P_0 = the atmospheric pressure in atmospheres absolute (1)

M_A = the molecular weight of air in g (28.7, per Method 1014)

M = the molecular weight of the tracer gas (helium) in g (4)

t_1 = the time of exposure to P_E in sec

t_2 = the dwell time between release of pressure and leak detection, in sec

V = the internal volume of the device package cavity in cc

The leak rate, R_1, is the same R used in the previous chapters of this book. Equation (6-1) assumes all flow to be molecular and that there is no helium sealed within the package. Tables relating values R_1 to L have been made available by Varian.[3] Graphs of R_1 versus L for a limited leak range have also been published.[4]

The first term to the right of the = sign,

$$\frac{LP_E}{P_0}\left(\frac{M_A}{M}\right)^{\frac{1}{2}}$$

contains the partial pressure difference (P_E) as the equation assumes no helium sealed in the package. L is based on a difference in pressure of one atmosphere of air between the inside and outside of the package, and is called the "equivalent standard leak rate" of air and equals the true leak rate of air. The leak rate of air is fictitious as the components of air leak at different rates. A RGA never reports air as a constituent. L for air is used here as a reference and leak rate limits have been set using this term. P_0 is always one when used in conjunction with L.

The term

$$\left(\frac{M_A}{M}\right)^{\frac{1}{2}}$$

converts the air leak rate to that of helium like the values in Table 3-1.

The second expression,

$$\left\{1-e^{-\left[\frac{Lt_1}{VP_0}\left(\frac{M_A}{M}\right)^{\frac{1}{2}}\right]}\right\}$$

is the same as the expression in Eqs. (4-7), (4-17), (4-21), and (4-22); namely:

$$\left(1-e^{-\frac{Lt}{V}}\right)$$

The third expression,

$$e^{-\left(\frac{Lt_1}{VP_0}\right)\left(\frac{M_A}{M}\right)^{\frac{1}{2}}}$$

is the same as that used in Ch. 3 to calculate the pressure remaining in a package as it is leaking out. In Eq. (6-1), it is the fraction of helium pressure remaining in the package at test time. The value of this term is usually one (except for very small packages, and large leaks, and long times).

Equation (6-1) can be rewritten as Eq. (6-2):

Eq. (6-2)

$$R_1 = 2.679LP_E\left(1-e^{-\frac{2.679Lt_1}{V}}\right)\left(e^{-\frac{2.679Lt_2}{V}}\right)$$

where $2.679 = \left(\frac{M_A}{M}\right)^{\frac{1}{2}}$ and $P_o = 1$

When the equation is written in terms of helium:

Eq. (6-3)

$$R = L_{HE}P_E\left(1-e^{-\frac{L_{HE}t_1}{V}}\right)\left(e^{-\frac{L_{HE}t_2}{V}}\right)$$

3.0 FIXED METHOD OF TESTING

The fixed method requires specific bombing pressures and times for specific package size ranges. The times and pressures are based on a Rome Air Development Center Report.[5] These conditions are delineated in Table 2 of Method 1014, and are repeated here as Table 6-1.

These fixed conditions "shall not be used if the maximum equivalent standard leak rate limit given in the acquisition document is less than the limits specified herein for the flexible method." The failure criteria for the flexible method is as specified in Table 6-2.

Table 6-1. Fixed Conditions for Helium Fine Leak Testing

Package Volume (cc)	PSIA ± 2 PSIA	Minimum Exposure Time t_1 (hr)	Maximum Dwell Time t_2 (hr)	R_1 Reject Limit (atm-cc/sec)
<0.05	75	2	1	5×10^{-8}
≥0.05 - <0.5	75	4	1	5×10^{-8}
≥0.5 - <1.0	45	2	1	1 × 10-7
≥1.0 - <10.0	45	5	1	5×10^{-8}
≥10.0 - <20.0	45	10	1	5×10^{-8}

Table 6-2. Failure Criteria for the Flexible Method

Package Volume, cc	L, Reject Limit (atm-cc/sec)
≤0.01	5×10^{-8}
$0.01 < V \leq 0.4$	1×10^{-7}
>0.4	1 × 10-6

Several calculations are necessary to understand the exact numerical relationship between the limits of the fixed method with that of the flexible method. The method of performing these calculations is to assign values of L in Eq. (6-1) and solve for R_1. When R_1 is plotted against L on a log-log graph, a straight line results if the time between bombing and test is short enough so that no appreciable helium has leaked out. Figures 6-1–6-5 are such plots.

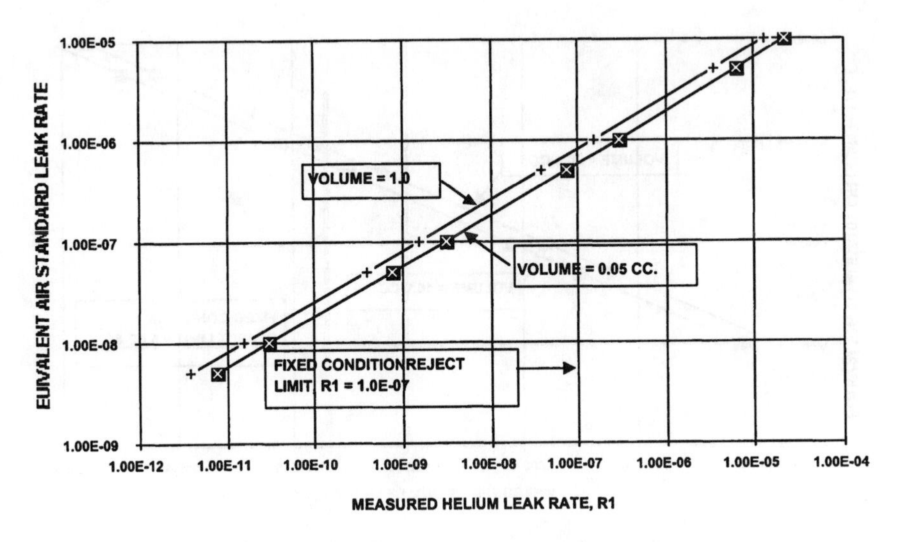

Figure 6-1. Measured helium leak rate versus the equivalent standard air leak rate, fixed condition 45 PSI, for 2 hr, 1 hr between bomb and test.

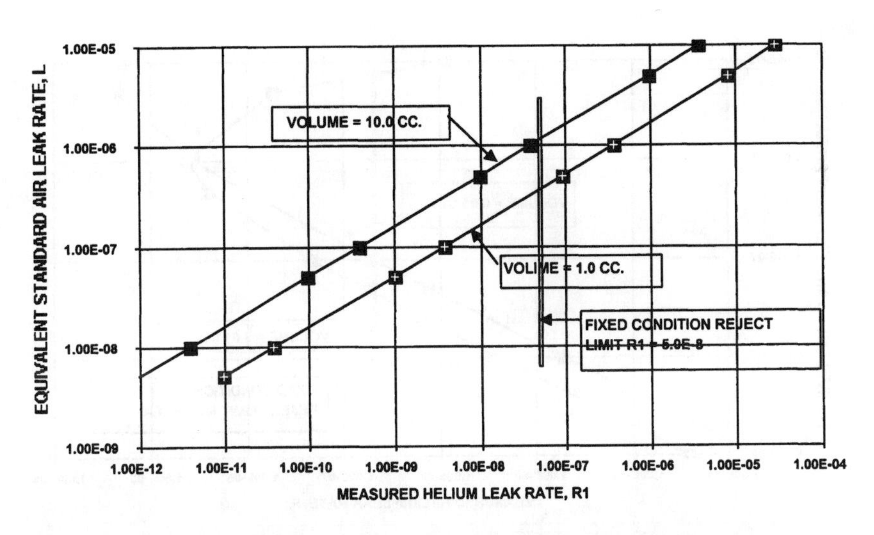

Figure 6-2. Measured helium leak rate versus the equivalent standard air leak rate, fixed condition 45 PSIA for 5 hr, 1 hr between bomb and test.

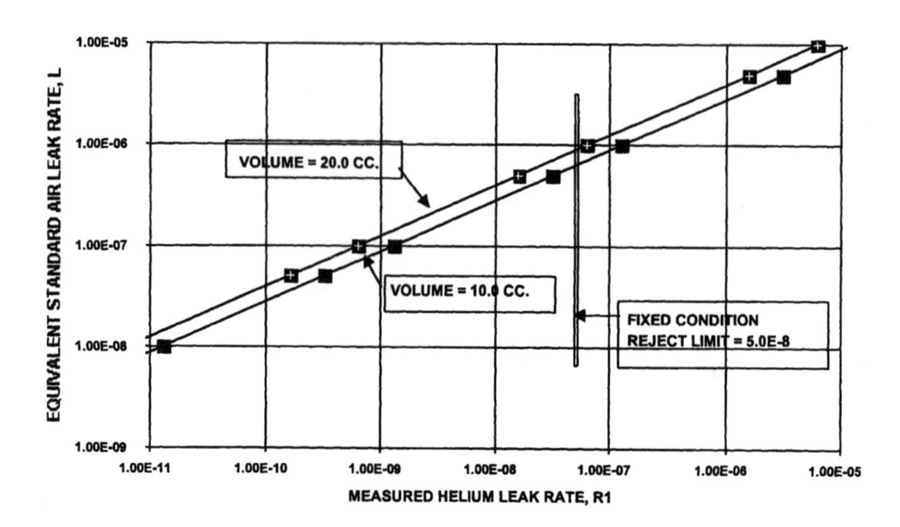

Figure 6-3. Measured helium leak rate versus the equivalent standard air leak rate, fixed condition 45 PSIA for 10 hr, 1 hr between bomb and test.

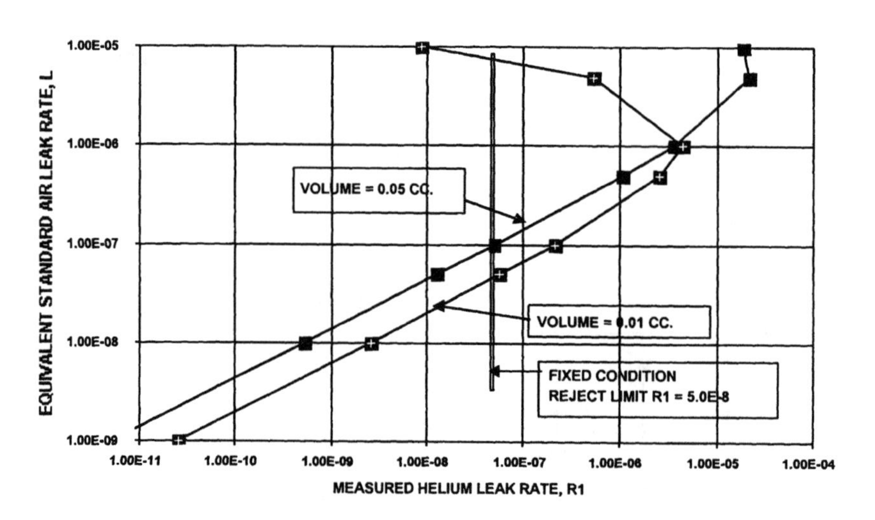

Figure 6-4. Measured helium leak rate versus the equivalent standard air leak rate, fixed condition 75 PSIA for 2 hr, 1 hr between bomb and test.

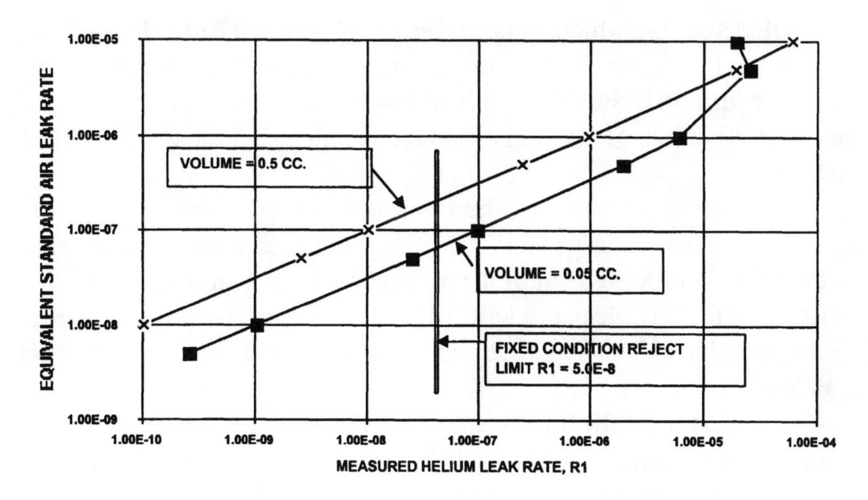

Figure 6-5. Measured helium leak rate versus the equivalent standard air leak rate, fixed condition 75 PSIA for 4 hr, 1 hr between bomb and test.

When the measured helium leak rate limit in Fig. 6-1 is 1×10^{-7} atm-cc/sec, it corresponds to an equivalent standard air leak rate of $L = 4 \times 10^{-7}$ atm-cc/sec for the volume = 0.5cc, and corresponds to an L of 7×10^{-7} atm-cc/sec for the 1.0 cc volume. For volumes between 0.5 cc and 1.0 cc, the L values will be between 4 and 7×10^{-7} atm-cc/sec.

When the measured helium leak rate limit in Fig. 6-2 is 5×10^{-8} atm-cc/sec it corresponds to an equivalent standard air leak rate (L) of 3.6×10^{-7} atm-cc/sec for the volume = 1.0 cc, and to an L of 1.1×10^{-6} atm-cc/sec for the 10.0 cc volume. For volumes between 1.0 cc and 10.0 cc, the L values will be between 3.6×10^{-7} and 1.1×10^{-6} atm-cc/sec.

When the measured helium leak rate limit in Fig.6-3 is 5×10^{-8} atm-cc/sec it corresponds to an equivalent standard air leak rate (L) of 6×10^{-7} atm-cc/sec for the volume = 10.0 cc, and to an L of 8×10^{-7} atm-cc/sec for the 20.0 cc volume. For volumes between 10.0 cc and 20.0 cc the L values will be between 6 and 8×10^{-7} atm-cc/sec.

The departure from straight lines in Fig. 6- 4 is due to an appreciable amount of helium leaking out of the package between removal from the bomb and the time of test. When this occurs, a dual value of L for a single value of R_1 results. Figure 6-4 shows that when a package with a volume of 0.01 cc has a measured value of 5×10^{-8} atm-cc/sec, the two L readings are 4×10^{-8} and 7×10^{-6} atm-cc/sec. The measured leak rate would indicate that this package is acceptable, but in fact it might not be. Retesting an hour later will determine which value is correct. If there is a large change in R_1 (to a lesser value), L is 7×10^{-6} atm-cc/sec. Such a package might also fail the gross leak test. Stanley Ruthberg has published curves showing the quantitative effect of the time between bomb and test, on the two L values.[6]

The primary reason for the gross leak test is to detect large leak rates. Packages with large leaks can pass the fine leak test because most of the helium has leaked out of the package between bombing and test. In most cases, the gross leak test will fail the part. The example above, where the larger value of $L = 7 \times 10^{-6}$ atm-cc/sec, would most likely fail the gross leak test.

The curves for the 0.05 cc volume in Figs.6- 4 and 6-5 also show the results when helium has leaked out between bombing and test. A dual value in the fine leak range does not occur in Fig.6-5 because the package contains more helium. The higher leak value is in the gross leak range and would fail that test.

Figure 6- 6 is a plot of the value of the last term in Eqs. (6-1) and (6-2), versus

$$\frac{Lt_2}{V}$$

where: L = the equivalent standard air leak rate in atm-cc/sec.

t_2 = the time between removal from the bomb and test, in sec

V = the internal volume of the package, in atm cc

$$\frac{Lt_2}{V} \text{ is in atm}$$

The maximum value of

$$\frac{Lt_2}{V}$$

for the package to retain at least 90% of the helium, is 0.039 atm.

Table 6-1 can be modified to include the equivalent standard air leak rate. The inclusion is in Table 6-3.

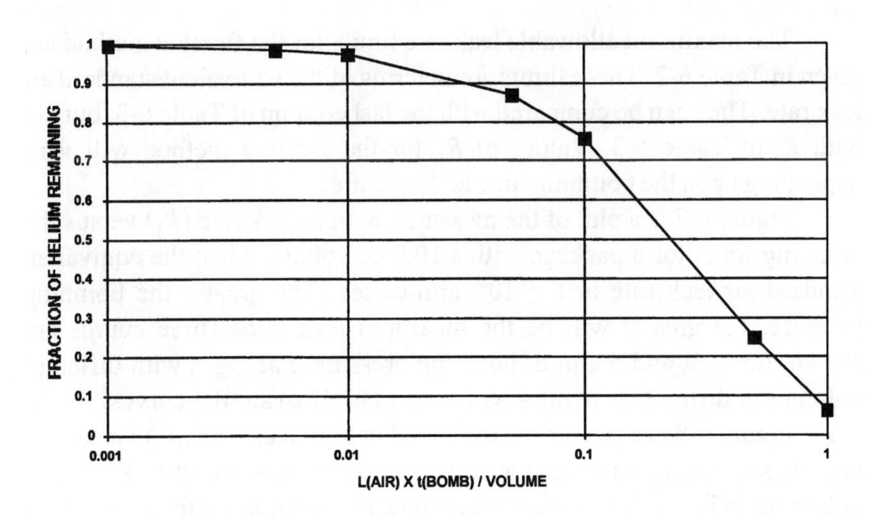

Figure 6-6. Fraction of helium remaining at test time as a function of equivalent standard air leak rate, time between bomb and test, and package volume.

Table 6-3. Fixed Conditions for Helium Fine Leak Testing with Equivalent Standard Air Leak Rates

Package Volumn (cc)	PSIA ±2 PSIA	Minimum Exposure Time t_1 (hr)	Maximum Dwell Time t_2 (hr)	R_1 Reject Limit (atm-cc/sec)	Equivalent Standard Air Leak Rate, L (atm-cc/sec)
<0.05	75	2	1	5×10^{-8}	$< 1 \times 10^{-7}$
≥0.05 - <0.5	75	4	1	5×10^{-8}	$7 \times 10^{-8} – 2.2 \times 10^{-7}$
≥0.5 – <1.0	45	2	1	1 × 10-7	$4 \times 10^{-7} – 7 \times 10^{-7}$
≥1.0 – <10.0	45	5	1	5×10^{-8}	$3.6 \times 10^{-7} – 1.1 \times 10^{-6}$
≥10.0– <20.0	45	10	1	5×10^{-8}	$6 \times 10^{-7} – 8 \times 10^{-7}$

4.0 FLEXIBLE METHOD OF TESTING

The maximum allowable leak rate limits for the flexible method are given in Table 6-2. These limits are in terms of the equivalent standard air leak rate. They can be compared with the last column of Table 6-3, but not with R_1 of Table 6-3. Values of R_1 for the flexible method will vary depending upon the bombing time and pressure.

Figure 6-7 is a plot of the measured helium leak rate (R_1) versus the bombing time, for a package with a 10.0 cc volume when the equivalent standard air leak rate is 1×10^{-6} atm-cc/sec. The greater the bombing pressure, the greater will be the measured leak rate. Three curves are shown, for 2, 3, and 5 atm of bombing pressure. Packages with different volumes or different leak rates will have a family of similar curves.

Figure 6-8 is a plot of the measured helium leak rate (R_1) versus the helium bombing pressure for the same equivalent standard air leak rate and volume as in Fig. 6-7. Five lines are plotted, representing different bombing times. The greater the bombing time, the larger the measured leak rate.

The value of R_1 for the flexible method can be calculated using Eq. (6-1) or (6-2). General solutions of R_1 can be achieved by plotting R_1 versus V for various values of L, $P_{E,}$ and the bombing time. Figure 6-9 is a plot for $L = 1 \times 10^{-6}$ atm-cc/sec (air). Figure 6-10 is for $L = 1 \times 10^{-7}$ atm-cc/sec (air) and Fig. 6-11 is for L = 5×10^{-8} atm-cc/sec (air).

Large amounts of helium are forced into the small packages of Fig. 6-11. The vertical line at 1.34×10^{-7} atm-cc/sec represents 1 atm of helium within the package for a total pressure of two atm. Points to the right of this line represent packages with more than two total atmospheres. It may be undesirable to have some packages at this high a pressure.

5.0 COMPARISON OF THE FIXED AND FLEXIBLE METHODS

The most frequent hermeticity requirement in a MCM/hybrid/IC acquisition document is "in accordance with Method 1014 of MIL-STD-883." The manufacturer has the option of the fixed or flexible method. He will select the option which will allow him to ship the most product. An example of making this choice is Example 6-1.

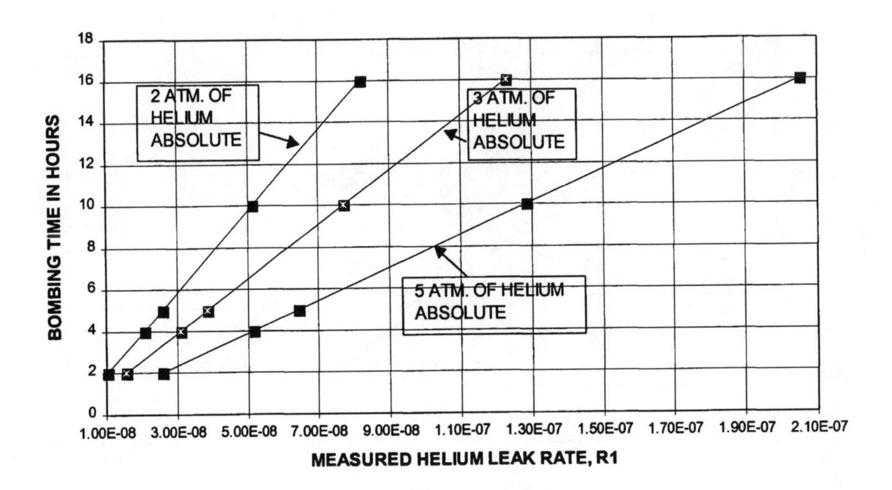

Figure 6-7. Measured helium leak rate versus bombing times for a 10.0 cc package volume and L=1 × 10^{-6} atm-cc/sec, at three different bombing pressures.

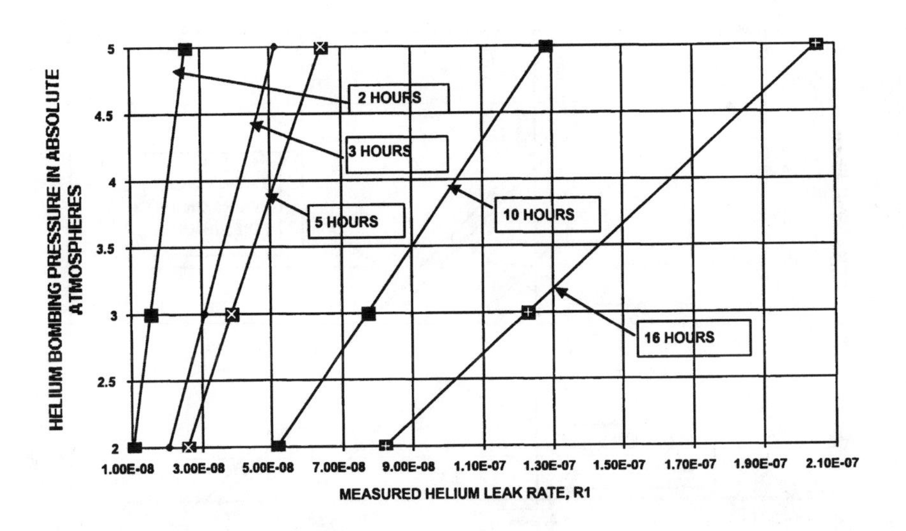

Figure 6-8. Measured helium leak rate versus the absolute helium bombing pressure for a 10.0 cc volume with L=1 × 10^{-6} atm-cc/sec; for five bombing times.

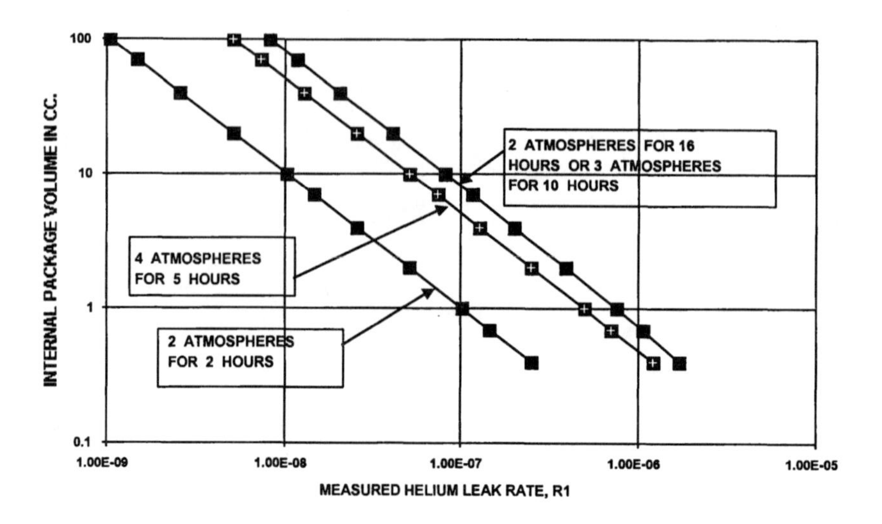

Figure 6-9. Measured helium leak rate versus the package volume for the flexible method, for various bombing times and pressures; L (air) = 1×10^{-6} atm-cc/sec.

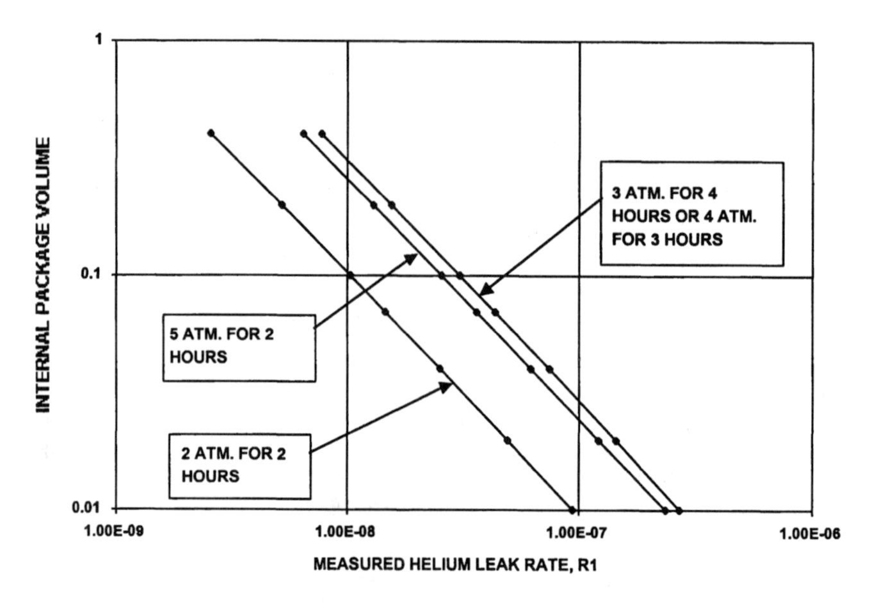

Figure 6-10. The measured helium leak rate versus the package volume for the flexible method, for various bombing times and pressures; L (air) = 1×10^{-7} atm-cc/sec.

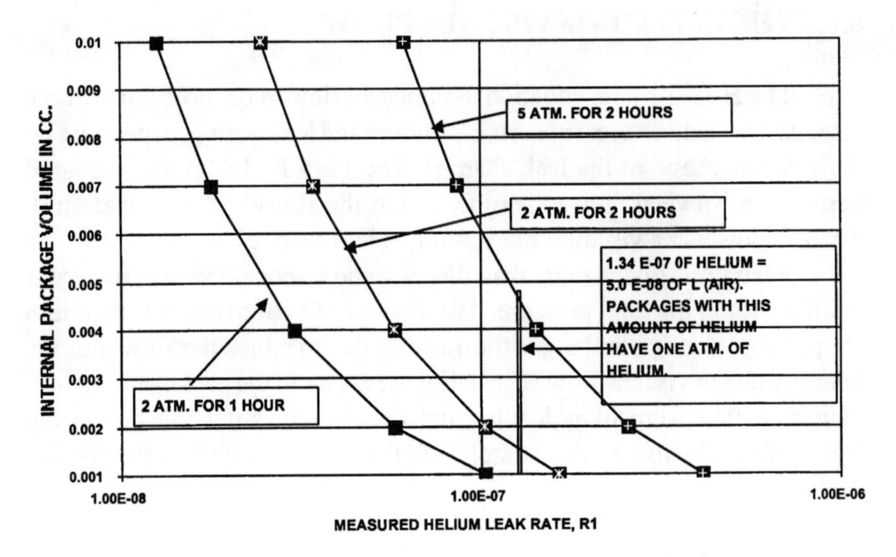

Figure 6-11. The measured helium leak rate versus the package volume for the flexible method, for various bombing times and pressures; L (air) = 5×10^{-8} atm-cc/sec.

Example 1. The internal volume of the package is 1.0 cc. Table 6-3 has an L value of 3.6×10^{-7} atm-cc/sec and a measured leak rate limit of 5×10^{-8} atm-cc/sec for the fixed method. Table 6-2 has an L value of 1×10^{-6} atm-cc/sec for the flexible method. The flexible method is much more lenient than the fixed method. Figure 6-9 shows three R_1 values for the 1.0 cc when using the flexible method. The values are 1×10^{-7} for bombing at 2 atm for 2 hr, 5×10^{-7} at 4 atm for 5 hr, and 7×10^{-7} for either 2 atm for 16 hr or 3 atm for 10 hr. Other combinations of pressures and time will give additional R_1 values. The purchaser of the device could specify the fixed method, thereby obtaining a device with a superior hermeticity. The selection of the bombing pressure and time is usually a trade off between the bombing time and the ease of measuring R_1. If the bombing was 4 atm for 5 hr, the measured limit would be about 5×10^{-7} atm-cc/sec.

6.0 THE EFFECT OF VISCOUS FLOW

The Howel-Mann equation assumes all flow to be molecular. Some viscous flow may be present during bombing and leak testing, depending on the size and shape of the leak channel. The 1975 RADC study had good correlation between their experiments and the Howel-Mann equation.[4] There is however a viscous contribution to leak testing.

Assume a leak channel that allows some viscous flow when there is a difference in the total pressure. The amount of helium being forced into the package during bombing is the quantity due to molecular flow plus the quantity due to viscous flow. The helium pressure in the package is greater than if the flow were all molecular, and more helium will come out during leak testing. The following example will illustrate the effect of the viscous contribution.

Example 2.

Assume, $L_{AIR} = 3.6 \times 10^{-7}$ atm-cc/sec

$L_{HE} = 2.679\, L_{AIR} = 2.679 \times 3.6 \times 10^{-7} = 9.64 \times 10^{-7}$ atm-cc/sec

$P_E = 3$ atm

$t = 5$ hr $= 18{,}000$ sec

$V = 1.0$ cc

all flow is molecular

The quantity of helium forced into the package due to molecular flow is:

$$Q_{HEMOL} = P_E\left(1 - e^{-L_{HE}t/V}\right)$$

$$Q_{HEMOL} = 3\left[1 - e^{\frac{-(9.64\times10^{-7})(18000)}{1.0}}\right]$$

$$Q_{HEMOL} = 3(1 - e^{-0.0173599}) = 3(1 - 0.98279) = 3(0.01721)$$

$$Q_{HEMOL} = 0.0516 \text{ atm of helium in the package}$$

$$R_{MOL} = 0.0516 \times L_{HE} = 0.0516 \times 9.64 \times 10^{-7}$$
$$= 4.97 \times 10^{-8} \text{ atm-cc/sec}$$

Now assume that 7% of the flow during leak testing is viscous (see Fig. 3-4)

$$L_{HEMOL} = (1 - 0.07) \times L_{HE} = 0.93 \times 9.64 \times 10^{-7}$$
$$= 8.96 \times 10^{-7} \text{ atm-cc/sec}$$

The 7% viscous is for a total pressure difference of 1 atm during leak testing and an average pressure difference of 0.5 atm. The average pressure difference during bombing is (3 + 1)/2 = 2 atm, so that the percentage viscous flow during bombing is (2/0.5) × 7% = 28%. Total helium in the package = helium due to molecular flow + 0.28 of total helium.

$$\text{Total helium} = 0.0516 + 0.28 \text{ Total helium}$$

$$\text{Total helium} = 0.0516/(1\text{-}0.28) = 0.0717 \text{ atm}$$

$$R_1 = R_{MOL+VIS} = 0.0717 \times L_{HE}$$
$$= 0.0717 \times 9.64 \times 10^{-7} = 6.91 \times 10^{-8} \text{ atm-cc/sec}$$

$$R_{MOL} = 0.93 \times R_1 = 0.93 \times 6.91 \times 10^{-8} = 6.43 \times 10^{-8} \text{ atm-cc/sec}$$

If there was only molecular flow, $R_1 = R_{MOL} = 4.97 \times 10^{-8}$ atm-cc/sec. If the flow during test was 7% (28% during bombing), $R_{MOL} = 6.43 \times 10^{-8}$ atm-cc/sec. If the flow was 7% viscous, but only molecular flow is assumed, then the percent error in the measurement of R_1 is $(6.43 - 4.97) \times 100/6.43 = 22.7\%$.

Some error may easily be hidden in the error of the measurement. Larger leak rates or greater bombing pressures will produce larger errors, but most leak rates are smaller. Good agreement in leak testing can be considered to be ±10%, so that the good agreement in the RADC study may not have seen the viscous contribution.

7.0 LEAK RATE LIMITS ARE TOO LENIENT

The leak rate limits in both the fixed and flexible procedures of Method 1014 are too lenient with respect to two criteria. The first criteria is a Qualification requirement in MIL-PRF-38534 and MIL-PRF-38535 that a hybrid, MCM, or integrated circuit must have less than 5,000 ppm of water after screening (tested only during Qualification). The second criteria is the desire to have less than three monolayers of water.

The amount of water that has leaked into a package can be calculated using Eq. 5-8:

$$Q_{H_2O} = \Delta p_{H_2O}\left(1 - e^{-\frac{L_{H_2O}t}{V}}\right)$$

$$L_{H_2O} = \sqrt{\frac{M_{AIR}}{M_{H_2O}}}L_{AIR} = \sqrt{\frac{28.96}{18}}L_{AIR} = 1.27L_{AIR}$$

Assuming 0.02 atm of water in the air, the time between seal and RGA test is 30 days, and that all the water comes from outside the package, then:

$$\Delta p_{H_2O} = 0.02 \text{ atm.} = 20{,}000 \text{ ppm}$$

$$t = 30 \text{ days} = 2{,}592{,}000 \text{ sec}$$

For a 1.0 cc package at the fixed leak rate limit of $L = 3.6 \times 10^{-7}$ atm-cc/sec: (See Table 6-3)

$$L_{H_2O} = 1.27 \times 3.6 \times 10^{-7} = 4.57 \times 10^{-7} \text{ atm-cc/sec}$$

$$Q_{H_2O} = 20{,}000\left[1 - e^{-\frac{(4.57\times10^{-7})(2{,}592{,}000)}{1.0}}\right]$$

$$Q_{H_2O} = 20{,}000(1 - e^{-1.184544})$$

$$= 20{,}000(1 - 0.305885) = 20{,}000(0.694114)$$

$$Q_{H_2O} = 13{,}882 \text{ ppm water}$$

Knowing the ppm of water in a package and the internal surface area of the package, the number of liquid monolayers of water can be calculated.

Eq. (6-4) Volume of water vapor = (ppm × package volume)/10^6

Eq. (6-5) Weight of water

= (volume of water vapor × 18 g)/22,400 cc

Substituting Eq. (6-4) into Eq. (6-5):

Eq. (6-6) $$\text{Weight of water} = \frac{\text{ppm} \times \text{package volume} \times 18\text{ g}}{10^6 \times 22{,}400\text{ cc}}$$

Volume of liquid water in cc = the weight of the water in grams.

Eq. (6-7) $$\text{Thickness of water} = \frac{\text{volume of liquid water}}{\text{internal surface area of package } (SA)}$$

One monolayer of water is 4×10^{-8} cm.

Eq. (6-8) $$\text{Number of monolayers} = \frac{\text{thickness (in cm)}}{4 \times 10^{-8}\text{ cm}}$$

Substituting Eq. (6-7) into Eq. (6-8):

Eq. (6-9) $$\text{Number of monolayers} = \frac{\text{volume of liquid water}}{4 \times 10^{-8} \times SA}$$

Substituting Eq. (6-6) into Eq. (6-9):

Eq. (6-10) $$\text{Number of monolayers} = \frac{\text{ppm} \times \text{package volume} \times 18}{10^6 \times 22{,}400 \times 4 \times 10^{-8} \times SA}$$

Eq. (6-11) $$\text{Number of monolayers} = \frac{\text{ppm} \times \text{package volume} \times 0.02}{SA}$$

Using Eqs. (6-8) and (6-11), the ppm of water and the number of monolayers of water have been calculated for selected volumes at their leak rate limits (*L*). The results are tabulated in Table 6-4.

Table 6-4. PPM of Water and the Number of Momolayers of Water in Selected Packages when their Leak Rate is at the MIL-SPEC Limit*

Volume (cc)	Method	L = L(air)	$L(H_2O)$	PPM of Water	Monolayers of Water
0.01	Flexible	5.0 E-08	6.35 E-08	20,000	14
0.05	Fixed	7.0 E-08	8.89 E-08	20,000	22
0.05	Flexible	1.0 E-07	1.27 E-07	20,000	22
0.4	Flexible	1.0 E-07	1.27 E-07	11,217	17
0.5	Fixed	4.0 E-07	5.08 E-07	18,563	30
0.5	Flexible	1.0 E-06	1.27 E-06	19,972	32
1.0	Fixed	3.6 E-07	4.57 E-07	13,882	24
1.0	Flexible	1.0 E-06	1.27 E-06	19,256	33
10.0	Fixed	6.0 E-07	7.62 E-07	3,585	10
10.0	Flexible	1.0 E-06	1.27 E-06	5,610	15
20.0	Fixed	8.0 E-07	1.02 E-06	2,476	7
20.0	Flexible	1.0 E-06	1.27 E-06	3,035	8

*Assumptions are:

0.02 atm of water in the air
30 days between seal and RGA test
the internal length and width of the package are equal
the inside height of packages 0.01 cc through 1.0 cc is 0.2 cm
the inside height of packages 10.0 cc and 20.0 cc is 0.3 cm
one monolayer of water is 4×10^{-8} cm thick

8.0 BACKFILLING THE PACKAGE WITH HELIUM

Backfilling with helium is sealing in an atmosphere which contains a percentage of helium. The detectability of a leak is directly proportional to the percentage of helium in the package at leak test time. The percentage sealed in is a trade off between the detectability and the cost of the helium. Ten percent is a common amount but higher percentages are often used for greater detectability.

It is important that the percentage of helium be controlled, so that the equivalent standard air leak rate or the true helium leak rate can be calculated. The control of the percentage should not be accomplished by controlling the two gases (helium and, for example, nitrogen) separately. A non-uniform flow in one gas will effect the helium percentage. The mixture should be controlled by having the helium be a fraction of the nitrogen. This control should be performed outside the sealing chamber, the controlled gas mixture entering the chamber. Any disruption in the nitrogen or helium flow will cause the other gas to change accordingly, thereby maintaining the desired ratio.

Method 1014 does not consider backfilling. MIL-PRC-38534 (General Specification for Hybrid Microcircuits) requires a minimum of 10% helium backfilling for Class K devices. Facilities that employ backfilling must guarantee a minimum helium percentage to satisfy the customer. The helium percentage can be controlled to within 20%, using the ratio control method. If the percentage is set to 12%, the mixture can be guaranteed to a minimum of 10%.

There are two advantages to backfilling. A leak test can be performed without bombing immediately after sealing. Such a test can give one confidence in the sealing operation, as well as a quantitative leak rate for the package. Bombing is still required, at a later time, to meet the requirements of Method 1014. If Method 1014 can be neglected, the leak rate can be verified by an additional non-bomb test at a future time, as the amount of helium in the package at a future time can be calculated.

The second advantage is a smaller leak rate detectability for all but the smallest packages. The measured leak rate with backfilling can be compared with the measured leak rate without backfilling for a given equivalent standard air leak rate. The calculation for the non-backfilled package is performed using Eq. (6-1). The measured leak rate when backfilled with 10% helium is simply 0.1 times the true helium leak rate, which equals 0.37 times the equivalent standard leak rate. Figure 6-12 shows this comparison for some selected leak rates for a 1.0 cc package. Figure 6-13 is a similar comparison for a package volume of 10.0 cc.

An example from Fig. 6-12 illustrates the comparison between backfilling and not backfilling. For an equivalent standard leak rate of 1.0×10^{-8}, the 2 hr at 5 atm would require a measured leak rate on the 10^{-11} range, not a realizable measurement. When bombed for 24 hr at 3 atm, the measured value is on the low 10^{-10} range, the limit of helium leak detectors and too small a value for production testing. The backfilled measured value for the same equivalent standard leak rate is between 2 and 3×10^{-9}, a value that can be used in production.

The disadvantage of backfilling is the cost of the helium, the cost of the control equipment, the time and labor to set up the process, and the cost of the RGAs to verify that the process is under control.

Figure 6-13 shows a similar comparison as in Fig. 6-12. Here, however, the difference between backfilling and not backfilling is even greater because of the larger package volume. The measured value for the backfilled package remains the same as in Fig. 6-12 (it is independent of package volume). The measured values for the bombed packages are a decade smaller. A tabular comparison of backfilling with no backfilling is shown in Table 6-5, where comparisons are made for other package volumes.

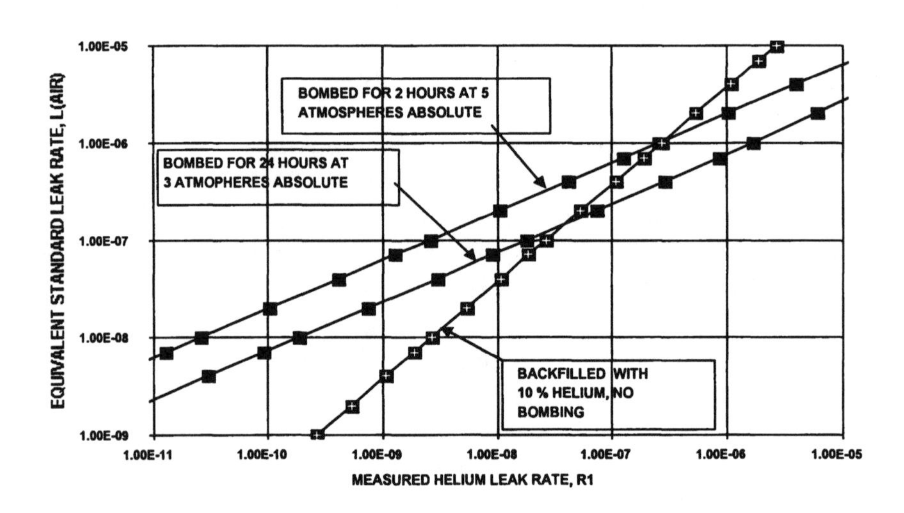

Figure 6-12. Comparison of measured leak rates of bombed packages, with packages backfilled with 10% helium but not bombed; the package volume = 1.0 cc.

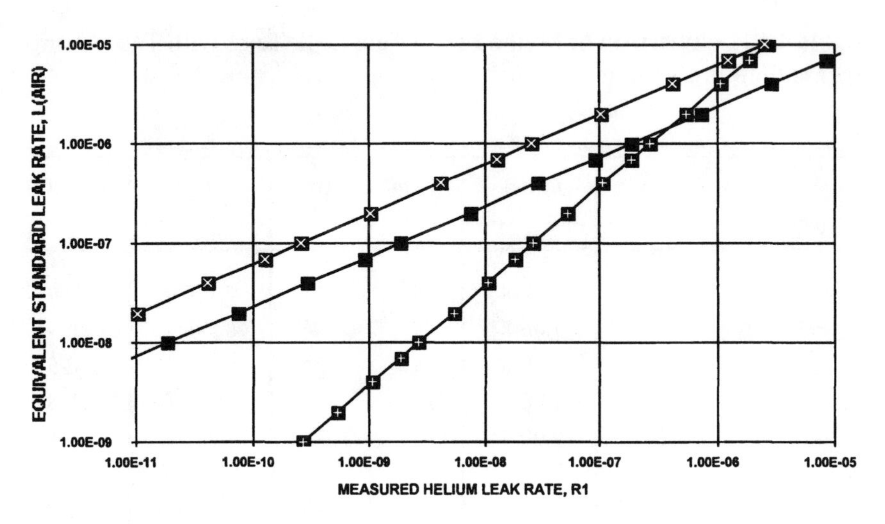

Figure 6-13. Comparison of measured leak rates of bombed packages, with packages backfilled with 10% helium but not bombed; the package volume = 10.0 cc.

9.0 BOMBING AFTER BACKFILLING

Military screening requirements include a bombing leak test even for packages that have been sealed with a percentage of helium. The measured leak rate, at a time subsequent to sealing and prior to bombing, will decrease with time as some helium will be leaking out. This has been discussed in Ch.3 and graphically presented in Fig. 3-2 in that chapter.

The amount of helium that is forced into the package during bombing depends upon the amount of helium in the package at that time, as well as the bombing pressure. The measured leak rate corresponding to an equivalent standard leak rate depends upon the atmospheres of helium in the package at the time of the leak test. An equation is needed to calculate the helium in the package at the test time.

The amount of helium forced into a package is Q_{HE}

Eq. (6-12) $$Q_{HE} = \Delta p_{HE}\left[1 - e^{-\frac{L_{HE}T}{V}}\right]$$

Table 6-5. Comparison of Method 1014 Fine Leak Tests with Tests Using 10% Helium Backfill

Volume (cc)	*L* (air)	*L* (he)	Atm	Hr	R_1	*R* (10%)
0.05	3.72E-9	1.0E-8	5	2	7.19E-11	1.0E-9
0.05	3.72E-9	1.0E-8	5	24	8.57E-10	1.0E-9
0.05	3.72E-9	1.0E-8	3	24	5.14E-10	1.0E-9
0.05	3.72E-8	1.0E-7	3	24	4.76E-8	1.0E-8
0.05	7.44E-8	2.0E-7	3	24	1.75E-7	2.0E-8
0.5	3.72E-9	1.0E-8	3	24	5.17E-11	1.0E-9
0.5	3.72E-8	1.0E-7	3	24	5.14E-9	1.0E-8
0.5	7.44E-8	2.0E-7	3	24	2.04E-8	2.0E-8
0.5	1.86E-7	5.0E-7	3	24	1.24E-7	5.0E-8
5.0	3.72E-9	1.0E-8	3	24	5.12E-12	1.0E-9
5.0	3.72E-8	1.0E-7	3	24	5.18E-10	1.0E-8
5.0	7.44E-8	2.0E-7	3	24	2.07E-9	2.0E-8
5.0	1.86E-7	5.0E-7	3	24	1.29E-8	5.0E-8
5.0	3.72E-7	1.0E-6	3	24	5.14E-8	1.0E-7
5.0	1.0E-6	2.69E-6	3	24	3.66E-7	2.69E-7
20.0	3.72E-9	1.0E-8	3	24	1.30E-12	1.0E-9
20.0	3.72E-8	1.0E-7	3	24	1.30E-10	1.0E-8
20.0	7.44E-8	2.0E-7	3	24	5.18E-10	2.0E-8
20.0	1.86E-7	5.0E-7	3	24	3.24E-9	5.0E-8
20.0	3.72E-7	1.0E-6	3	24	6.41E-8	1.0E-7

where: Q_{HE} = atmospheres of helium forced into the package

Δp_{HE} = the difference in the helium partial pressure between the bomb and that inside the package = $P_E - p_t$

L_{HE} = the true helium leak rate

T = the bombing time in sec

V = the internal volume of the package in cc

P_E = bombing pressure in absolute atmospheres

p_t = atmospheres of helium in the package at time t

The value of L_{HE} is easily calculated from the measured leak rate immediately after sealing without bombing.

$$L_{HE} = \frac{R\text{ (measured value)}}{\text{atmospheres of helium in the package}}$$

For a 10% backfill, $L_{HE} = \dfrac{R}{0.1}$

Δp_{HE} will vary with the time between sealing and bombing.

Eq. (6-13) $$\Delta p_{HE} = P_E - p_t$$

where

Eq. (6-14) $$p_t = p_i\left(e^{-\frac{L_{HE}t}{V}}\right)$$

where: p_t = the helium remaining at test time.

t = the time between seal and test

p_i = the atmospheres of helium sealed within the package

P_E = the bombing pressure in absolute atmospheres

Substituting Eq. (6-14) into Eq. (6-13):

Eq. (6-15) $$\Delta p_{\text{HE}} = P_E - \left[p_i \left(e^{-\frac{L_{\text{HE}} t}{V}} \right) \right]$$

Substituting Eq. (6-15) into Eq. (6-12):

Eq. (6-16) $$Q_{\text{HE}} = \left\{ P_E - \left[p_i \left(e^{-\frac{L_{\text{HE}} t}{V}} \right) \right] \right\} \left(1 - e^{-\frac{L_{\text{HE}} T}{V}} \right)$$

The total helium in the package at test time is the amount forced in plus the amount remaining at test time due to backfilling.

$$Q_{\text{HE-TOTAL}} = Q_{\text{HE}} + p_t$$

Eq. (6-17) $$Q_{\text{HE-TOTAL}} = Q_{\text{HE}} + p_i \left(e^{-\frac{L_{\text{HE}} t}{V}} \right)$$

Example:

Let: the bombing pressure $P_E = 3$ atm absolute of helium

the helium sealed into the package, $p_i = 0.1$ atm

the volume of the package = 1.0 cc

the bombing time (T) = 10 hr = 36,000 sec

the true helium leak rate, $L_{\text{HE}} = 1 \times 10^{-7}$ atm-cc/sec

the time between seal and test (t) = 30 days = 2,592,000 sec

$$Q_{\text{HE}} = \left\{ 3 - \left[0.1 \left(e^{-\frac{(1 \times 10^{-7})(2{,}592{,}000)}{1.0}} \right) \right] \right\} \left[1 - e^{-\frac{1 \times 10^{-7}(36{,}000)}{1.0}} \right]$$

$$Q_{HE} = (3 - 0.077166)(1 - 0.99964) = (2.9228)(0.0003599)$$

$$Q_{HE} = 0.00105 \text{ atm bombed in}$$

$$Q_{HE\text{-}TOTAL} = Q_{HE} + \text{the amount remaining from bombing}$$

$$Q_{HE\text{-}TOTAL} = 0.00105 + 0.077166 = 0.07822 \text{ atm of helium}$$

Figure 6-14 shows the amount of helium in the package when backfilled with 10% helium and subsequently bombed at 2 atm absolute. Each curve is for a different L_{HE}/V value and bombing time.

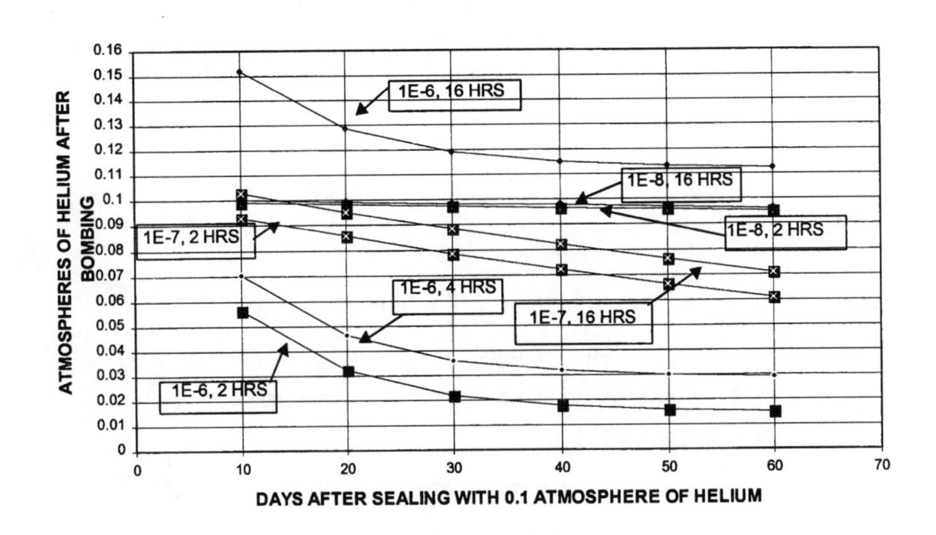

Figure 6-14. Atmospheres of helium in the package for 10% backfilling and then bombing at 2 atm absolute for various $L_{(He)}/V$ values and specific bombing times.

10.0 PROBLEMS AND THEIR SOLUTIONS

Problem 1. An acquisition document for a microwave hybrid, with an internal volume of 0.2 cc, allows either the fixed or flexible fine leak method. Which method allows the largest leak?

Solution. The equivalent standard air leak rate limit (L) when using the flexible method is given in Table 6-2, and is 1×10^{-7} atm-cc/sec. The fixed method gives only a measured leak rate limit (R_1) of 5×10^{-8} atm-cc/sec (see Table 6-1). A true leak rate limit corresponding to the fixed measured limit must be calculated so that a comparison can be made. The package volume, the bombing conditions, and the fixed leak rate limit determine the corresponding true leak rate limit.

Figure 6-5 shows the L for the fixed method to be approximately 1×10^{-7} atm-cc/sec. A calculation is necessary to answer which method allows the larger leak. L cannot be solved directly, but, by using Eq. (6-2), various values of L can be assigned and R_1 calculated. Equation (6-2) is:

$$R1 = (2.679)LP_E\left[1 - e^{-\frac{(2.679)LT_B}{V}}\right]\left[e^{-\frac{(2.679)LT_T}{V}}\right]$$

where: R_1 = the measured helium leak rate in atm-cc/sec

L = the equivalent standard air leak rate (true air leak rate)

P_E = the bombing pressure in atmospheres absolute

T_B = the bombing time in sec

T_T = the time between removal from the bomb and the time of test

V = the internal volume of the package, allowing for the volume of the parts inside the package

2.679 = the conversion factor between helium and air (between R_1 and L)

The fixed method for a volume of 0.2 cc uses 5 atm of helium for 4 hr (14,400 sec). The maximum time between removal from the bomb, to the time of leak testing is 1 hr (3,600 sec), for $L = 1 \times 10^{-7}$ atm-cc/sec.

$$R_1 = (2.679)(1\times10^{-7})(5)\left[1-e^{-\frac{(2.679)(10^{-7})(14{,}400)}{0.2}}\right]\left[e^{-\frac{(2.679)(10^{-7})(3600)}{0.2}}\right]$$

$$R_1 = 1.3395\times10^{-6}(1-e^{-0.0192888})(e^{-0.00096})$$

$$R_1 = 1.3395\times10^{-6}(1-0.980896)(0.99987)$$

$$R_1 = 1.3395\times10^{-6}(0.0191)(1) = 2.56\times10^{-8}\text{ atm-cc/sec, helium}$$

This value of R_1 is smaller than the 5×10^{-8} limit, so that L is larger than 1×10^{-7}, for $L = 1.5 \times 10^{-7}$ atm-cc/sec.

$$R_1 = (2.679)(1.5\times10^{-7})(5)\left[1-e^{-\frac{(2.679)(1.5\times10^{-7})}{0.2}}\right]$$

$$R_1 = 2.01\times10^{-6}(1-e^{-0.0289332})$$

$$R_1 = 2.01\times10^{-6}(1-0.9714813) = 2.01\times10^{-6}(0.0285)$$

$$R_1 = 5.73\times10^{-8}\text{ atm-cc/sec}$$

This a little larger than the 5×10^{-8} limit. This limit corresponds to approximately 1.4×10^{-7} atm-cc/sec. *The fixed method allows a larger leak.*

Problem 2. The equivalent standard air leak rate limit (L) for a 6.0 cc package is 1×10^{-7} atm-cc/sec. The package is sealed with 100% nitrogen and the leak test is performed 10 min after removal from the bomb. What is the measured leak rate limit (R_1) if the package is bombed in three atm absolute of helium for 10 hr?

Solution. Equation 6-2 can be used to solve this problem.

$$R_1 = (2.679)LP_E\left[1-e^{-\frac{(2.679)LT_B}{V}}\right]\left[e^{-\frac{(2.679)LT_T}{V}}\right]$$

where: R_1 = the measured helium leak rate

L = the equivalent standard leak rate = 1×10^{-7} atm-cc/sec

P_E = the bombing pressure in atm = 3

T_B = the bombing time in sec = 36,000

T_T = the time between removal from the bomb and the time of test = 600 sec

V = the volume of the package = 6.0 cc

$$R_1 = 2.679\left(1 \times 10^{-7}\right)(3)\left[1 - e^{-\frac{2.679\left(1\times10^{-7}\right)(36,000)}{6.0}}\right]\left[e^{-\frac{2.679\left(1\times10^{-7}\right)(600)}{6.0}}\right]$$

$$R_1 = 8.04 \times 10^{-7}(1 - e^{-0.0016074})(e^{-0.00002679})$$

$$R_1 = 8.04 \times 10^{-7}(1 - 0.998394)(1) = 8.04 \times 10^{-7}(0.001606)$$

$$R_1 = 1.29 \times 10^{-9} \text{ atm-cc/sec}$$

Problem 3. An aquisition document specifies the equivalent standard leak rate requirement for a 1.1 cc package, sealed with 100% nitrogen, as 5×10^{-8} atm-cc/sec. The maximum pressure that the package can withstand is 2 atm absolute. How long do you have to bomb the package at that pressure so the measured leak rate is 1×10^{-8} atm-cc/sec?

Solution. First the L of 5×10^{-8} atm-cc/sec is converted to the true leak rate of helium (L_{HE})

$$L_{HE} = 2.679 \times 5 \times 10^{-8} = 1.34 \times 10^{-7} \text{ atm-cc/sec}$$

The atmospheres in the package needed to yield $R_1 = 1 \times 10^{-8}$ atm-cc/sec should next be calculated.

$$\text{Atmospheres of helium in the package} = R_1/L_{HE}$$

$$= 1 \times 10^{-8}/1.34 \times 10^{-7} = 0.075$$

Using Eq. (4-9):

$$t = -\frac{V}{L}\left[\ln\left(1 - \frac{Q_{inp}}{\Delta p_i}\right)\right]$$

where: t = the bombing time in sec

V = the package volume = 1.1 cc

$L = L_{HE} = 1.34 \times 10^{-7}$ atm-cc/sec

Q_{inp} = 0.075 atm of helium in the package

Δp_i = 2 atm of helium

$$t = -\frac{1.1}{1.34 \times 10^{-7}}\left[\ln\left(1 - \frac{0.075}{2}\right)\right]$$

$$t = -8.21 \times 10^{-6}[\ln(0.96250)]$$

$$t = -8.21 \times 10^{-6}(-0.03822)$$

$$t = 313{,}796 \text{ sec} = 87.16 \text{ hr}$$

Problem 4. What is the measured helium leak rate limit (R_1) when tested 1 hr after bombing, for:

(a) A 4.0 cc package using the flexible method when bombed at 5 atm absolute for 4 hr?

(b) A 0.02 cc package bombed at 5 atm absolute for 2 hr?

Solution.

(a) The equivalent standard leak rate for this size package, when using the flexible method, is 1×10^{-6} atm-cc/sec. Using Eq. (2):

$$R_1 = (2.679)LP_E\left[1 - e^{-\frac{(2.679)LT_B}{V}}\right]\left[e^{-\frac{(2.679)LT_T}{V}}\right]$$

$$R_1 = 2.679\left(1\times10^{-6}\right)(5)\left[1-e^{-\frac{2.679\left(1\times10^{-6}\right)(14{,}400)}{4.0}}\right]\left[e^{-\frac{2.679\left(10^{-6}\right)(3{,}600)}{4.0}}\right]$$

$$R_1 = 1.34 \times 10^{-5}(1 - e^{-0.009644})(e^{-0.0024})$$

$$R_1 = 1.34 \times 10^{-5}(1 - 0.99759)(1) = 1.34 \times 10^{-5}(0.009575)$$

$$R_1 = 1.28 \times 10^{-7}\text{atm-cc/sec}$$

(b) The equivalent standard leak rate for this size package is 1×10^{-7} atm.-cc/sec when using the flexible method.

$$R_1 = 2.679\left(1\times10^{-7}\right)(5)\left[1-e^{-\frac{2.679\left(1\times10^{-7}\right)(7200)}{0.02}}\right]\left[e^{-\frac{2.579\left(10^{-7}\right)(3{,}600)}{0.02}}\right]$$

$$R_1 = 1.34 \times 10^{-6}(1 - e^{-0.096444})(e^{-0.04822})$$

$$R_1 = 1.34 \times 10^{-6}(0.09194)(0.095292)$$

$$R_1 = 1.17 \times 10^{-7} \text{ atm-cc/sec}$$

Problem 5. A package with an internal volume of 0.03 cc is sealed with 100% nitrogen, and is tested in accordance with the fixed method two hr after removal from the bomb. What is the R_1 limit?

Solution. The fixed method measured limit for this size package is 5 × 10^{-8} atm-cc/sec when bombed at 5 atm for 2 hr. This assumes that the test occurs within one hr after removal from the bomb. Testing the package two hr after removal from the bomb will result in a smaller R_1 limit as some of the helium has leaked out.

The equivalent standard leak rate (L) or the true helium leak rate (L_{HE}) must be known in order to calculate R_1. A direct calculation of L and L_{HE} cannot be made as they are not known. A log-log plot of L_{HE} versus R_1 will yield a straight line similar to Figs. 6-1–6-5. The equation for R_1 in terms of L_{HE} is:

$$R_1 = L_{HE} P_E \left[1 - e^{-\frac{L_{HE} t_1}{V}}\right]\left[e^{-\frac{L_{HE} t_2}{V}}\right]$$

where: R_1 = the measured helium leak rate

L_{HE} = the true helium leak rate

P_E = the bombing pressure in absolute atmospheres = 5

t_1 = the bombing time in sec = 2 hr = 7200 sec

V = the internal volume of the package = 0.03 cc

t_2 = the time between removal from the bomb and leak testing = 7200 sec

First L_{HE} must be determined. If the test is performed immediately after removal from the bomb, the term in the last bracket = 1.

Table 6-3 indicates that the value of L is less than 1×10^{-7} atm.-cc/sec, and the value of L_{HE} is less than 2.679×10^{-7} atm-cc/sec for this size package. Assigning values of L_{HE}, we can calculate R_1.

Let $L_{HE} = 1 \times 10^{-7}$ atm-cc/sec

$$R_1 = \left(1 \times 10^{-7}\right)(5)\left[1 - e^{-\frac{\left(1\times10^{-7}\right)(7200)}{0.03}}\right]$$

$$R_1 = 5 \times 10^{-7}(1 - e^{-0.024}) = 5 \times 10^{-7}(1 - 0.97628)$$

$$R_1 = 5 \times 10^{-7}(0.0237) = 1.18 \times 10^{-8} \text{ atm-cc/sec}$$

This is less than 5×10^{-8} so that L_{HE} must be larger than 1×10^{-7}.

Let $L_{HE} = 3 \times 10^{-7}$ atm-cc/sec

$$R_1 = \left(3 \times 10^{-7}\right)(5)\left[1 - e^{-\frac{\left(3\times10^{-7}\right)(7200)}{0.03}}\right]$$

$$R_1 = 1.5 \times 10^{-6}(1 - e^{-0.072}) = 1.5 \times 10^{-6}(1 - 0.93053)$$

$$R_1 = 1.5 \times 10^{-6}(0.069469) = 1.04 \times 10^{-7} \text{ atm-cc/sec}$$

This is greater than 5×10^{-8} so that the value of L_{HE} is between 1 and 3 $\times 10^{-7}$.

Let $L_{HE} = 2 \times 10^{-7}$ atm-cc/sec

$$R_1 = \left(2 \times 10^{-7}\right)\left(5\right)\left[1 - e^{-\frac{\left(2\times10^{-7}\right)\left(7200\right)}{0.03}}\right]$$

$$R_1 = 1 \times 10^{-6}(1 - e^{-0.048}) = 1 \times 10^{-6}(1 - 0.95313)$$

$$R_1 = 1 \times 10^{-6}(0.046866) = 4.69 \times 10^{-8} \text{ atm-cc/sec}$$

Let $L_{HE} = 2.07 \times 10^{-7}$ atm-cc/sec

$$R_1 = \left(2.07 \times 10^{-7}\right)\left(5\right)\left[1 - e^{-\frac{\left(2.07\times10^{-7}\right)\left(7200\right)}{0.03}}\right]$$

$$R_1 = 1.035 \times 10^{-6}(1 - e^{-0.04968}) = 1.035 \times 10^{-6}(1 - 0.95153)$$

$$R_1 = 1.035 \times 10^{-6}(0.048466) = 5.016 \times 10^{-7} \text{ atm-cc/sec}$$

Let $L_{HE} = 2.06 \times 10^{-7}$ atm-cc/sec

$$R_1 = \left(2.06 \times 10^{-7}\right)\left(5\right)\left[1 - e^{-\frac{\left(2.06\times10^{-7}\right)\left(7200\right)}{0.03}}\right]$$

$$R_1 = 1.03 \times 10^{-6}(1 - e^{-0.04944}) = 1.03 \times 10^{-6}(1 - 0.95176)$$

$$R_1 = 1.03 \times 10^{-6}(0.048238) = 4.97 \times 10^{-7} \text{ atm-cc/sec}$$

Interpolating, $L_{HE} = 2.067 \times 10^{-7}$ atm-cc/sec.

The leak rate, R_1, can now be calculated using this value for L_{HE} and including the second bracket with a value of 2 hr for t_2 (7200 sec).

$$R_1 = (2.067 \times 10^{-7})(5)\left[1 - e^{-\frac{(2.067\times10^{-7})(7200)}{0.03}}\right]\left[e^{-\frac{(2.067\times10^{-7})(7200)}{0.03}}\right]$$

$$R_1 = 1.0335 \times 10^{-6}(1 - e^{-0.049608})(e^{-0.049608})$$

$$R_1 = 1.0335 \times 10^{-6}(1 - 0.951602)(0.9516)$$

$$R_1 = 1.0335 \times 10^{-6}(0.0483976)(0.9516)$$

$$R_1 = 4.76 \times 10^{-8} \text{ atm-cc/sec}$$

Problem 6. Two devices, one with an internal volume of 0.036 cc and the other with an internal volume of 0.876 cc, are hermetically tested using the fixed method. Both have the maximum allowable leak rate. How much water will be in each device after one year in an ambient of 65% relative humidity at 25°C?

Solution. The key missing information is the true water leak rate (L_{H_2O}). When the true helium leak rate (L_{HE}) is known, L_{H_2O} can be calculated. The bombing is 2 hr at 75 psia (5 atm). The true helium leak rate is found using the approach used in Problem 5. In that problem, the volume was 0.03 cc. The first volume in this problem is 20% larger, so that L_{HE} will be a little greater than 2.067×10^{-7} atm-cc/sec. The 0.036 cc package has a maximum measured leak rate limit of 5×10^{-8} atm-cc/sec.

$$R_1 = L_{HE} P_E \left[1 - e^{-\frac{L_{HE} t}{V}}\right]$$

Let $L_{HE} = 2.5 \times 10^{-7}$ atm-cc/sec.

$$R_1 = (2.5 \times 10^{-7})(5)\left[1 - e^{-\frac{(2.5\times10^{-7})(7200)}{0.036}}\right]$$

$$R_1 = 1.25 \times 10^{-6}(1 - e^{-0.05}) = 1.25 \times 10^{-6}(1 - 0.951229)$$

$$R_1 = 1.25 \times 10^{-6}(0.04877)$$

$$R_1 = 6.09 \times 10^{-8} \text{ atm-cc/sec, too large}$$

Let $L_{HE} = 2.2 \times 10^{-7}$ atm-cc/sec.

$$R_1 = (2.2 \times 10^{-7})(5)\left[1 - e^{-\frac{(2.2\times10^{-7})(7200)}{0.036}}\right]$$

$$R_1 = 1.1 \times 10^{-6}(1 - e^{-0.044}) = 1.1 \times 10^{-6}(1 - 0.95695)$$

$$\mathrm{R}_1 = 1.1 \times 10^{-6}(0.043046)$$

$$\mathrm{R}_1 = 4.73 \times 10^{-8} \text{ atm-cc/sec, too small}$$

Interpolating, $L_{HE} = 2.26 \times 10^{-7}$ atm-cc/sec.

Let $L_{HE} = 2.26 \times 10^{-7}$ atm-cc/sec.

$$R_1 = (2.26 \times 10^{-7})(5)\left[1 - e^{-\frac{(2.26\times10^{-7})(7200)}{0.036}}\right]$$

$$R_1 = 1.13 \times 10^{-6}(1 - e^{-0.0452}) = 1.13 \times 10^{-6}(1 - 0.955806)$$

$$R_1 = 1.13 \times 10^{-7}(0.044194)$$

$$R_1 = 4.99 \times 10^{-8} \text{ atm-cc/sec, close enough}$$

$$\mathrm{L}_{H_2O} = 0.471\ L_{HE} = 0.471 \times 2.26 \times 10^{-7} \text{ (see Table 3-1)}$$

$$\mathrm{L}_{H_2O} = 1.06 \times 10^{-7} \text{ atm-cc/sec}$$

The amount of water leaking into the package in one year can now be calculated.

$$Q_{H_2O} = \Delta p \left[1 - e^{-\frac{L_{H_2O}t}{V}} \right]$$

where: Q_{H_2O} = the amount of water that has leaked in, in atm

Δp = the partial pressure of water in an ambient of 65% relative humidity at 25°C = 0.0203 atm

L_{H_2O} = 1.06×10^{-7} atm-cc/sec

t = 1 yr = 3.154×10^{7} sec

V = 0.036 cc

$$Q_{H_2O} = 0.0203 \left[1 - e^{-\frac{(1.06\times10^{-7})(3.154\times10^{7})}{0.036}} \right]$$

$$Q_{H_2O} = 0.0203(1 - e^{-93.258}) = 0.0203(1 - 0)$$

$$Q_{H_2O} = 0.0203 \text{ atm} = 2.03\% \text{ water}$$

For the package having a volume of 0.876 cc, the fixed method has an R_1 limit of 1×10^{-7} atm-cc/sec.

P_E = 3 atm, bombing time is 2 hr

P_E = 7200 sec

Figure 6-1 indicates that the L value is between 5 and 8×10^{-7} atm-cc/sec. L_{HE} is between 1.3 and 2.1×10^{-6} atm-cc/sec

Let $L_{HE} = 1.9 \times 10^{-6}$ atm-cc/sec.

$$R_1 = (1.9 \times 10^{-6})(3) \left[1 - e^{-\frac{(1.9\times10^{-6})(7200)}{0.876}} \right]$$

$$R_1 = 5.7 \times 10^{-6}(1 - e^{-0.015616}) = 5.7 \times 10^{-6}(1 - 0.9845)$$

$$R_1 = 5.7 \times 10^{-6}(0.015495) = 8.83 \times 10^{-8} \text{ atm-cc/sec}$$

L_{HE} is greater than 1.9×10^{-6} atm-cc/sec because R1 is too small.

Let $L_{HE} = 2.0 \times 10^{-6}$ atm-cc/sec.

$$R_1 = \left(2.0 \times 10^{-6}\right)(3)\left[1 - e^{-\frac{\left(2.0\times10^{-6}\right)(7200)}{0.876}}\right]$$

$$R_1 = 6 \times 10^{-6}(1 - e^{-0.016438}) = 6 \times 10^{-6}(1 - 0.983696)$$

$$R_1 = 6 \times 10^{-6}(0.0163) = 9.78 \times 10^{-8} \text{ atm-cc/sec}$$

R_1 is still too small.

Let $L_{HE} = 2.03 \times 10^{-6}$ atm-cc/sec.

$$R_1 = \left(2.03 \times 10^{-6}\right)(3)\left[1 - e^{-\frac{\left(2.03\times10^{-6}\right)(7200)}{0.876}}\right]$$

$$R_1 = 6.09 \times 10^{-6}(1 - e^{-0.0166849}) = 6.09 \times 10^{-6}(1 - 0.98345)$$

$$R_1 = 6.09 \times 10^{-6}(0.016546)$$

$$R_1 = 1.008 \times 10^{-7} \text{ atm-cc/sec, just a little high}$$

Let $L_{HE} = 2.02 \times 10^{-6}$ atm-cc/sec

$$R_1 = \left(2.02 \times 10^{-6}\right)(3)\left[1 - e^{-\frac{\left(2.02\times10^{-6}\right)(7200)}{0.876}}\right]$$

$$R_1 = 6.06 \times 10^{-6}(1 - e^{-0.0166027}) = 6.06 \times 10^{-6}(1 - 0.983534)$$

$$R_1 = 6.06 \times 10^{-6}(0.016465)$$

$$R_1 = 9.978 \times 10^{-8} \text{ atm-cc/sec, a little low}$$

Assume $L_{HE} = 2.025 \times 10^{-6}$ atm-cc/sec, then $L_{H_2O} = 0.471 \text{ L}_{HE}$

$$L_{H2O} = 9.54 \times 10^{-7} \text{ atm-cc/sec}$$

$$Q_{H_2O} = 0.0203\left[1 - e^{-\frac{(9.54\times10^{-7})(3.154\times10^{7})}{0.876}}\right]$$

$$Q_{H_2O} = 0.0203(1 - e^{-34.35}) = 0.0203(1)$$

$$Q_{H_2O} - 0.0203 \text{ atm or } 2.03\% \text{ water}$$

Both packages will have reached equilibrium with the ambient having 2.03% water

Problem 7. A 1.0 cc package is sealed with 100% nitrogen, and bombed in 3 absolute atmospheres of helium for 10 hr. It is immediately leak tested and the measured value is 5×10^{-9} atm-cc/sec. Assume a rectangular leak channel 0.1 cm. long, with an aspect ratio of 100:1. What is the true molecular and true viscous leak rates?

Solution. The true helium leak rate is unknown because the amount of helium in the package is unknown. The percent viscous flow can therefore not be determined from Fig. 3-4. To obtain an approximate true helium leak rate, assume all flow is molecular. Then the amount of helium forced into the package is Q_{HE}.

$$Q_{HE} = \Delta p_{HE}\left(1 - e^{-\frac{L_{HE}t}{V}}\right)$$

where: Δp_{HE} = 3 atm

L_{HE} is unknown

t = 10 hr = 36,000 sec

V = 1.0 cc

The value of R_1 can be calculated for different values of L_{HE} .

$$L_{HE} = \frac{R_1}{Q_{HE}} = \frac{R_1}{\Delta p_{HE}\left(1 - e^{-\frac{L_{HE}t}{V}}\right)}$$

The following table is prepared using this equation:

L_{HE}	Lt/V	$e^{-Lt/V}$	$(1 - e^{-Lt/V})$	$L\Delta p_{HE}$	R_1
1×10^{-7}	0.0036	0.996406	0.003593	3×10^{-7}	1.078×10^{-9}
1×10^{-6}	0.036	0.96464	0.035359	3×10^{-6}	1.061×10^{-7}
1×10^{-5}	0.36	0.697676	0.30232	3×10^{-5}	9.070×10^{-6}
5×10^{-7}	0.018	0.98216	0.017839	1.5×10^{-6}	2.676×10^{-8}
2×10^{-7}	0.0072	0.992826	0.007174	6×10^{-7}	4.30×10^{-9}
2.1×10^{-7}	0.00756	0.992468	0.007531	6.3×10^{-7}	4.74×10^{-9}
2.2×10^{-7}	0.00792	0.992111	0.0078887	6.6×10^{-7}	5.21×10^{-9}

The above table indicates a value for L_{HE} to be approximately 2.15×10^{-7} atm-cc/sec. Figure 3-4 shows a viscous contribution of about 5%.

True molecular helium leak rate (L_{HE}) = $0.95 \times 2.15 \times 10^{-7}$

$$L_{HE} = 2.04 \times 10^{-7} \text{ atm-cc/sec}$$

Viscous helium leak rate = $0.05 \times 2.15 \times 10^{-7} = 1.08 \times 10^{-8}$ atm-cc/sec

Problem 8. The mechanical design of an MCM limits the ambient pressure external to the package to one atmosphere. The volume of the package is 5.0 cc and is sealed with 100% nitrogen. The L leak rate limit is 1×10^{-6} atm-cc/sec (air). How do you leak test the package?

Solution. The package can not be bombed at a pressure greater than one atmosphere absolute. Bombing at one atmosphere of helium will still cause helium to enter the package because of the difference in partial helium pressures. First, an $R1$ value that is easily measurable must be assumed. Then, the amount of helium that has to be in the package to yield that $R1$ value is calculated, and lastly, the time needed for bombing at one atmosphere is calculated.

$$L_{\text{HE}} = 2.679\, L = 2.679 \times 10^{-6} \text{ atm-cc/sec}$$

Equation 4-9, can be used to calculate the required bombing time.

$$t = -\frac{V}{L_{\text{HE}}}\left[\ln\left(1 - \frac{Q}{\Delta p_i}\right)\right]$$

where: t = time in sec

V = package volume = 5.0 cc

$L_{\text{HE}} = 2.679 \times 10^{-6}$

Q = the atmospheres of helium that have leaked into the package

Δp_i = the helium partial pressure difference =1 atm

$$\text{atm helium in pkg.} = \frac{R_1}{L_{\text{HE}}} = Q$$

The larger the R_1 value, the longer the bomb time.

If $R_1 = 1 \times 10^{-7}$, then

$$Q = \frac{1 \times 10^{-7}}{2.679 \times 10^{-6}} = 0.037 \text{ atm helium}$$

$$t = -\frac{5.0}{2.679 \times 10^{-6}}\left[\ln\left(1 - \frac{0.037}{1}\right)\right]$$

$$t = -1{,}866{,}368 \ln(0.963) = -1{,}866{,}368(-0.0377)$$

$$t = 70{,}366 \text{ sec} = 19.5 \text{ hr}$$

If $R_1 = 1 \times 10^{-8}$, then $Q = 0.0037$ atm helium

$$t = -1{,}866{,}368 \ln(0.9963) = -1{,}866{,}368(-0.0037)$$

$$t = 6{,}918 \text{ sec} = 1.93 \text{ hr}$$

Bombing for 2 hr will result in an R_1 limit of 1×10^{-8} atm-cc/sec helium.

Problem 9. A 2.5 cc package is backfilled with 10% helium. The R_1 immediately after sealing is 5×10^{-9} atm-cc/sec. Forty five days later, the package is bombed for 16 hr at 3 atm absolute of helium. If the L limit is 1×10^{-7} atm-cc/sec, what is the R_1 limit after bombing?

Solution.

$$L_{\text{HE}} = \frac{R_1}{\text{atm helium in package}} = \frac{5 \times 10^{-9}}{0.1} = 5 \times 10^{-8}$$

Using Eq. (6-16):

$$Q_{\text{HE}} = \left[P_E - \left(p_i e^{-\frac{L_{\text{HE}} t}{V}}\right)_i\right]\left[1 - e^{-\frac{L_{\text{HE}} T}{V}}\right]$$

where: Q_{HE} = the amount of helium forced into the package during bombing, 45 days after sealing

P_E = helium bombing pressure = 3 atm absolute

p_I = atmospheres of helium backfill = 0.1

V = internal volume of the package = 2.5 cc

$$L_{HE} = 5 \times 10^{-8} \text{ atm-cc/sec}$$

$$t = 45 \text{ days} = 3{,}888{,}000 \text{ sec}$$

$$T = \text{helium bombing time} = 16 \text{ hr} = 57{,}600 \text{ sec}$$

$$p_i\left(e^{-\frac{L_{HE}t}{V}}\right) = \text{helium in package after 45 days, prior to bombing}$$

$$Q_{HE} = \left\{3 - \left[0.1e^{-\frac{(5\times10^{-8})(3{,}888{,}000)}{2.5}}\right]\right\}\left[1 - e^{-\frac{(5\times10^{-8})(57{,}600)}{2.5}}\right]$$

$$Q_{HE} = \{3 - [0.1(0.92519)]\}(1 - 0.998848)$$

$$Q_{HE} = (2.0748)(0.001151)$$

$$Q_{HE} = 0.00239 \text{ atm of helium due to bombing}$$

To this must be added the helium in the package just before bombing. This is:

$$p_i\left(e^{-\frac{L_{HE}t}{V}}\right) = 0.1\ (0.92519) = 0.09252$$

Total helium in the package = 0.09252 + 0.00239=0.09491 atm.

$$R_1 \text{ limit} = \text{atm helium in package} \times L_{HE} \text{ limit}$$

$$L_{HE} \text{ limit} = 2.679\, L \text{ limit} = 2.679 \times 1 \times 10^{-7} = 2.679 \times 10^{-7}$$

$$R_1 \text{ limit} = 0.09491 \times 2.679 \times 10^{-7} = 2.54 \times 10^{-8} \text{ atm-cc/sec}$$

Problem 10. A new MCM, having an internal volume of 2.0 cc, is being qualified to MIL-PRF-38534. The leak rate is to be performed using the fixed method. The RGA requirement is in accordance with Method 1018 of MIL-STD-883 and is to be performed thirty days or more after sealing. The RGA sample shall be taken from devices that have been previously subjected to screening and Subgroup 1 of Group C of MIL-PRF-38534. Is the Method 1014 leak rate requirement adequate to meet the 5,000 ppm water vapor requirement of Method 1018? If not, what measured leak rate limit is needed to keep the moisture below the 5,000 ppm limit? Assume the MCM is in an atmosphere containing 0.02 atm of water.

Solution. To determine if the specified leak rate limit will or will not allow 5,000 ppm of water to enter the MCM in 30 days, Eq. (5-8) can be used:

$$Q_{H_2O} = \Delta p_{H_2O}\left(1 - e^{-\frac{L_{H_2O}t}{V}}\right)$$

This equation requires a value for L_{H_2O}, the true water leak rate. L_{H_2O} is known if any other true leak rate is known. Table 6-3 shows the equivalent standard leak rate (L_{AIR}) for a 1.0 cc volume to be 3.6×10^{-7}, and for a 10.0 cc volume to be 1.1×10^{-6} atm-cc/sec. A 2.0 cc volume will be between these two values, close to the 3.6×10^{-7} value. A log-log plot similar to Fig. 6-2 could be made to arrive at the equivalent standard leak rate. Another approach is to perform several calculations of $R1$ for selected values of L_{HE}. When L_{HE} is known, L_{H_2O} can be calculated. The fixed R_1 value for the 2.0 cc package is 5.0×10^{-8} atm-cc/sec of helium, when bombed for 5 hr at 3 atm absolute.

Using Eq. 6-2:

$$R_1 = 2.679 \times L \times P_E\left(1 - e^{-\frac{2.679 \times Lt_1}{V}}\right)\left(e^{-\frac{2.679 \times Lt_2}{V}}\right)$$

where: $L_{HE} = 2.679 \times L$

$P_E = 3$ atm

$t_1 = 5$ hr $= 18{,}000$ sec

$t_2 = 10$ min $= 600$ sec (typical time from bomb to test)

$V = 2.0$ cc

Let $L = 4 \times 10^{-7}$ atm-cc/sec

$$R_1 = (2.679)(4 \times 10^{-7})(3)\left[1 - e^{-\frac{(2.679)(4\times10^{-7})(18{,}000)}{2.0}}\right]\left[e^{-\frac{(2.679)(4\times10^{-7})(600)}{2.0}}\right]$$

$$R_1 = 3.21 \times 10^{-6}(1 - e^{-0.0096444})(e^{-0.00032})$$

$$R_1 = 3.21 \times 10^{-6}(1 - 0.99040)(0.9907)$$

$$R_1 = 3.21 \times 10^{-6}(0.0096)$$

$R_1 = 3.08 \times 10^{-8}$ atm-cc/sec, too small

Let $L = 5 \times 10^{-7}$ atm-cc/sec

$$R_1 = (2.679)(5 \times 10^{-7})(3)\left[1 - e^{\frac{(2.679)(5\times10^{-7})(18{,}000)}{2.0}}\right]$$

$$R_1 = 4.02 \times 10^{-6}(1 - e^{-0.01205}) = 4.02 \times 10^{-6}(1 - 0.988017)$$

$$R_1 = 4.02 \times 10^{-6}(0.01198)$$

$R_1 = 4.82 \times 10^{-8}$ atm-cc/sec, too small

Let $L = 5.1 \times 10^{-7}$ atm-cc/sec

$$R_1 = (2.679)(5.1 \times 10^{-7})(3)\left[1 - e^{\frac{(2.679)(5.1\times10^{-7})(18,000)}{2.0}}\right]$$

$$R_1 = 4.10 \times 10^{-6}(1 - e^{-0.0122966}) = 4.10 \times 10^{-6}(1 - 0.987779)$$

$$R_1 = 4.10 \times 10^{-6}(0.01222)$$

$$R_1 = 5.01 \times 10^{-8} \text{ atm-cc/sec, close enough}$$

If $L = 5.1 \times 10^{-7}$, then $L_{H_2O} = 1.263\, L_{AIR}$

$$L_{H_2O} = 1.263 \times 5.1 \times 10^{-7} = 6.44 \times 10^{-7} \text{ atm-cc/sec}$$

Using Eq. (5-8) for the ingress of water,

$$Q_{H_2O} = \Delta p_{H_2O}\left(1 - e^{-\frac{L_{H_2O}t}{V}}\right)$$

where: Δp_{H_2O} = 0.02 atm = 20,000 ppm of water
t = 30 days = 2,592,000 sec
L_{H_2O} = 6.44×10^{-7} atm-cc/sec
V = 2.0 cc

$$Q_{H_2O} = 20,000\left[1 - e^{-\frac{(6.44\times10^{-7})(2,592,000)}{2.0}}\right]$$

$$Q_{H_2O} = 20,000(1 - e^{-0.8346}) = 20,000(1 - 0.43404)$$

$$Q_{H_2O} = 20,000(0.56596) = 11,319 \text{ ppm}$$

The 11,319 ppm of water is the amount of water that would enter the package if the fixed limit of Method 1014 were imposed.

To determine the leak rate that would preclude a water content greater than 5,000 ppm, Eq. (4-10) is used to calculate L_{H_2O}.

$$L_{H_2O} = -\frac{V}{t}\left[\ln\left(1 - \frac{Q_{H_2O}}{\Delta p_{H_2O}}\right)\right]$$

$$L_{H_2O} = -\frac{2.0}{2{,}592{,}000}\left[\ln\left(1 - \frac{5{,}000}{20{,}000}\right)\right]$$

$$L_{H_2O} = -7.72 \times 10^{-7}[\ln(1 - 0.25)] = -7.72 \times 10^{-7}(\ln 0.75)$$

$$L_{H_2O} = -7.72 \times 10^{-7}(-0.28768) = 2.22 \times 10^{-7} \text{ atm-cc/sec}$$

$$L = L_{AIR} = 1/1.263(L_{H_2O}) = 2.22 \times 10^{-7}/1.263$$

$$L = 1.76 \times 10^{-7} \text{ atm-cc/sec}$$

If the bombing is at 3 atm for 5 hr, then:

$$R_1 = (2.679)(1.76 \times 10^{-7})(3)\left[1 - e^{\frac{(2.679)(1.76\times10^{-7})(18{,}000)}{2.0}}\right]$$

$$R_1 = 1.41 \times 10^{-6}(1 - e^{-0.004224354}) = 1.41 \times 10^{-6}(1 - 0.99576)$$

$$R_1 = 1.41 \times 10^{-6}(0.004234)$$

$$R_1 = 5.97 \times 10^{-9} \text{ atm-cc/sec}$$

This is the measured leak rate limit necessary to meet 5,000 ppm of water in the 2.0 cc package when bombed for 5 hr at 3 atm of helium.

Problem 11. A microwave hybrid having a volume of 0.036 cc has a maximum measured leak rate limit of 5×10^{-8}, in accordance with Method 1014. If the package is backfilled with 10% helium and not bombed, what is the measured leak rate limit corresponding to the true leak rate limit?

Solution. The true helium leak rate limit for this package was calculated in Problem 6 and is 2.26×10^{-7} atm-cc/sec. For 10% backfill:

$$R_1 = 0.1 \times L_{HE} = 0.1 \times 2.26 \times 10^{-7} = 2.26 \times 10^{-8} \text{ atm-cc/sec}$$

Problem 12. MIL-PRF-38534 is being revised to include rewelding of packages that fail the fine leak test when performed to the fixed test conditions. The resealing must be performed within 24 hours of sealing, and only if the measured leak rate is sufficiently small so that no more than 1,000 ppm of water can leak into the package during the 24 hours. What is the maximum measured leak rate for a 1.1 cc package that can be reworked? Assume the ambient atmosphere contains 20,000 ppm of water.

Solution. First, the true water leak rate which would allow 1,000 ppm to enter the package in 24 hours is calculated.

$$L_{H_2O} = -\frac{V}{t}\left[\ln\left(1 - \frac{Q_{H_2O}}{\Delta p_{H_2O}}\right)\right]$$

where:

$V = 1.1$ cc

$t = 24 \text{ hr} = 86{,}400 \text{ sec}$

$Q_{H_2O} = 1{,}000$ ppm

$\Delta p_{H_2O} = 20{,}000$

$$L_{H_2O} = -\frac{1.1}{86{,}400}\left[\ln\left(1 - \frac{1{,}000}{20{,}000}\right)\right]$$

$$L_{H_2O} = -1.27 \times 10^{-5}(\ln 0.95) = -1.27 \times 10^{-5}(-0.05129)$$

$$L_{H_2O} = 6.51 \times 10^{-7} \text{ atm-cc/sec}$$

$$L_{HE} = 2.12 \times L_{H2O} = 2.12 \times 6.51 \times 10^{-7}$$

$$L_{HE} = 1.38 \times 10^{-6} \text{ atm-cc/sec}$$

Next, R_1 is calculated.

$$R_1 = L_{HE} P_E \left(1 - e^{-\frac{L_{HE} t}{V}} \right)$$

where: L_{HE} = 1.38 × 10^{-6} atm-cc/sec

P_E (the bombing pressure) = 3 atm absolute

t = bombing time = 5 hr = 18,000 sec

V = 1.1 cc

$$R_1 = \left(1.38 \times 10^{-6}\right)(3)\left[1 - e^{-\frac{\left(1.38\times10^{-6}\right)(18,000)}{1.1}}\right]$$

$$R_1 = 4.14 \times 10^{-6}(1 - e^{-0.02258}) = 4.14 \times 10^{-6}(1 - 0.97767)$$

$$R_1 = 4.14 \times 10^{-6}(0.02232) = 9.24 \times 10^{-8} \text{ atm-cc/sec}$$

Packages with a measured leak rate between 5 and 9.24 × 10^{-8} can be reworked within 24 hr.

Problem 13. A particular equipment contains several MCMs having various internal volumes. The MCMs are backfilled with 10% helium and have a measured leak rate limit of 5 × 10^{-9} atm-cc/sec. The packages can be rewelded if the leak rate is sufficiently small so that no more than 1,000 ppm of water can leak into the package within 24 hr. What is the maximum measured leak rate for all package sizes that can be reworked?

Solution. Dividing both sides of Eq. 4-10, by V yields:

$$\frac{L_{H_2O}}{V} = -\frac{1}{t}\left[\ln\left(1 - \frac{Q_{H_2O}}{\Delta p_{H_2O}}\right)\right]$$

$$\frac{L_{H_2O}}{V} = -\frac{1}{86,400}\left[\ln\left(1 - \frac{1,000}{20,000}\right)\right]$$

$$\frac{L_{H_2O}}{V} = -1.16 \times 10^{-5}\left[\ln(0.95)\right] = 1.16 \times 10^{-5}(-0.05129)$$

$$\frac{L_{H_2O}}{V} = 5.94 \times 10^{-7} \text{ atm/sec}$$

$$L_{HE} = 2.12\, L_{H_2O}$$

$$\frac{L_{HE}}{V} = 2.12\frac{L_{H_2O}}{V} = 2.12(5.94 \times 10^{-7}) = 1.26 \times 10^{-6} \text{ atm/sec}$$

$$R_1 = 0.1\, L_{HE}$$

$$\frac{R_1}{V} = 0.1\frac{L_{HE}}{V} = 1.26 \times 10^{-7} \text{ atm/sec}$$

$$R_1 = 1.26 \times 10^{-7}\ V \text{ atm-cc/sec}$$

The maximum measured leak rate that can be reworked in 24 hr is equal to 1.26×10^{-7} times the internal volume of the package.

REFERENCES

1. Howel, D. A., and Mann, C. P., *Vacuum*, 15(7):347–352, Pergamon Press Ltd., Great Britain (1965)
2. MIL-STD 883, Method 1014, *Test Methods And Procedures For Microelectronics*, U.S. Department Of Defense (March, 1995)
3. Briggs, W. E., and Burnett, S. G., "Helium Mass Spectrometer Leak Testing Of Pressure Bombed Sealed Enclosures," *Varian Vacuum Division/NRC Operation*, Newton Highlands, MA (Oct., 1968)
4. ASTM Designation F 134, *1979 Annual Book Of ASTM Standards*, pp. 545–552 (1979)
5. Banks, S. B., McCullough, R. E., and Roberts, E. G., "Investigation Of Microcircuit Seal Testing," *RADC-TR-75-89* (1975)
6. Ruthberg, S., "Graphical Solution For The Back Pressurization Method Of Hermetic Test," *IEEE Transactions On Components, Hybrids And Manufacturing Technology*, CHMT-4(2) (June, 1981)

7

Fine Leak Measurements Using a Helium Leak Detector

1.0 PRINCIPLE OF OPERATION

A helium leak detector is a mass spectrometer tuned to analyze the helium gas. The leak detecting system consists of a mass spectrometer tube, a high vacuum system, a means for helium tuning, a voltage supply, an ion current amplifier, a display system and an inlet system for the sample gas.

The mass spectrometer tube, located in the high vacuum environment, receives the gas and ionizes it by means of a hot filament. The ionized gas is then separated into ion beams of equal mass to charge ratios. The ion beams are directed to an ion collector, producing an ion current. The ion current is amplified and read by the display system.

In a helium leak detector, the mass spectrometer is tuned so that only helium ions strike the collector. A fine tuning control is available to the operator to peak the helium response. The helium ion current is proportional to the partial pressure of the helium. The display system consists of an analog meter, or a chart recorder, or a bar meter, or a digital readout. An attenuating system establishes several sensitivity ranges. The successful operation of a helium detector requires the understanding of the sources of measurement errors and their elimination.

2.0 DEFINITIONS

- *Residual Helium Background* is a signal due to helium which has been previously adsorbed on to the internal walls of the leak detector, and now released.
- *Injected Helium Background* is a signal due to helium other then from the sample being tested. An example of this is the helium in the ambient air.
- *Non-helium Background* is a signal due to an electronic circuit malfunction or due to a gas other than helium.
- *Zero* is the nulling of all signals when there is no helium present.
- *Drift* is the slow change of one or more of the backgrounds or a change in the zero.
- *Noise* is a rapid change of the background or zero.
- *Minimum Detectable Signal* is the sum of the drift and noise, or 2% of the full scale reading which ever is greater.
- *Sensitivity* is the value of the signal due to a calibrated leak, divided by the value of the calibrated leak.
- *Minimum Detectable Leak* is the minimum detectable signal divided by the sensitivity.

3.0 CALIBRATION USING A STANDARD LEAK

Calibration consists of using a standard leak together with the careful use of the leak detector. A standard leak consists of a sealed container filled with 100% helium at a pressure from 1–3 atm. The helium leaks out through a porous material such as glass. The mechanism is permeation rather than a leak channel.

The National Institute of Standards and Technology (NIST) has primary leak rate standards for each decade range. Laboratories that sell, make, or calibrate leak standards use a secondary standard that NIST calibrates. NIST guarantees the secondary standards to an absolute accuracy of ±7%. Using this 7% standard, the calibration laboratory calibrates standards to an additional accuracy of ±3% absolute. Therefore, the standards that industries use has an absolute accuracy of ±10%. These standards are effected by temperature and aging.

The change in the leak rate due to temperature is 3.3% per degree C. The higher the temperature, the more helium leaks out and the greater the leak rate. Standards are calibrated at 22°C (71.6°F). MIL-STD-883 allows the ambient temperature to be 25°C (+3°C, –5°C). Ambient temperatures at different facilities are often near this limit of 20–28°C (68–82.4°F) range.

The value of the standard leak decreases with time because helium is leaking out of the standard leak. The smaller the leak, the more linear the decrease with time. The amount leaking out the first year after calibration can vary between 1 and 3%, depending upon the size of the leak, the helium pressure inside the standard, and the volume of helium. This depletion rate is available from the laboratory that calibrates the standard.

Summarizing the errors in the standard leak:

- ±10% of the stated value
- +3.3% per +1°C
- –3.3% per –1°C
- –1% to –3% per year due to aging (helium escaping)

The actual calibration of the leak detector consists of the following steps:

1. Set the mechanical zero of the display with the leak detector off.
2. Connect the standard to the leak detector, standard valve open.
3. Set the electronic zero with the filament off, using the zero adjust control. This should be done on the scale that will be used to read the standard.
4. Turn the filament on and peak the helium response using the tuning control.
5. Set the gain to read the value of the standard.
6. Check the zero with the filament off.
7. Turn the filament back on.
8. Close the standard valve and read the background.
9. Open the standard valve and set the gain to read the standard plus the background.

10. Close the standard valve and read the background.
11. Set the gain to read the standard value plus the average of the two background readings.

The sensitivity of the leak detector can be measured at this time. Turn the gain to its maximum value and record the value. Divide this value by the value of the standard plus its background. This is the sensitivity of the machine.

Eq. (7-1) $$\text{Sensitivity} = \frac{\text{value at maximum gain}}{\text{standard} + \text{background}}$$

Example 1. The standard is labeled 5.1×10^{-8} atm-cc/sec helium. The zero is balanced on the 10×10^{-8} range. The background average is 0.3×10^{-8}. The reading is 6.6×10^{-8} at the maximum gain setting. The sensitivity is:

$$\frac{6.6}{5.4} = 1.22$$

4.0 MEASUREMENT ERRORS, NOT INCLUDING BACKGROUND ERRORS

The errors in the standard leak have been considered, but the error in making a measurement has not. There can be an error in setting the zero (when the filament is on) or in measuring the background. Assuming no error in these two parameters, there are two other errors that cannot be canceled. These are the repeatability of the measurement and the error introduced when there is a scale change.

The repeatability of a measurement is the sum of the drift and noise, but never less than ±1% of the full scale reading. When on the 10×10^{-8} range, the best repeatability on that range is $\pm 0.01 \times 10 \times 10^{-8} = \pm 0.1 \times 10^{-8}$. The background value in Example 1 is really $0.3 \times 10^{-8} \pm 0.1 \times 10^{-8}$. When the gain is adjusted to read the value of the standard plus the value of the background, an additional error of ±1% is introduced. The 5.4×10^{-8} reading has the two errors totaling $\pm 0.02 \times 10 \times 10^{-8}$, so that the correct reading is $(5.4 \pm 0.2) \times 10^{-8}$.

Measurements can be made by canceling the background instead of including it in the measurements. If the background is canceled with the zero adjust, and the filament on, the ±1% error is now in the zero setting instead of the background reading. The total error is the same for both methods.

Changing decade scales in the display of the leak detector is done electronically by multiplying or usually dividing by 10. This introduces an error of at least ±1% of the full scale reading of the largest range. The error could be as large as ±5% for leak detectors that have not been well maintained. In the following example, a leak detector with an error of ±1% will be used. Assume the standard has no error.

Example 2. A standard has a value of 9.0×10^{-8} atm-cc/sec helium, and the background is 0.1×10^{-8}, both read on the 10×10^{-8} range. The calibration sets the gain so that the standard reads 9.1×10^{-8}. A sealed package reads 5.3×10^{-9} including 0.3×10^{-9} for the sample's background read on the 10×10^{-9} range. The 5.3×10^{-9} reading has the following errors:

- Error in reading the standard with its background = $\pm 0.02 \times 10 \times 10^{-8} = \pm 2 \times 10^{-9}$ ($\pm 1 \times 10^{-9}$ for the standard, $\pm 1 \times 10^{-9}$ for the background)
- Error in changing scales = $\pm 0.01 \times 10 \times 10^{-8} = \pm 1 \times 10^{-9}$
- Error in reading the sample with its background = $\pm 0.02 \times 10 \times 10^{-9} = \pm 0.2 \times 10^{-9}$ ($\pm 0.1 \times 10^{-9}$ for the sample, $\pm 0.1 \times 10^{-9}$ for the background)
- Total error = $\pm 3.2 \times 10^{-9}$

The sample with its background has a value of $(5.3 \pm 3.2) \times 10^{-9}$. The leak rate of the sample = $(5.0 \pm 3.2) \times 10^{-9}$ atm-cc/sec helium. This large percentage error is due to the standard being on a different range then the sample. Consider the same problem, but with the standard = 9.0×10^{-9}.

Example 3. The standard equals 9.0×10^{-9} read on the 10×10^{-9} range. The background of the standard equals 0.1×10^{-9} read on the 10×10^{-9} range. The sample and its background reads 5.3×10^{-9} read on the 10×10^{-9} range. The background of the sample is also read on the 10×10^{-9} range. The errors are now:

- Error in the standard and its background = $\pm 0.02 \times 10 \times 10^{-9} = \pm 0.2 \times 10^{-9}$
- Error in reading the sample with its background = $\pm 0.02 \times 10 \times 10^{-9} = \pm 0.2 \times 10^{-9}$

- Total error = $\pm 0.4 \times 10^{-9}$

The sample and its background have a value of $(5.3 \pm 0.4) \times 10^{-9}$. The leak rate of the sample is $(5.0 \pm 0.4) \times 10^{-9}$ atm-cc/sec helium.

Examples 2 and 3 show that the sample and the standard should be on the same range.

5.0 BACKGROUND ERRORS

The background can be separated into a background due to helium, and a background due to non-helium. The source of the helium background is due to the sum of the injected helium and the residual helium The injected helium is due to the helium in the ambient air leaking into the leak detector. This is a function of the size of the leak, the pumping speed of the high vacuum system and the pressure at the detector. The leak rate of the injected helium is generally different when the standard is connected than when the sample chamber is connected. The connection of the standard consists of one seal between the standard and the leak detector. The connection of the sample chamber consists of two seals, one that connects the chamber to the leak detector, and the seal that is made when the chamber is closed. The injected helium background is generally greater for the sample chamber because of the additional seal and total greater cross-sectional area of the chamber seal. The injected helium background can be different every time the seal is made.

The source of residual helium is the desorption of helium from the walls of the detector. This is caused by a change in the pressure at the detector, usually after a large ingress of helium due to a gross leak in a sample.

A background signal due to a gas other than helium is caused by the leak detector not being optimally tuned to helium. The peaking of the helium is sensitive to the tuning adjustment and is non symmetrical. There is also a false peak but it is much smaller than the helium peak. Tuning to the false peak will result in a sensitivity of less than one. A background signal due to an electronic circuit malfunction is short in duration and is considered noise.

A change in the residual and injected helium, as well as a change in the tuning will cause a change in the background. This change in the background with time is called drift. A method for correcting the drift is to measure the background before and after the measurement, and then use the average of the background.

Example 4. The background for a sample reads 0.8×10^{-9} on the 10×10^{-9} range. The sample with the background reads 5.9×10^{-9}. The background is again measured, and reads 0.4×10^{-9}. Using the average background value of 0.6×10^{-9}, the corrected value of the sample is $(5.9 - 0.6) \times 10^{-9} = 5.3 \times 10^{-9}$.

For accurate measurements when the drift is changing rapidly, the background should be measured before and after each sample. Background measurements are required less frequently when the drift is less.

6.0 ERRORS DUE TO HELIUM ON THE EXTERNAL SURFACE OF THE PACKAGE

There is a small amount of helium attached the external surface of a bombed package. If this external helium is not removed, the leak rate will read greater than its correct value. The amount of helium attached to the external surface will vary with the size and construction of the package as well as the bombing pressure. The external helium can be removed by heating the package for 30 min at 125°C.

Packages with *porous glass* to metal seals, can have a considerable amount of helium attached to the external surface. These seals may be hermetic if the porosity is confined to the surface of the glass. The following experiment shows the quantitative effect of porous glass on the measured leak rate.

Twelve plug-in type packages were backfilled and sealed with 10% helium. Each gold-plated package had an external surface area of 6.13 in^2 and contained 23 glass to metal seals. The external surface area of the total 23 glass beads was 0.0306 in^2. Immediately after sealing the 12 packages, they were separately leak tested without bombing. The average value of the leak rates was 0.5×10^{-9} atm-cc/sec helium. The packages were then bombed for 16 hr at 3 atm of helium. After removal from the bomb, the packages were individually leak tested, and then tested again in 30 min intervals for a total time of eight hours. The average leak rates for these 12 packages for this eight hour time is shown in Fig. 7-1.

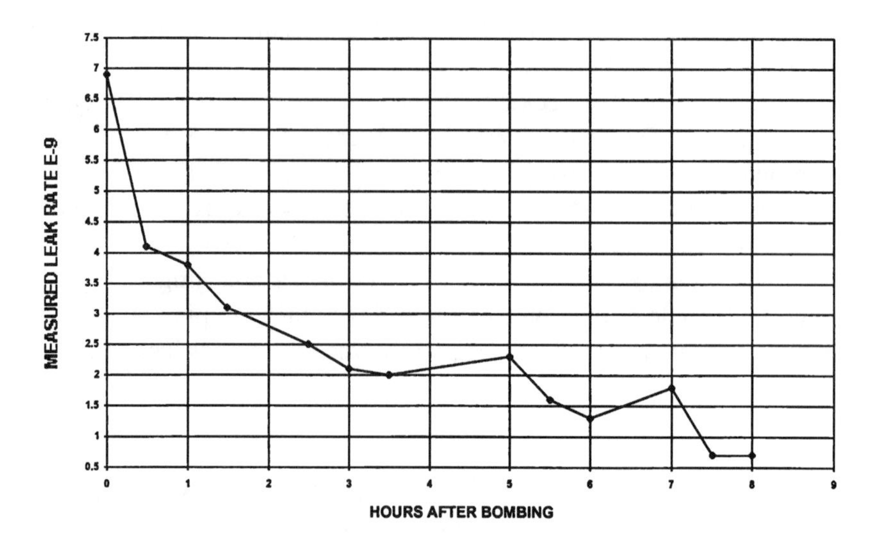

Figure 7-1. Leak rate decrease due to desorption of helium from the external package surface.

All readings of the standard and the packages were corrected for background drift. The readings are accurate to $\pm 0.2 \times 10^{-9}$. The value after eight hours (0.7×10^{-9}) is about the same as the value before bombing. The adsorbed helium contributed about 6×0^{-9}. If the measured leak rate limit was 1×10^{-8} or greater, the adsorped helium may not have been a concern. The leak rate limit in this instance was 5×10^{-9}, so that good packages read as rejects.

The leak rates in Fig. 7-1 are measured helium leak rates. The true helium leak rate is ten time larger because the packages are backfilled with 10% helium. The decrease in the measured leak rate is not due to helium escaping from the package. The decrease in leak rate due to helium escaping from this size package when the leak rate is this small, is less than two percent after 1000 hr.

Seven packages of the type described above were sealed with 10% helium and measured four times in the sequence as indicated in Table 7-1. The values in the table are the average for the seven packages. The data in Table 7-1 show the effectiveness of a 30 min bake at 125°C for removing adsorbed helium from the exterior surface of a bombed package. There is no typical error due to adsorbed helium on the surface of a package.

Table 7-1. The Effect of Heating after Bombing

After Original Bombing	After Gross Leak, 1 hr at 125°C Rebombed	After 125°C for 30 min	After Being Bombed Again
3.5×10^{-9}	3.0×10^{-9}	2.1×10^{-10}	5.1×10^{-9}

A plug-in type, gold-plated package with four glass to metal feedthroughs and four brazed pins, showed large amounts of helium adsorbed on its surface. The surface area of the package was 1.73 in^2. The glass was not known to be porous. The seven packages were sealed with 90% nitrogen and 10% helium. The measured leak rate before bombing was 1.6×10^{-9} for each package. The average adsorbed helium after bombing for four hours at 3 atm absolute was 2.32×10^{-9}, with a standard deviation of 0.32×10^{-9}. The calculated amount of helium forced into the package due to the bombing is 0.00048 atm, or an increase of 0.48% to the 0.1 atm originally there. This small increase is not measurable.

A larger package with more feedthroughs showed less helium adsorption. Twenty five gold-plated kovar plug-in packages were bombed for four hours at 3 atm absolute. The packages were sealed with 90% nitrogen and 10% helium, and had pre-bombed values from 1.4×10^{-9} to 3.4×10^{-9}. The packages had 23 glass to metal feedthroughs and 1 brazed pin. The surface of the package was 4.125 in^2. The adsorbed helium was 0.04×10^{-9} with a standard deviation of 0.71×10^{-9}. The variation of the adsorbed helium may not only vary with the type of package, but could also vary from lot to lot. Leak rates in the low 10^{-9} range in numerous packages have shown to have zero leaks as indicated by RGA.

7.0 MINIMUM DETECTABLE LEAK (MDL)

The minimum detectable leak has been defined as:

Eq. (7-2) $$\text{MDL} = \frac{\text{Minimum Detectable Signal}}{\text{Sensitivity}}$$

The minimum detectable signal on a well designed and maintained leak detector is 2% of the full scale reading on the most sensitive scale. The sensitivity is calculated using Eq. (7-1).

Example 5. The most sensitive scale of a particular leak detector is 10×10^{-9} atm-cc/sec helium. The minimum detectable signal is 2% × 10×10^{-9} atm-cc/sec = 2×10^{-10} atm-cc/sec. Using the sensitivity calculated in Example 1, and Eq. (7-2):

$$\text{MDL} = \frac{2 \times 10^{-10}}{2.2} = 9.1 \times 10^{-11} \text{ atm-cc/sec helium}$$

The minimum detectable signal is often greater than the 2% of full scale, of the most sensitive scale, because the leak detector system's properties do not support the 2%. In other words, adding an additional scale does not necessarily decrease the minimum detectable signal. The MDL can be calculated using the properties of the leak detector. The normal helium background is due to the helium in the air which is one part in 200,000 or 5×10^{-6} atm. helium per atmosphere of air. If the pumping speed and pressure at the detector are known, they can be used to calculate the MDL. This type of calculation is based on the assumption that the MDL is equal to the helium background at the detector (HeD), i.e., a signal to noise ratio of one.

Eq. (7-3) $$\text{He}D = (5 \times 10^{-6})(\text{pressure})(\text{pumping speed})$$

Example 6. The pressure at the detector is 1×10^{-6} Torr. The pumping speed is 20 l/sec.

$$\text{He}D = (5 \times 10^{-6})(10^{-6} \text{ Torr})(20 \text{ l/sec})$$

$$\text{He}D = (5 \times 10^{-6}) \left(\frac{\dfrac{10^{-6}\ \text{torr}}{760\ \text{torr}}}{\text{atm}} \right) \left(\frac{20\,1}{\text{sec}} \frac{1000\ \text{cc}}{1} \right)$$

$$\text{He}D = (5 \times 10^{-6})(1.3 \times 10^{-9}\ \text{atm})(20{,}000\ \text{cc/sec})$$

$$\text{He}D = 1.3 \times 10^{-10}\ \text{atm-cc/sec helium}$$

or,

$$MDL = 1.3 \times 10^{-10}\ \text{atm-cc/sec}$$

Equation (3) requires some clarification. One might falsely surmise that increasing the pumping speed would result in a smaller MDL, but Eq. (7-3) states just the opposite. If it takes a larger pumping speed to maintain the same pressure, the pump will force more helium to strike the detector in a unit time. This results in a larger helium background and a larger MDL.

8.0 CORRELATION OF STANDARD LEAKS

Suppose a manufacturer and buyer of MCMs each have a standard leak on the 10^{-9} range. Each standard leak is accurate to ±10%. It is prudent for the manufacturer and buyer to assume that the standard at the manufacturer is 10% low. If the manufacturer institutes a 10% guard band, then he can be sure that the buyer will not reject MCMs because of an excessive leak rate. The buyer is satisfied that all parts meet the leak rate.

The manufacturer can argue that his standard is not 10% low, and indeed he may be correct. It may be high and the buyer's standard may be 10% low. A correlation standard may be used to settle the question. The manufacturer could seal a pinless package with 10% helium. The leak rate of the package could be measured at both facilities without bombing, thereby determining which standard reads the lowest, relative to the true value.

Example 7. The standard leak at the manufacturer is labeled 3.0×10^{-9} atm-cc/sec helium. The standard at the buyer is labeled 5.0×10^{-9} atm-cc/sec helium. The pinless package correlation standard measures 2.0×10^{-9} at the manufacturer and 1.8×10^{-9} at the buyer. The standard at the buyer must be a lower value, relative to its labeled value, than the standard at the manufacturer, relative to its labeled value. The buyer's standard should be considered to be 10% low. Assuming this, the correlation standard is then $(1.8 + 0.18) \times 10^{-9} = 2.0 \times 10^{-9}$. The manufacturer need not guard band because of the standard.

9.0 LOCATING LEAKS IN PACKAGES

Package related problems can often be illuminated by knowing the exact location of the leak path. Experimental verification of the leak being at the cover-header interface, will lead to an investigation of the sealing process. Repetitive leaks at the same pin location may be an indication of an inadequate package design. Cracks in a ceramic package may also indicate a package design fault or excessive mechanical stresses. Knowing the exact location of the leak can be a major factor in a failure analysis.

The method of finding the location of the leak consists of connecting the leak detector to the inside of the package, and supplying helium to the outside of the package. This is similar to the leak test for header integrity (A4 of Method 1014 of MIL-STD-883). The A4 test floods the entire package; this is not suitable for determining a specific leak location. Instead of flooding the entire package, a hypodermic needle is used with a slow flow of helium. The end of the needle can be placed next to each pin or lead, or at the header-cover interface. There are two ways to pump on the inside of the package. The first method requires removing the cover and then placing the open package side against a fixture connected to the leak detector. This is a simple fixture of the type used to leak test headers, but this method will of course not test the header-cover interface. The second method consists of punching a hole in the cover and connecting the hole to the leak detector. This involves a more complicated fixture, with an "o-ring" making contact with the cover.

10.0 PROBLEMS AND THEIR SOLUTIONS

Problem 1. A package and standard leak are measured at 28°C. The value stamped on the standard leak is 3.50×10^{-9} atm-cc/sec helium. The package measures 8.4×10^{-9} atm-cc/sec. The background measures 1×10^{-9} atm-cc/sec. The gain of the leak detector is set so that the standard reads 4.50×10^{-9} atm-cc/sec (standard value plus the background). What is the correct value for the package?

Solution. The standard leak is calibrated at 22°C, and has a temperature correction of 3.3% per degree C. The standard leak at 28°C has increased helium escaping by (28–22)(3.3%) = 19.8%. The package will therefore read 19.8% high. The correct leak rate for the package $= 8.4 \times 10^{-9} - 19.8\,\% = (8.4 - 1.66) \times 10^{-9} = 6.74 \times 10^{-9}$ atm-cc/sec.

Problem 2. What is the leak rate and error of the package due to only changing scales when the standard is 1×10^{-7} atm-cc/sec helium read on the 10×10^{-8} scale, and the package is 7×10^{-9} atm-cc/sec helium read on the 10×10^{-9} scale?

Solution. The error due changing scales is at least ±1% full scale of the largest scale. If 1% is applicable, the error is $\pm 0.01 \times 10 \times 10^{-8} = \pm 1 \times 10^{-9}$. The leak rate of the package is $7 \times 10^{-9} \pm 1 \times 10^{-9}$ atm-cc/sec helium.

Problem 3. What is the leak rate of the package, with its measurement error only due to changing scales, when the standard is 1×10^{-7} atm-cc/sec helium read on the 10×10^{-8} scale and the package measures 8.0×10^{-10} atm-cc/sec helium read on the 10×10^{-10} scale?

Solution. There are two scale changes (two electronic divisions) from 10×10^{-8} to 10×10^{-9} and from 10×10^{-9} to 10×10^{-10}. The error due to the first scale change is $\pm 0.01 \times 10 \times 10^{-8} = 1 \times 10^{-9}$. The error due to the second change is $\pm 0.01 \times 10 \times 10^{-9} = \pm 1 \times 10^{-10}$. The total change due to changing scales is 1.1×10^{-9}. The leak rate of the package is $8.0 \times 10^{-10} \pm 1.1 \times 10^{-9}$ atm-cc/sec helium. The standard being two scales from the package measurement results in an error larger than the leak rate.

Problem 4. Five packages are measured on a leak detector that has an appreciable background drift. The measurements of the five packages and their backgrounds are shown in the following table.

Assuming errors only due to the background of the packages, what is the average of the five packages after correcting for the background?

Item Tested	Time of Test	Measured Value
Background	1:01	1×10^{-9}
Package #1	1:03	8×10^{-9}
Background	1:05	2×10^{-9}
Package #2	1:07	8.5×10^{-9}
Background	1:09	2.5×10^{-9}
Package #3	1:11	7×10^{-9}
Background	1:13	3×10^{-9}
Package #4	1:15	8×10^{-9}
Background	1:17	3.5×10^{-9}
Package #5	1:19	9×10^{-9}
Background	1:21	4×10^{-9}

Solution. The average of the backgrounds immediately before and after the measurement of the package should be subtracted from the package measurement. The table above is repeated with an additional correction column.

Item Tested	Time of Test	Measured Value	Corrected Value
Background	1:01	1×10^{-9}	
Package #1	1:03	8×10^{-9}	$(8 - 1.5) \times 10^{-9} = 6.5 \times 10^{-9}$
Background	1:05	2×10^{-9}	
Package #2	1:07	8.5×10^{-9}	$(8.5 - 2.25) \times 10^{-9} = 6.25 \times 10^{-9}$
Background	1:09	2.5×10^{-9}	
Package #3	1:11	7×10^{-9}	$(7 - 2.75) \times 10^{-9} = 4.25 \times 10^{-9}$
Background	1:13	3×10^{-9}	
Package #4	1:15	8×10^{-9}	$(8 - 3.25) \times 10^{-9} = 4.75 \times 10^{-9}$
Background	1:17	3.5×10^{-9}	
Package #5	1:19	9×10^{-9}	$(9 - 3.75) \times 10^{-9} = 5.25 \times 10^{-9}$
Background	1:21	4×10^{-9}	

The average of the corrected values is 5.4×10^{-9}.

Problem 5. A standard leak and package are both measured on the 10×10^{-9} scale. The standard is labeled 5.0×10^{-9} atm-cc/sec helium. The background is drifting. The background, the standard, and a package are measured three times, as shown in the following table.

Item Tested	Time of Test	Measured Value
Background	1:01	1.0×10^{-9}
Standard	1:04	6.0×10^{-9}
Background	1:07	1.5×10^{-9}
Package	1:11	6.0×10^{-9}
Background	1:15	2×10^{-9}
Standard	1:19	7.1×10^{-9}
Background	1:23	2.5×10^{-9}
Package	1:27	7.0×10^{-9}
Background	1:31	2.7×10^{-9}
Standard	1:35	8.0×10^{-9}
Background	1:39	3.0×10^{-9}
Package	1:43	8.0×10^{-9}
Background	1:47	3.2×10^{-9}

What is the corrected leak rate value of the package?

Solution. Both the standard and the package readings have to be corrected for the background. The following table is a copy of the above table with the addition of three correction columns.

When the standard reads below the calibrated value, the difference must be added to the value of the package, when high it must be subtracted. The average of the corrected values is $(4.5 + 4.55 + 4.75)/3 = 4.60 \times 10^{-9}$.

Item Tested	Time of Test	Measured Value	Background Correction	Error in Standard	Total Correction
Background	1:01	1.0×10^{-9}			
Standard	1:04	6.0×10^{-9}	$(6.0 - 1.25) = 4.75 \times 10^{-9}$	-0.15×10^{-9}	
Background	1:07	1.5×10^{-9}			
Package	1:11	6.0×10^{-9}	$(6.0 - 1.75) = 4.25 \times 10^{-9}$		$(4.25 + 0.25) = 4.5 \times 10^{-9}$
Background	1:15	2×10^{-9}			
Standard	1:19	7.1×10^{-9}	$(7.1 - 2.25) = 4.85 \times 10^{-9}$	-0.15×10^{-9}	
Background	1:23	2.5×10^{-9}			
Package	1:27	7.0×10^{-9}	$(7.0 - 2.6) = 4.4 \times 10^{-9}$		$(4.4 + 0.15) = 4.55 \times 10^{-9}$
Background	1:31	2.7×10^{-9}			
Standard	1:35	8.0×10^{-9}	$(8.0 - 2.85) = 5.15 \times 10^{-9}$	$+0.15 \times 10^{-9}$	
Background	1:39	3.0×10^{-9}			
Package	1:43	8.0×10^{-9}	$(8.0 - 3.1) = 4.9 \times 10^{-9}$		$(4.9 - 0.15) = 4.75 \times 10^{-9}$
Background	1:47	3.2×10^{-9}			

Problem 6. An external standard for a leak detector is labeled 2.0×10^{-7} atm-cc/sec helium. The background on the 10×10^{-7} scale is 0.2×10^{-7}. The gain of the leak detector is adjusted so the standard reads 2.2×10^{-7} atm-cc/sec helium. A package is leak tested and measures 4.0×10^{-8} atm-cc/sec helium on the 10×10^{-8} range. The background on the 10×10^{-8} scale is 1×10^{-8}.

(a) Assuming the value of the standard is correct, what is the error in measuring the standard with its background?

(b) What is the leak rate of the package including the measurement errors?

Solution.

(a) There are two errors associated with measuring the standard and its background. The ±1% of full scale reading of the standard, and ±1% of full scale reading of the background.

$$\text{standard} = 2.0 \times 10^{-7} \pm 0.01 \times 10 \times 10^{-7} = 2.0 \times 10^{-7} \pm 1 \times 10^{-8}$$

$$\text{background} = 0.2 \times 10^{-7} \pm 0.01 \times 10 \times 10^{-7} = 0.2 \times 10^{-7} \pm 1 \times 10^{-8}$$

$$\text{standard with its background} = 2.2 \times 10^{-7} \pm 2 \times 10^{-8}$$

(b) The errors in measuring the package are:

- The error in measuring the standard = $\pm 2.0 \times 10^{-8}$
- The ±1% of full scale when reading the package with its background = $\pm 0.01 \times 10 \times 10^{-8} = \pm 1.0 \times 10^{-9}$
- The ±1% of full scale when reading the background to the package = $\pm 0.01 \times 10 \times 10^{-8} = \pm 1.0 \times 10^{-9}$
- The ±1% of full scale because of changing scales between the standard and the package. $\pm 0.01 \times 10 \times 10^{-7} = \pm 1 \times 10^{-8}$
- The package plus the background = 4.0×10^{-8}, $\pm 2.0 \times 10^{-8}$, $\pm 1 \times 10^{-9}$, $\pm 1 \times 10^{-9}$, $\pm 1 \times 10^{-8} = 4.0 \times 10^{-8} \pm 3.2 \times 10^{-8}$ atm-cc/sec helium.

The leak rate of the package = the leak rate of the package with its background – the background = $(4.0 \times 10^{-8} \pm 3.2 \times 10^{-8}) - 1 \times 10^{-8} = 3.0 \times 10^{-8} \pm 3.2 \times 10^{-8}$ atm-cc/sec helium.

Problem 7. An operator calibrates a leak detector at 22°C using a standard leak labeled 3.0×10^{-8} atm-cc/sec helium. The background for the standard measures 0.5×10^{-8} on the 10×10^{-8} scale. He sets the gain of the leak detector so the standard with its background reads 3.5×10^{-8} atm-cc/sec helium. A package is leak tested and measures 5×10^{-9} on the 10×10^{-9} scale, including the background of 1×10^{-9}. The operator rechecks the standard and finds that the background to the standard is really 1×10^{-8}. What is the correct leak rate of the package, including all errors?

Solution. There are the following errors:

- Error in reading the standard with background = ±1% of full scale = $\pm 0.01 \times 10 \times 10^{-8} = \pm 1.0 \times 10^{-9}$

- Error in reading the background of the standard = ±1% of full scale = $\pm 0.01 \times 10 \times 10^{-8} = \pm 1.0 \times 10^{-9}$
- The percent error due to measuring the incorrect background of the standard = $(3.5 - 4.0) \times 100/4.0 = -12.5\%$
- Error in reading the package with its background = ±1% of full scale = $\pm 0.01 \times 10 \times 10^{-9} = \pm 1.0 \times 10^{-10}$
- Error in reading the background to the package = $\pm 1.0 \times 10^{-10}$
- Error due to changing scales = $\pm 0.01 \times 10 \times 10^{-8} = \pm 1.0 \times 10^{-9}$

Package plus its background = 5×10^{-9}, +12.5%, $\pm 2.0 \times 10^{-9}$ (std + bg), $\pm 2.0 \times 10^{-10}$ (pkg + bg), $\pm 1.0 \times 10^{-9}$ (scale) = $5.62 \times 10^{-9} \pm 3.2 \times 10^{-9}$. The leak rate of the package = the package + background – background = $5.62 \times 10^{-9} \pm 3.2 \times 10^{-9} - 1 \times 10^{-9} = 4.62 \times 10^{-9} \pm 3.2 \times 10^{-9}$ atm-cc/sec helium.

Problem 8. Assuming a signal to noise ratio of one, what is the minimum detectable leak (MDL) for the following leak detecting systems?

(a) Pressure at the detector = 1×10^{-7} atm, pumping speed = 20 l/sec

(b) Pressure at the detector = 1×10^{-8} atm, pumping speed = 60 l/sec

(c) Pressure at the detector = 1×10^{-9} atm, pumping speed = 100 l/sec

(d) Pressure at the detector = 1×10^{-6} torr, pumping speed = 40 l/sec

Solution. The helium in the air = 5×10^{-6} atm of helium per atm of air. Using Eq. (7-3):

MDL = HeD = (5×10^{-6})(pressure at the detector)(pumping speed)

(a) MDL = $(5 \times 10^{-6})(1 \times 10^{-7}$ atm)(20 l/sec)(1000 cc/l) = 1×10^{-8} atm-cc/sec helium

(b) MDL = $(5 \times 10^{-6})(1 \times 10^{-8}$ atm)(60 l/sec)(1000 cc/l) = 3×10^{-9} atm-cc/sec helium

(c) MDL = $(5 \times 10^{-6})(1 \times 10^{-9}$ atm)(100 l/sec) (1000 cc/sec) = 5×10^{-10} atm-cc/sec helium

(d) MDL = $(5 \times 10^{-6})(1 \times 10^{-6}$ torr)(1 atm/760 torr)(40 l/sec)(1000 cc/l) = 2.6×10^{-10} atm-cc/sec

Problem 9. The company wants to purchase a leak detector and you are to recommend the one that can measure the smallest MDL. Assume that all leak detectors have a sensitivity of one, and you have the following four choices:

(a) Pressure at the detector = 1×10^{-9} atm, pumping speed = 100 l/sec, most sensitive scale = 10×10^{-9}

(b) Pressure at the detector = 1×10^{-9} atm, pumping speed = 20 l/sec, most sensitive scale = 10×10^{-10}

(c) Pressure at the detector = 1×10^{-8} atm, pumping speed = 30 l/sec, most sensitive scale = 10×10^{-10}

(d) Pressure at the detector = 1×10^{-7} torr, pumping speed = 40 l/sec, most sensitive scale = 10×10^{-10}

Solution. The MDL of a leak detector is the greater of the value found in Eq. (7-3), or the 2% of full scale of the most sensitive scale divided by the sensitivity.

(a) From Eq. (7-3):

MDL = (5×10^{-6})(pressure at the detector)(pumping speed)

MDL = $(5 \times 10^{-6})(1 \times 10^{-9}$ atm)(100 l/sec)(1000 cc/l)

MDL = 5×10^{-10} atm-cc/sec helium

2% of full scale = $0.02 \times 10 \times 10^{-9} = 2 \times 10^{-10}$ atm-cc/sec helium

The MDL of leak detector (a) is the greater of the two and equals 5×10^{-10} atm-cc/sec helium.

For leak detector (b):

MDL = $(5 \times 10^{-6})(1 \times 10^{-9}$ atm)(20 l/sec)(1000 cc/l)

MDL = 1×10^{-10} atm-cc/sec helium

2% of full scale = $0.02 \times 10 \times 10^{-10} = 2 \times 10^{-11}$ atm-cc/sec helium

The MDL of leak detector (b) is 1×10^{-10} atm-cc/sec helium.

For leak detector (c):

$$MDL = (5 \times 10^{-6})(1 \times 10^{-8}\ \text{atm})(30\ \text{l/sec})(1000\ \text{cc/l})$$

$$MDL = 1.5 \times 10^{-9}\ \text{atm-cc/sec helium}$$

$$2\%\ \text{of full scale} = 0.02 \times 10 \times 10^{-10} = 2 \times 10^{-11}\ \text{atm-cc/sec helium}$$

The MDL of leak detector (c) is 1.5×10^{-9} atm-cc/sec helium.

For leak detector (d):

$$MDL = (5 \times 10^{-6})(1 \times 10^{-7}\ \text{torr})(1\ \text{atm}/760\ \text{torr})(40\ \text{l/sec})(1000\ \text{cc/l})$$

$$MDL = 2.63 \times 10^{-11}\ \text{atm-cc/sec helium}$$

$$2\%\ \text{of full scale} = 0.02 \times 10 \times 10^{-10} = 2 \times 10^{-11}\ \text{atm-cc/sec helium}$$

The MDL of leak detector (d) is 2.63×10^{-11} atm-cc/sec helium. *Leak detector (d) will measure the smallest MDL.*

Problem 10. A subcontractor has a leak detector which uses an internal standard leak = 1×10^{-7} atm-cc/sec helium. The detector's most sensitive scale is 10×10^{-9}. The subcontractor measures a package to be 5×10^{-9} atm-cc/sec helium. It is not known if the standard was measured on the 10×10^{-7} or on the 10×10^{-8} range. It is also not known if the package was measured on the 10×10^{-8} or on the 10×10^{-9} range. Assuming no background errors, what are the measurement errors for the four possible measurement options?

Solution. The four possible measurement options are:

(a) Leak standard on the 10×10^{-7} range, package on the 10×10^{-8} range.

(b) Leak standard on the 10×10^{-7} range, package on the 10×10^{-9} range.

(c) Leak standard on the 10×10^{-8} range, package on the 10×10^{-8} range.

(d) Leak standard on the 10×10^{-8} range, package on the 10×10^{-9} range.

Error for (a):

- The error in reading the standard = $\pm 1\% \times 10 \times 10^{-7} = \pm 1.0 \times 10^{-8}$
- The error in reading the package = $\pm 1\% \times 10 \times 10^{-8} = \pm 1.0 \times 10^{-9}$
- The error in changing scales = $\pm 1.0 \times 10 \times 10^{-7} = \pm 1.0 \times 10^{-8}$
- The total error is 2.1×10^{-8}

The leak rate of the package is $5 \times 10^{-9} \pm 2.1 \times 10^{-8}$ atm-cc/sec helium.

Error for (b):

- The error in reading the standard = $\pm 1\% \times 10 \times 10^{-7} = \pm 1.0 \times 10^{-8}$
- The error in reading the package = $\pm 1\% \times 10 \times 10^{-9} = \pm 1.0 \times 10^{-10}$
- The error in changing the scale from 10×10^{-7} to $10 \times 10^{-8} = \pm 1\% \times 10 \times 10^{-7} = \pm 1.0 \times 10^{-8}$
- The error in changing the scale from 10×10^{-8} to $10 \times 10^{-9} = \pm 1\% \times 10 \times 10^{-8} = \pm 1.0 \times 10^{-9}$
- The total error = $\pm 2.11 \times 10^{-8}$

The leak rate of the package = $5 \times 10^{-9} \pm 2.11 \times 10^{-8}$ atm- cc/sec helium.

Error for (c):

- The error in reading the standard = $\pm 1\% \times 10 \times 10^{-8} = \pm 1.0 \times 10^{-9}$
- The error in reading the package = $\pm 1\% \times 10 \times 10^{-8} = \pm 1.0 \times 10^{-9}$
- No scale change
- Total error = $\pm 2.0 \times 10^{-9}$

The leak rate of the package = $5 \times 10^{-9} \pm 2.0 \times 10^{-9}$ atm-cc/sec helium.

Error for d):

- The error in reading the standard = $\pm 1\% \times 10 \times 10^{-8} = \pm 1.0 \times 10^{-9}$
- The error in reading the package = $\pm 1\% \times 10 \times 10^{-9} = \pm 1.0 \times 10^{-10}$
- The error in changing the scale from 10×10^{-8} to $10 \times 10^{-9} = \pm 1\% \times 10 \times 10^{-8} = \pm 1.0 \times 10^{-9}$

- The total error = $\pm 2.1 \times 10^{-9}$

The leak rate of the package = $5 \times 10^{-9} \pm 2.1 \times 10^{-9}$ atm-cc/sec helium.

Problem 11. A vendor has their leak detector modified so that it reads the equivalent water leak rate instead of helium. They have an external standard that is labeled 2.5×10^{-9} atm-cc/sec helium. Using this standard leak, they set the gain of the detector so it reads 2.5×10^{-9} on the water scale. A package containing 10% helium measures 5×10^{-9} on the water scale. Assuming no background errors, what is the true helium leak rate of the package?

Solution. There are two errors: first in setting the gain, and second in using a water scale instead of the helium scale.

$$L_{HE} = L_{H_2O}\,(M_{H_2O}/M_{HE})^{1/2} = L_{H_2O}\,(18/4)^{1/2} = 2.12 \times L_{H_2O}$$

so that the gain was really set to $2.12 \times 2.5 \times 10^{-9} = 5.3 \times 10^{-9}$ atm-cc/sec helium. When the package read 5×10^{-9} on the water scale, this was $2.12 \times 5 \times 10^{-9}$ on the helium scale = 1.06×10^{-8} atm-cc/sec helium. But since the gain was set high by a factor of 2.12, this helium rate must be divided by 2.12:

$$1.06 \times 10^{-8}/2.12 = 5.0 \times 10^{-9}$$

Therefore the leak rate of the package = 5×10^{-9} atm-cc/sec helium. The two errors were self canceling.

Problem 12. A lot of headers is suspected of having porous glass that could adsorb helium when bombed. A sample of five unsealed headers are subjected to the same bombing conditions as the sealed packages. Immediately after bombing, the five headers are leak tested as a group. The leak detector measures 4×10^{-8} atm-cc/sec helium. What is the expected helium adsorption of a single sealed package?

Solution. The five open headers have ten times the glass bead area as a sealed package. Five times because there are five headers, and a factor of two because both the inside and the outside of the headers are exposed to the helium. A sealed package is expected to have 4×10^{-9} atm-cc/sec helium adsorbed on its surface.

Problem 13. A customer purchased some MCMs from a vendor, with a maximum measured leak rate of 5.0×10^{-9} atm-cc/sec helium. The customer and vendor agreed that an external leak standard of 3.0×10^{-9} atm-cc/sec helium would be used, all measurements were to be made on the 10×10^{-9} scale. The vendor would guard band for the errors under these conditions. What is the correct guard band and maximum leak rate limit?

Solution. The guard band is the total measurement errors for the conditions.

- Error in measuring the standard and its background = $\pm 1\% \times 10 \times 10^{-9} = \pm 1.0 \times 10^{-10}$
- Error in measuring the background of the standard = $\pm 1\% \times 10 \times 10^{-9} = \pm 1.0 \times 10^{-10}$
- Error in measuring the package and its background = $\pm 1\% \times 10 \times 10^{-9} = \pm 1.0 \times 10^{-10}$
- Error in measuring the background of the package = $\pm 1\% \times 10 \times 10^{-9} = \pm 1.0 \times 10^{-10}$

The total error = 4×10^{-10} atm-cc/sec helium and this is equal to the guard band. The maximum leak rate limit = $5 \times 10^{-9} - 4 \times 10^{-10} = 4.6 \times 10^{-9}$ atm-cc/sec helium.

Problem 14. The customer received several lots of MCMs from the vendor as presented in Problem 13. The vendor subsequently sent a letter stating that the last lot shipped was measured incorrectly. The conditions were:

- The maximum allowed measured limit was 4.6×10^{-9} atm-cc/sec helium
- All MCMs were measured on the 10×10^{-9} range
- A 1×10^{-7} leak standard was used and measured on the 10×10^{-8} range instead of the 3.0×10^{-9} standard measured on the 10×10^{-9} range

The vendor has recorded leak rate data by serial numbers of the MCMs. What units should be returned to the vendor?

Solution. Here again, the guard band is equal to the total measurement error.

- Error in measuring the standard and its background = $\pm 1\% \times 10\text{x}0^{-8} = \pm 1.0 \times 10^{-9}$
- Error in measuring the background to the standard = $\pm 1\% \times 10 \times 0^{-8} = \pm 1.0 \times 10^{-9}$
- Error in measuring the MCM and its background = $\pm 1 \times 10 \times 10^{-9} = \pm 1.0 \times 10^{-10}$

- Error in measuring the background of the MCM $= \pm 1 \times 10 \times 10^{-9} = \pm 1.0 \times 10^{-10}$
- Error in changing scales from 10×10^{-8} to $10 \times 10^{-9} = \pm 1\% \times 10x0^{-8} = \pm 1.0 \times 10^{-9}$

The total error and guard band $= 3.2 \times 10^{-9}$ atm-cc/sec helium. The limit is $(5 - 3.2) \times 10^{-9} = 1.8 \times 10^{-9}$ atm-cc/sec helium. MCMs that measured greater than 1.8×10^{-9} should be returned to the vendor.

Problem 15. An external standard reads 4.0×10^{-9} on the 10×10^{-9} range. A sample reads 1×10^{-9} on the 10×10^{-9} range. What is the leak rate of the sample including the repeatability errors and an estimated helium adsorption of 1×10^{-9} atm-cc/sec?

Solution.

- The error in reading the standard $= \pm 1\%$ of full scale $= \pm 0.1 \times 10^{-9}$
- The error in reading the background of the standard $= \pm 1\%$ of full scale $= \pm 0.1 \times 10^{-9}$
- The error in reading the sample $= \pm 1\%$ of full scale $= \pm 0.1 \times 10^{-9}$
- The error in reading the background of the sample $= \pm 1\%$ of full scale $= \pm 0.1 \times 10^{-9}$
- The error due to the adsorbed helium $= -1 \times 10^{-9}$

The leak rate of the sample $= 1 \times 10^{-9} \pm 0.4 \times 10^{-9}, -1 \times 10^{-9} =$ a maximum value of 4×10^{-10} and a minimum value of zero.

8

Gross Leaks

1.0 INTRODUCTION

Gross leaks can be defined in three ways: by the leak rate of the package, by detection of a liquid or its byproducts in the package, or by the size of the leak channel. Standard air equivalent leak rates greater than 1×10^{-5} atm-cc/sec are considered gross leak rates. Gross leaks are detected rather than measured quantitatively. A quantitative value of a gross leak rate is usually of no interest, as any package having a gross leak is unacceptable. An exception to this is a failure analysis investigation into the cause of the gross leak. If a gross leak test method is positive (there is a gross leak), the package is a gross leaker. A negative response to a particular gross leak method does not eliminate a possible gross leak, as different test methods have different sensitivities. Packages which contain liquids, gases derived from liquids, or a large influx of gases, are gross leakers. As discussed later in this chapter, leak channels with all cross-sectional dimensions greater than 1×10^{-4} cm are gross leakers.

There are generally four methods of ascertaining a gross leak:

1. Forcing a liquid into the package through the gross leak channel, vaporizing or decomposing this liquid in the package thereby forcing the resultant gas out through the same leak channel, and then detecting the gas.
2. Forcing a liquid into the package, then detecting its presence by a change in weight or by the deflection of the lid due to an increase in pressure.

3. Forcing a dye penetrant into the package, opening the package and observing the presence of the dye.
4. Performing a residual gas analysis (RGA).

2.0 FORCING A LIQUID INTO A PACKAGE

Liquids enter a package by viscous flow. This can only occur when there is a difference in total pressure between the inside and outside of the package. Packages which have large holes, such as when a pin falls out, can have liquids enter the package by displacing some of the gas. This type of liquid injection is not considered in this chapter.

The viscous conductance of a cylindrical leak channel, with slip and end correction, has been given in Ch. 2 as Eq. (2-4), and is repeated here as Eq. (8-1):

Eq. (8-1)
$$F_{VC} = \frac{\pi D^4 P_a Y}{128\eta\left[\ell + \left(\frac{Y}{Z}\right)D\right]}$$

where F_{VC} = viscous conductance in cc/sec

D = the diameter of the leak channel in cm

P_a = average pressure in dynes/cm^2 $(P_1+P_2)/2$

η = the viscosity in poise [(dyne-sec)/cm^2]

ℓ = length of the leak channel in cm

Y and Z are calculated using Eqs. (8-2) and (8-3):

Eq. (8-2)
$$Y = 1 + \left[\frac{4.4445}{\left(\frac{D}{\text{mfp}}\right) + 1.1949}\right]$$

Eq. (8-3)
$$Z = 1.6977 + \left[\frac{4.6742}{\left(\frac{D}{\text{mfp}}\right) + 2.0444} \right]$$

The equations in Ch. 2 were applied to the conductance of viscous *gases* but the equation is also valid for the viscous conductance of *liquids*. The viscosity of liquids is much greater than the viscosity of gases. The typical viscosity of the detector liquid forced into packages during the gross leak test is 0.012 poise, compared to the viscosity of helium being 194×10^{-6} poise. For the same size cylindrical leak channel, the viscosity of the detector liquid is:

$$\frac{0.012}{194 \times 10^{-6}} = 61.856 \text{ times the viscosity of helium}$$

The measured viscous flow rate is equal to the viscous conductance multiplied by the initial difference in total pressure. The true or standard viscous flow rate can be defined as when the pressure difference is one atmosphere and the average pressure is one half an atmosphere.

The viscous flow of gases was discussed in Ch. 4. The viscous equations in Ch. 4 also apply to liquids. Equation (4-16a) from Ch. 4 is repeated here as Eq. (8-4):

Eq. (8-4)
$$Q_{inv} = -V \Delta P_i \left[\left(e^{-\frac{L_V t}{V}} \right) - 1 \right]$$

where: Q_{inv} = quantity of fluid entering a package by viscous flow in units of volume × pressure

V = the internal volume of the package

ΔP_i = the initial total pressure difference in atm

L_V = the true viscous leak rate in atm-cc/sec

t = time in sec

To get Q in terms of atmospheres, Eq. (8-4) is divided by the volume, resulting in Eq. (8-5):

Eq. (8-5) $$Q_{invP} = \Delta P_i \left(1 - e^{-\frac{L_V t}{V}} \right)$$

where Q_{invP} = quantity of fluid entering the package in atm

When the difference in total pressure is not one atmosphere, Eq. (8-5) becomes:

Eq. (8-6) $$Q_{invP} = \Delta P_i \left(1 - e^{-\frac{R_V t}{V \Delta P_i}} \right)$$

where R_v = the measured or calculated viscous leak rate

As in the case for gases, the shape of the leak channel is unknown. The shape having the greatest conductance is a cylinder. Calculation of liquid injection through cylindrical shaped channels represents the maximum liquid injection for that cross-sectional area. Viscous flow through rectangular channels is more restrictive.

The relationship between the viscous conductance of the detector liquid and the viscous conductance of the helium, for cylindrical leak channels, is a function of the differences in their viscosity (η), their mean free path (mfp), their average pressure and their correction factor (Y). This relationship is presented as Eq. (8-7):

$$F_{CDET} = \frac{(P_{AVEDET})(Y_{DET})}{\eta_{DET}}$$

$$F_{VCHE} = \frac{(P_{AVEHE})(Y_{HE})}{\eta_{HE}}$$

Dividing F_{CDET} by F_{VCHE},

$$\frac{F_{CDET}}{F_{VCHE}} = \frac{(P_{AVDET})(Y_{DET})(\eta_{HE})}{(\eta_{DET})(P_{AVEHE})(Y_{HE})}$$

Eq. (8-7) $$F_{CDET} = F_{VCHE} \times \frac{\eta_{HE}}{\eta_{DET}} \times \frac{P_{AVEDET}}{P_{AVEHE}} \times \frac{Y_{DET}}{Y_{HE}}$$

The mfp of the detector liquid is approximately = 1×10^{-7} cm so that *D*/mfp is very large (100–1000). Therefore Y_{DET} is very close to 1.00, Z_{DET} is very close to 1.70 and *Y*/*Z* is very close to 0.59.

The following calculations can be made for a cylindrical leak channel:

1. The viscous leak rate of helium during leak testing.
2. The molecular leak rate of helium during leak testing.
3. Adding the above two, gives the total helium leak rate during testing.
4. The leak rate of the gross leak detector liquid being forced into a package during bombing.
5. The volume and weight of the detector liquid forced into a package for a particular bombing condition.

Example 1. Following is an example of such calculations. The parameters are:

- Channel diameter $D = 1 \times 10^{-4}$ cm
- Channel length = 0.1 cm
- Package volume = 4.0 cc
- MFP of helium = 37×10^{-6} cm
- *D*/mfp for helium = $(1 \times 10^{-4})/(37 \times 10^{-6}) = 2.703$
- MFP of detector liquid = 1×10^{-7} cm
- Viscosity of helium = 194×10^{-6} poise

- Viscosity of the detector fluid = 0.012 poise
- Specific gravity of detector fluid = 1.7
- Average pressure during leak testing = 0.5 atm = 0.5 × 1.01 × 10^6 dynes/cm^2
- Bombing pressure of detector liquid = 3 atm absolute for 4 hr
- Initial pressure in the package = 1 atm of helium
- Average pressure during detector bombing = (3 + 1)/2 × 1.01 × 10^6 dynes/cm

1. To calculate the average leak rate of helium during leak testing:

Using Eq. (8-1), the viscous conductance of helium during leak testing is:

$$F_{VCHE} = \frac{\pi D^4 P_a Y}{128\eta\left[\ell + \left(\frac{Y}{Z}\right)D\right]}$$

From Eq. (8-2):

$$Y = 1 + \frac{4.4445}{2.703 + 1.1949} = 1 + 1.140 = 2.140$$

From Eq. (8-3):

$$Z = 1.6977 + \frac{4.6742}{2.703 + 2.0444} = 1.6977 + 0.9846 = 2.682$$

$$Y/Z = 2.140/2.682 = 0.798$$

$$F_{VCHE} = \frac{\pi\left(1\times10^{-4}\right)^4 \times 0.5 \times \left(1.01\times10^6\right)\times 2.140}{128\left(194\times10^{-6}\right)\left[0.1 + \left(0.798\times10^{-4}\right)\right]} = \frac{1.367\times10^{-8}}{0.1}$$

$F_{VCHE} = 1.367 \times 10^{-7}$ cc/sec, and for $\Delta P = 1$ atm

The true helium viscous leak rate $L_{VHE} = 1.367 \times 10^{-7}$ atm-cc/sec.

2. To calculate the molecular leak rate of helium during leak testing:

Equation (2-23) is for molecular conduction.

$$F_{MCHE} = 4.961 \times 10^4 \left\{ \frac{D^3}{\left[\ell + \left(\frac{4D}{3}\right)\right]\left[\frac{D}{\text{mfp}} + 1.509\right]} \right\}$$

$$F_{MCHE} = 4.961 \times 10^4 \left[\frac{10^{-12}}{\left(0.1 + \frac{4 \times 10^{-4}}{3}\right)(2.703 + 1.509)} \right]$$

$F_{MCHE} = 4.961 \times 10^4 \, (2.373 \times 10^{-12}) = 1.178 \times 10^{-7}$ cc/sec, and for $\Delta p = 1$ atm

The molecular true helium leak rate $L_{MHE} = 1.178 \times 10^{-7}$ atm-cc/sec.

3. To calculate the total true helium leak rate:

$L_{HE} = 1.367 \times 10^{-7} + 1.178 \times 10^{-7} = 2.545 \times 10^{-7}$ atm- cc/sec

4. To calculate the leak rate of the gross leak detector liquid being forced into a package during bombing:

The viscous conductance of the detector liquid F_{VCDET} is:

$$F_{VCDET} = \frac{\pi D^4 P_a Y}{128\eta\left[\ell + \left(\frac{Y}{Z}\right)D\right]}$$

$$D/\text{mfp} = 10^{-4}/10^{-7} = 1000$$

$$P_a = (3 \text{ atm} + 1 \text{ atm})/2 = 2 \text{ atm}$$

$$Y = 1 + \frac{4.4445}{1000 + 1.1949} = 1 + 0.004 = 1.004$$

$$Z = 1.6977 + \frac{4.6742}{1000 + 2.0444} = 1.6977 + 0.0047 = 1.702$$

$$Y/Z = 0.598$$

$$F_{CDET} = \frac{\pi\left(10^{-4}\right)^4 \times 2 \times \left(1.01 \times 10^6\right) \times 1.004}{128 \times 0.012\left[0.1 + \left(0.590 \times 10^{-4}\right)\right]}$$

$$F_{CDET} = 4.146 \times 10^{-9} \text{ cc/sec}$$

Or using Eq. (8-7):

$$F_{CDET} = 1.367 \times 10^{-7} \times \frac{194 \times 10^{-6}}{0.012} \times \frac{2}{0.5} \times \frac{1.00}{2.15}$$

$F_{CDET} = 4.11 \times 10^{-9}$ cc/sec (less than 1% difference from the 4.146×10^{-9})

The leak rate of the detector liquid is:

$$R_{CDET} = \Delta P_i \times F_{DET} = 2 \times 4.146 \times 10^{-9} = 8.292 \times 10^{-9} \text{ atm-cc/sec.}$$

5. The quantity of detector liquid forced into a package is calculated using Eq. (8-6):

$$Q_{DET} = \Delta P_i\left(1 - e^{-\frac{R_V t}{V \Delta P_i}}\right)$$

where: Q_{DET} = the atm of detector liquid entering the package

ΔP_i = the liquid bombing pressure – the pressure in the package and = 3 – 1 atm = 2 atm in this example

R_V = the viscous leak rate of the detector liquid = 8.292×10^{-9} atm-cc/sec

t = bombing time in sec, typically 4 hr = 14,400 sec

V = 4.0 cc

$$Q_{DET} = 2\left[1 - e^{-\frac{(8.292\times10^{-9})(14,400)}{4.0\times2}}\right] = 2\left(1 - e^{-0.0000149}\right) = 2(1 - 0.9999851)$$

$$Q_{DET} = 2(1.4925 \times 10^{-5}) = 2.99 \times 10^{-5} \text{ atm}$$

This is the liquid pressure inside the package. To get the liquid volume, the atmospheres are multiplied by the volume of the package, i.e.:

$$\frac{\text{atm liquid}}{\text{total atm in package}} = \frac{\text{volume of liquid}}{\text{volume of package}}$$

$$\text{Volume of detector liquid} = \frac{2.99\times10^{-5}\,\text{atm}\times4.0\,\text{ml}}{1\,\text{atm}}$$

$$= 1.19 \times 10^{-4} \text{ ml} = 0.119\ \mu\text{l}$$

The weight in grams = volume in ml × specific gravity = $1.19 \times 10^{-4} \times 1.7 = 2.03 \times 10^{-4}$ grams = 0.20 mg

Table 8-1 shows the results of several calculations similar to those in Ex. 1. The leak channels are 0.1 cm long and the package volume is 4.0 cc. The first seven columns of the table are independent of the package volume. If the true helium leak rate is plotted versus the microliters of detector liquid in the package, a straight line results. The log-log curves for the above data are shown in Fig. 8-1.

Table 8-1. This shows the results of detector liquid in a 4.0 cc package, sealed with 100% helium, for cylindrical leak channels 0.1 cm long, when bombed for 4 hours at 3 atms absolute.* #

D (cm)	L_{VHE} Helium Viscous Leak Rate**	L_{MHE} Helium Molecular Leak Rate**	L_{HE} Total Helium Leak Rate**	R_{DET} Liquid Leak Rate	Q_{DET} Atm In Pkg. $\times 10^{-3}$	Volume Detector Liquid In Pkg. (μl)	Weight Detector In Pkg. (mg)
1.0E-3	7.35E-4	1.72E-5	7.52E-4	8.24E-5	275	1,100	1,870
9.0E-4	4.89E-4	1.38E-5	5.03E-4	5.40E-5	195.0	780.1	1,326
8.0E-4	3.11E-4	1.09E-5	3.22E-4	3.36E-5	121.3	485.1	824.7
7.0E-4	1.86E-4	8.25E-6	1.95E-4	1.96E-5	69.4	277.5	471.7
6.0E-4	1.04E-4	6.00E-6	1.10E-4	1.06E-5	37.8	107.2	257.0
5.0E-4	5.18E-5	4.13E-6	5.60E-5	5.16E-6	18.5	73.8	125.0
4.0E-4	2.24E-5	2.58E-6	2.50E-5	2.12E-6	7.58	30.3	51.5
3.0E-4	7.64E-6	1.39E-6	9.03E-6	6.70E-7	2.41	9.62	16.4
2.0E-4	1.71E-6	5.72E-7	2.28E-6	1.32E-7	0.47	1.89	3.21
1.6E-4	7.55E-7	3.48E-7	1.10E-6	5.40E-8	0.196	0.76	1.28
1.5E-4	5.91E-7	3.01E-7	9.98E-7	4.18E-8	0.152	0.60	1.00
1.09E-4	1.85E-7	1.52E-7	3.37E-7	1.16E-8	0.044	0.167	0.28
1.0E-4	1.37E-7	1.18E-7	2.55E-7	8.30E-9	0.030	0.119	0.20
9.0E-5	9.32E-8	9.18E-8	1.85E-7	5.44E-9	0.020	0.080	0.13
8.0E-5	6.12E-8	6.92E-8	1.30E-7	3.38E-9	0.012	0.049	0.08
7.0E-5	3.74E-8	5.01E-8	8.75E-8	1.98E-9	0.007	0.029	0.05
6.0E-5	1.85E-8	3.42E-8	5.27E-8	1.07E-9	0.004	0.015	0.03
5.0E-5	9.37E-9	2.17E-8	3.11E-8	5.15E-10	0.0018	0.0074	0.013

*All leak rates are in atm-cc/sec.

**Leak rates are for the leak testing condition (i.e., in vacuum) $P_{AVE} = 0.5$ atm.

#The first seven columns are independent of package volume.

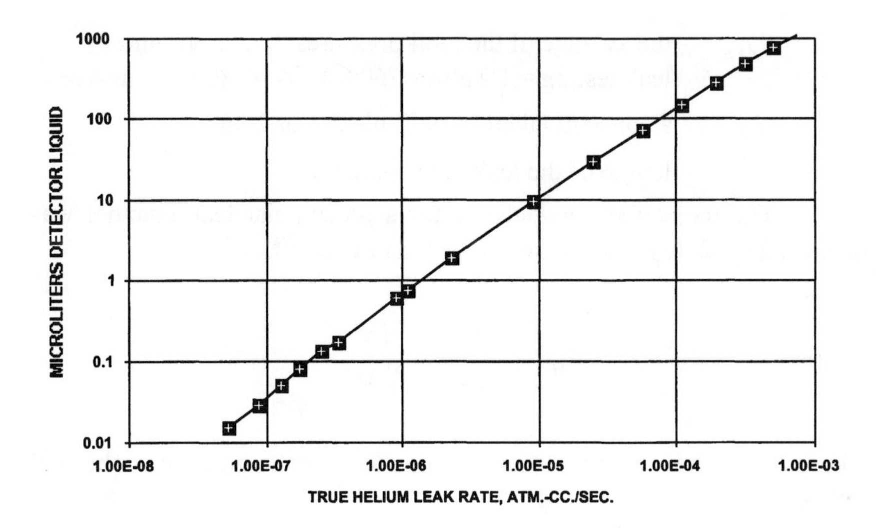

Figure 8-1. Microliters detector liquid in a 4.0 cc package when bombed for 4 hr at 3 atm absolute, as a function of the true helium leak rate in cylindrical leak channels.

The calculated values of the liquid forced into a package agree with the experimental data by William H. Hedley et al.[1] These authors found that the smallest diameter cylindrical channel that water could be forced through with a pressure difference of one atm, was one micron (1×10^{-4} cm). In the above table, when the diameter is 1×10^{-4} and a pressure difference of two atm, the volume of liquid is 0.119 microliters.

A similar analysis as performed for cylindrical leak channels can be performed for rectangular leak channels. The viscous conductance for a rectangular leak channel was presented in Eq. (2-6), and is repeated here as Eq. (8-8):

Eq. (8-8) $$F_{VR} = \frac{0.24a^2b^2P_{AVE}G}{\ell}$$

where F_{VR} = viscous conductance of a rectangular leak channel in l/sec. The constant for helium at room temperature = 0.24 (0.26 for air), and has the dimensions of 1/micron-sec, *a* and *b* are the cross-sectional dimensions in cm

P_{AVE} = the average of the total pressures in microns and for leak testing = 1/2 atm = $760 \times 10^3/2 = 380 \times 10^3$ microns

G = a constant taken from Table 2-1 or 2-1a

ℓ = length of the leak channel in cm

The molecular conductance for a rectangular leak channel was given as Eq. (2-18) and is repeated here as Eq. (8-9).

Eq. (8-9)
$$F_{MV} = \frac{9.7a^2b^2}{(a+b)\ell}\sqrt{\frac{T}{M}}$$

where F_{MV} = the molecular conductance for a rectangular leak channel in l/sec.

The constant 9.7 is in units of cm/sec

$\sqrt{\frac{T}{M}}$ is dimensionless

a, b and ℓ are as in Eq. (8-8)

T = degrees Kelvin

M = molecular weight of the gas

For helium at 20°C, = $\sqrt{\frac{T}{M}} = \sqrt{\frac{293}{4}} = 8.56$

Example 2 is an example of a calculation for a rectangular leak channel.

Example 2.

Let $a = 1 \times 10^{-4}$ cm

$b = 1 \times 10^{-2}$ cm

$\ell = 0.1$ cm

$G = 0.038$ (Table 2-1a)

Using Eq. (8-8) to calculate the viscous conductance:

$$F_{VRHE} = \frac{0.24(10^{-4})^2(10^{-2})^2(380\times10^3)(0.038)}{0.1}\ \text{l/sec}$$

$$F_{VRHE} = 3.466 \times 10^{-8}\ \text{l/sec}$$

When the difference in total pressure is one atmosphere, the true viscous leak rate for helium is:

$$L_{VRHE} = 3.466 \times 10^{-5}\ \text{atm-cc/sec}$$

The molecular conductance is calculated using Eq. (8-9):

$$F_{MRHE} = \frac{9.7(10^{-4})^2(10^{-2})^2(8.56)}{(10^{-4}+10^{-2})(0.1)} = \frac{8.303\times10^{-11}}{0.00101} = 8.221\times10^{-8}\ \text{l/sec}$$

When the difference in partial helium pressure is 1 atm, the true molecular leak rate is:

$$L_{MRHE} = 8.221 \times 10^{-5}\ \text{atm-cc/sec}$$

The total true helium leak rate (L_{RHE}) is the viscous plus the molecular leak rates:

$$L_{RHE} = 3.466 \times 10^{-5} + 8.221 \times 10^{-5} = 1.169 \times 10^{-4}\ \text{atm-cc/sec}$$

The viscous conductance of the detector liquid is the viscous conductance of helium, corrected for the difference in viscosity and the average pressure. In this case, the average pressure = 2.

$$\begin{aligned} F_{DET} &= F_{HE} \times \frac{\eta_{HE}}{\eta_{DET}} \times \frac{P_{AVEDET}}{P_{AVEHE}} \\ &= 3.466\times10^{-5} \times \frac{194\times10^{-6}}{0.012} \times \frac{2}{0.5} \\ &= 2.24\times10^{-6}\ \text{cc/sec} \end{aligned}$$

$$R_{DET} = F_{DET} \times \Delta P = 2.24 \times 10^{-6}\ \text{cc/sec} \times 2\ \text{atm} = 4.88 \times 10^{-6}\ \text{atm-cc/sec}$$

The calculation of the amount of liquid entering the package when bombed at 3 atm absolute for 4 hr, is done the same as for the cylindrical leak channel.

$$Q_{DET} = 2\left[1 - e^{-\frac{(4.48\times10^{-6})(14400)}{4\times2}}\right]$$

$$Q_{DET} = 2(1 \times e^{-0.008068}) = 2(1 \times 0.991964) = 2(0.0080358)$$

$$Q_{DET} = 0.01607 \text{ atm}$$

This corresponds to a volume of 0.01607 × 4 cc = 0.06429 cc (ml) which equals 64.3 µl. The weight of this volume in milligrams is equal to 64.3 × 1.7 = 109.3 mg.

Table 8-2 shows the results of several calculations similar to those in Example 2. The first eight columns of the table are independent of package volume.

Figure 8-2 is a log-log plot of the true helium leak rate versus the volume of detector liquid in the package, using the data in Table 8-2.

Tables 8-1 and 8-2, and Figs. 8-1 and 8-2, are for an internal volume of 4.0 cc. The amount of detector liquid forced into a package is generally dependent upon the internal volume. This is because the pressure difference between the inside and outside of the package is decreasing as the liquid is being forced into the package. Many gross leaks only allow a small amount of liquid to enter the package. When this is so, the volume of liquid relative to the volume of the package is insignificant. Therefore, the pressure inside the package remains essentially unchanged for any size package. If a very small leak channel in a 1,000 cc package allows 0.167 ml to be forced in under a particular bombing condition, 0.167 microliters will also be forced into a 0.01 cc package that has the same small leak channel for the same bombing condition.

Table 8-2. This shows the results of detector liquid in a 4 cc package, sealed with 100% helium, for rectangular leak channels 0.1 cm long, when bombed for 4 hr at 3 atm absolute.*#

a (cm)	b (cm)	L_{VRHE} Helium Viscous Leak Rate**	L_{MRHE} Helium Molecular Leak Rate**	L_{RHE} Total Helium Leak Rate**	R_{DET} Liquid Leak Rate	Q_{DET} Atm In Pkg. × 10^{-3}	Volume Detector Liquid In Pkg. (μl)	Weight Detector Liquid In Pkg. (mg)
1.0E-4	1.0E-2	3.46E-5	8.22E-5	1.17E-4	4.48E-6	16.05	64.2	109.1
2.0E-4	2.0E-3	3.36E-5	6.04E-5	9.39E-5	4.34E-6	15.60	62.3	105.9
1.0E-4	1.0E-3	2.10E-6	7.55E-6	9.66E-6	2.72E-7	0.976	3.90	6.6
1.0E-4	9.0E-4	1.77E-6	6.72E-6	8.49E-6	2.28E-7	0.824	3.28	5.6
1.0E-4	8.0E-4	1.46E-6	5.90E-6	7.36E-6	1.89E-7	0.680	2.72	4.8
1.0E-4	7.0E-4	1.16E-6	5.09E-6	6.25E-6	1.50E-7	0.540	2.16	3.68
1.0E-4	6.0E-4	9.85E-7	4.27E-6	5.26E-6	1.27E-7	0.456	1.84	3.12
1.0E-4	5.0E-4	9.58E-7	3.46E-6	4.42E-6	1.24E-7	0.448	1.80	3.08
1.0E-4	4.0E-4	7.30E-7	2.66E-6	3.39E-6	9.44E-8	0.340	1.36	2.32
1.0E-4	3.0E-4	4.92E-7	1.87E-6	2.36E-6	6.37E-8	0.228	0.92	1.56
1.0E-4	2.0E-4	2.99E-7	1.11E-6	1.41E-6	3.87E-8	0.140	0.56	0.96
1.0E-4	1.0E-4	9.12E-8	4.15E-7	5.06E-7	1.18E-8	0.042	0.170	0.289
1.0E-4	9.9E-5	8.94E-8	4.09E-7	4.98E-7	1.16E-8	0.042	0.167	0.284
1.0E-4	9.0E-5	7.39E-8	3.54E-7	4.28E-7	9.55E-9	0.034	0.138	0.235
1.0E-4	8.0E-5	5.72E-8	2.95E-7	3.53E-7	7.40E-9	0.027	0.106	0.184
1.0E-4	7.0E-5	4.25E-8	2.39E-7	2.82E-7	5.50E-9	0.020	0.079	0.135
1.0E-4	6.0E-5	2.95E-8	1.87E-7	2.16E-7	1.91E-9	0.014	0.055	0.093

*All leak rates are in atm-cc/sec

**Leak rates are for the leak testing condition (i.e., in vacuum) P_{AVE} = 0.5 atm

#The first eight columns are independent of package volume

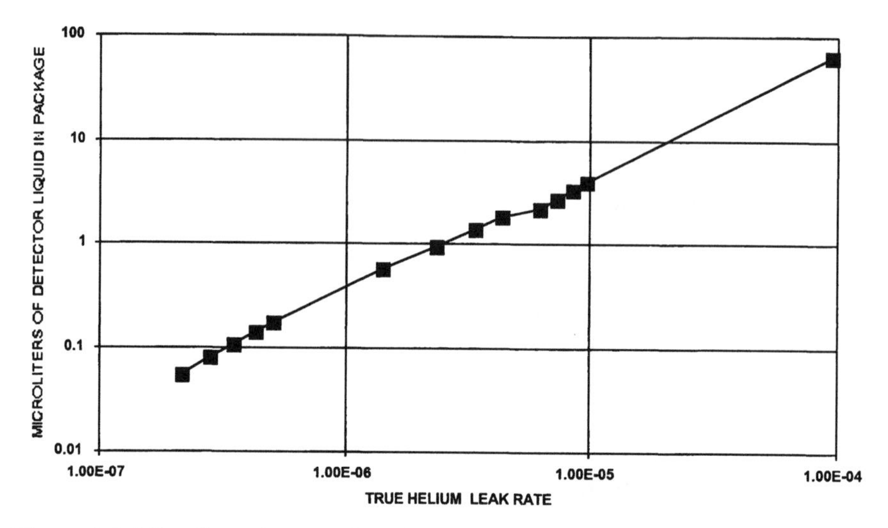

Figure 8-2. Microliters detector liquid in a 4.0 cc package when bombed for 4 hr at 3 atm absolute, as a function of rhe true helium leak rate in rectangular leak channels.

3.0 FLUOROCARBON VAPOR EXITING A PACKAGE

The gross leak detector fluid usually has the chemical formula of:

$$\begin{array}{l} CF_3\text{—}O\text{—}\underset{\displaystyle CF_3}{\underset{|}{CF}}\text{—}CF_2\text{—}O\text{—}CF_2\text{—}OCF_3 \end{array}$$

This liquid decomposes when heated above its boiling temperature, each liquid molecule giving off three molecules of CF_3, two molecules of CF_2 and one molecule of CF. CF_3, CF_2 and CF are gases so that their individual volumes per molecule is many times greater than the volume of the liquid molecule. The exact volumetric ratios are calculated as follows.

The specific gravity of the detector liquid is 1.7, so that 1 ml of liquid weighs 1.7 g and 1 micro-liter (μl) weighs 1.7 mg. To calculate the number of μl of CF_3 gas that there will be for 1 μl of detector liquid, first the number of molecules in 1 μl of the detector liquid must be calculated.

One molecular weight of the detector liquid weighs 386 g and contains 6×10^{23} molecules.

$$\frac{6\times10^{23}\,\text{molecules}}{386\text{g}}=\frac{\text{molecules}/\mu\text{l}}{1.7\times10\ ^{3}\text{g}/\mu\text{l}}$$

$$\text{molecules in 1 }\mu\text{l of liquid}=2.64\times10^{18}$$

The number of CF_3 molecules in 1 μl of liquid = 3 × 2.64 × 10^{18} = 7.93 × 10^{18}. One molecular weight of a gas, or 6 × 10^{23} molecules, occupy 22.4 l.

$$\frac{6\times10^{23}\,\text{molecules}}{22.4\text{ liters}}=\frac{7.93\times10^{18}\,\text{molecules}}{\text{volume of }CF_3}$$

$$\text{volume of }CF_3=2.96\times10^{-4}\text{ l}=0.296\text{ ml}=296\text{ ml}$$

One ml of detector liquid produces 296 ml of CF_3.

The ratios of CF_3 to CF_2 and to CF are 3:2 and 3:1 respectively. Therefore:

One μl of detector liquid produces 197 μl of CF_2

One μl of detector liquid produces 98.6 μl of CF

The total fluorocarbon gaseous volume = 592 μl

These volumes are at the boiling temperature of the liquid, 91°C (364°K). At 125°C (398°K), the volume will increase by the factor 398/364.

Although the relationship between the amount of detector liquid in a package and the volume of gas produced is known, the amount of gas exiting the package is not directly known. This latter amount depends upon the leak rate of the gases, which depend on the total and partial pressure differences between the inside and outside of the package and the conductance of the leak channel. The mechanisms and equations of Ch. 3 are applicable for determining the amount and rate of these gases that leave the package.

Assume a cylindrical leak channel with a diameter of 3 × 10^{-4} cm as in Table 8-1. The amount of liquid forced into the 4.0 cc package is 9.62 μl. Multiplying this volume by the CF_3/liquid ratio of 296, the volume of CF_3 is 2.85 ml. Doing the same for CF_2 and CF, the volume of CF_2 is 1.90 ml and for CF is 0.95 ml. The total volume for the three gasses is 5.70 ml. These are the volumes at 91°C. Correcting for the temperature increase from 91°C to 125°C, the values are multiplied by 398/364.

The new volumes are:

$CF_3 = 3.12$ ml

$CF_2 = 2.08$ ml

$CF = 1.04$ ml

Total = 6.24 ml

These are the volumes if the 4.0 cc package expanded to maintain the pressure inside at 1 atm. If the package is rigid, the volume remains at 4.0 cc, and the total fluorocarbon pressure is (6.24/4) × 1 atm = 1.56 atm.

The pressure of the 4 ml of helium that was originally sealed in the package at 22°C (295°K) will increase when the temperature rises to 125°C. The helium pressure at 125°C = (398/295) × 1 atm = 1.35 atm. The total pressure at 125°C = 1.56 + 1.35 = 2.91 atm. There will be viscous as well as molecular flow of CF_3, CF_2, CF, and helium out of the package if the package is heated to 125°C in air or vacuum.

In Table 8-1 for the 9.62 μl of detector liquid, the true helium viscous leak rate is $L_{VHE} = 7.64 \times 10^{-6}$ atm-cc/sec and the true molecular leak rate (L_{MHE}) equals 1.39×10^{-6} atm-cc/sec. This helium viscous leak rate is based on an average pressure of 0.5 atm. The average pressure for the viscous flow out of the 4 cc package when the package is in a vacuum at 125°C is (2.91 + 0)/2 = 1.455 atm. The true helium viscous leak rate of 7.64×10^{-6} must be increased for the increase in average pressure. The viscous leak rate of the gases leaving the 4 cc package containing 9.62 μl is:

$$L_{V9.62} = (7.64 \times 10^{-6})(1.455/0.5) = 2.22 \times 10^{-5} \text{ atm-cc/sec}$$

This leak rate must now be corrected for the fact that the total pressure difference is not 1 atm but 2.91 atm when leaking into the vacuum. The corrected viscous leak rate is:

$$L_{V9.62C} = 2.91 \times 2.22 \times 10^{-5} = 6.47 \times 10^{-5} \text{ atm-cc/sec}$$

All the gases will flow out at this rate when the package is at 125°C.

When the detector liquid is at 125°C, the composition of the gases inside the package in terms of atmospheres is:

Helium = (5.4 ml/4.0 ml) × 1 atm = 1.35 atm

Fluorocarbons = 1.56 atm, consisting of:

CF_3 =(1.56/2) = 0.78 atm

CF_2 = (2/3) × 0.78 = 0.52 atm

CF = (1/3) × 0.78 = 0.26 atm

The effective viscous leak rate of CF_3 is:

$$L_{VCF_3} = \text{viscous leak rate} \times \text{fraction of } CF_3 \text{ in the package}$$
$$= 6.47 \times 10^{-5}(3.12/11.64) = 1.73 \times 10^{-5} \text{ atm-cc/sec}$$

The effective viscous leak rate of CF_2 is:

$$L_{VCF_2} = (2/3) \times 1.73 \times 10^{-5} = 1.16 \times 10^{-5} \text{ atm-cc/sec}$$

The effective viscous leak rate of CF is:

$$L_{VCF} = (1/3) \times 1.73 \times 10^{-5} = 5.78 \times 10^{-6} \text{ atm-cc/sec}$$

There will be additional amounts of fluorocarbon gases exiting the package due to molecular flow. The molecular flow rates are based on the true helium molecular leak rate L_{MHE} = 1.39 × 10^{-6} atm-cc/sec (see Table 8-1). The relationship between the true molecular helium leak rate and the true molecular leak rate for the fluorocarbons is: the molecular weight of CF_3 = 69, so that the molecular leak rate of CF_3 with respect to helium is:

$$L_{MCF_3} = L_{MHE}\sqrt{\frac{4}{69}} = 0.24\, L_{HE}$$

The molecular weight of CF_2 = 50, the molecular weight of CF = 31.

$$L_{MCF_2} = L_{MHE}\sqrt{\frac{4}{50}} = 0.283\, L_{HE}$$

$$L_{MCF} = L_{MHE}\sqrt{\frac{4}{31}} = 0.36\,L_{HE}$$

The pressure difference of the fluorocarbon gases from inside to outside the package is not 1 atm (true leak rates) so that the leak rates must be corrected in accordance with the actual partial pressure difference. Therefore:

$$L_{MCF_3} = 0.24 \times 1.39 \times 10^{-6} \times 0.78 = 2.60 \times 10^{-7} \text{ atm-cc/sec}$$

$$L_{MCF_2} = 0.283 \times 1.39 \times 10^{-6} \times 0.52 = 2.05 \times 10^{-7} \text{ atm-cc/sec}$$

$$L_{MCF} = 0.36 \times 1.39 \times 10^{-6} \times 0.26 = 1.30 \times 10^{-7} \text{ atm-cc/sec}$$

The total leak rates for the fluorocarbon gases exiting the 4.0 cc package at 125°C that contained 9.62 μl of detector liquid are:

$$L_{CF_3T} = (1.73 \times 10^{-5}) + (2.60 \times 10^{-7}) = 1.76 \times 10^{-5} \text{ atm-cc/sec}$$

$$L_{CF_2T} = (1.16 \times 10^{-5}) + (2.05 \times 10^{-7}) = 1.18 \times 10^{-5} \text{ atm-cc/sec}$$

$$L_{CFT} = (5.78 \times 10^{-6}) + (1.30 \times 10^{-7}) = 5.91 \times 10^{-6} \text{ atm-cc/sec}$$

Total fluorocarbon gas leak rate = 3.53×10^{-5} atm-cc/sec.

Table 8-3 gives the volume and leak rates of fluorocarbon gases at 125°C into a vacuum for some rectangular leak channels in a 4.0 cc package taken from Table 8-2.

4.0 THE BUBBLE TEST

An early gross leak test was the bubble test. It is still in present use and is an approved method in standard military test methods, such as test condition C1 of MIL-STD-883, Method 1014. The test consists of bombing the package with a fluorocarbon liquid (detector liquid) that has a boiling temperature from 50°C to 95°C for a length of time and pressure. Lengths of time and pressure combinations are prescribed in the Military

Table 8-3. The Volumes and Leak Rates of Fluorocarbon Gases at 125°C for Rectangular Leak Channels in a 4.0 cc Package*

Leak Channel Number	L_{VHE} True Viscous Helium Leak Rate	L_{MHE} True Molecular Helium Leak Rate	L_{HE} True Total Helium Leak Rate	Detector Liquid Volume μl	Pressure of CF_3 (atm) @ 125°C**	Pressure of CF_2 (atm) @ 125°C**	Pressure of CF (atm) @ 125°C**	Total Pressure in the Package (atm) @ 125°C ***	L_{VPKG} Viscous Leak Rate From the Package @ 125°C
1	1.70E-4	2.04E-4	3.74E-4	310.6	25.12	16.75	8.375	51.6	0.453
2	3.36E-5	6.04E-5	9.38E-5	62.3	5.04	3.360	1.680	11.43	4.39E-3
3	2.10E-6	7.55E-6	9.66E-6	3.90	0.316	0.211	0.106	1.981	8.24E-6
4	1.77E-6	6.72E-6	8.49E-6	3.28	0.266	0.177	0.089	1.881	6.26E-6
5	1.46E-6	5.90E-6	7.36E-6	2.72	0.220	0.147	0.074	1.790	4.68E-6
6	1.16E-6	5.09E-6	6.25E-6	2.16	0.175	0.117	0.089	1.700	3.35E-6
7	9.85E-7	4.27E-6	5.26E-6	1.84	0.149	0.099	0.050	1.648	2.68E-6
8	7.30E-7	2.66E-6	3.39E-6	1.36	0.110	0.073	0.037	1.570	1.80E-6
9	4.92E-7	1.87E-6	2.66E-6	0.92	0.074	0.049	0.025	1.498	1.11E-6
10	2.99E-7	1.11E-6	1.14E-6	0.56	0.0453	0.030	0.015	1.441	6.21E-7
11	8.94E-8	4.09E-7	4.98E-7	0.167	0.0135	0.009	0.005	1.377	1.69E-7

Table 8-3. (Cont'd.)

Leak Channel Number	L_{VCF_3} Viscous Leak Rate of CF_3	L_{MCF_3} Molecular Leak Rate of CF_3	L_{CF_3T} Total Leak Rate of CF_3	L_{VCF_2} Viscous Leak Rate of CF_2	L_{MCF_2} Molecular Leak Rate of CF_2	L_{CF_2T} Total Leak Rate of CF_2	L_{VCF} Viscous Leak Rate of CF	L_{MCF} Molecular Leak Rate of CF	L_{CFT} Total Leak Rate of CF	L_{FLC} Total Fluorocarbon Leak Rate
1	0.221	1.23E-3	0.222	0.147	9.67E-4	0.148	7.35E-2	6.15E-4	7.41E-2	0.444
2	1.94E-3	7.03E-5	2.01E-3	1.29E-3	5.74E-5	1.84E-3	6.45E-4	3.66E-5	6.82E-4	4.03E-3
3	1.31E-6	5.72E-7	1.88E-6	8.73E-7	4.49E-7	1.32E-6	4.37E-7	2.86E-7	7.23E-7	3.92E-6
4	8.84E-7	4.28E-7	1.31E-6	5.89E-7	3.36E-7	9.25E-7	3.95E-7	2.14E-7	5.09E-7	2.74E-6
5	5.75E-7	3.11E-7	8.86E-7	3.83E-7	2.11E-7	5.94E-7	1.42E-7	1.56E-7	2.98E-7	1.78E-6
6	3.45E-7	2.13E-7	5.58E-7	2.30E-7	1.68E-7	3.98E-7	1.15E-7	1.07E-7	2.22E-7	1.18E-6
7	2.42E-7	1.53E-7	3.95E-7	1.61E-7	1.20E-7	2.81E-7	8.05E-8	7.65E-8	1.57E-7	8.33E-7
8	1.26E-7	7.02E-8	1.96E-7	8.40E-8	5.51E-8	1.39E-7	4.20E-8	3.52E-8	7.72E-8	4.12E-7
9	5.52E-8	3.24E-8	8.76E-8	3.68E-8	2.63E-8	6.31E-8	1.84E-8	1.68E-8	3.52E-8	1.86E-7
10	1.95E-8	1.21E-8	3.16E-8	1.30E-8	9.50E-9	2.25E-8	6.50E-9	6.09E-9	1.26E-8	6.67E-8
11	1.66E-9	1.32E-9	2.98E-8	1.10E-9	1.03E-9	2.13E-9	8.30E-9	6.63E-10	1.49E-9	6.60E-9

*All leak rates are in atm-cc/sec

**Based on one microliter of detector liquid producing 296 microliters of CF_3 gas at the boiling point of the liquid (91°C), and increasing by (398/364) at 125°C

***Includes the helium pressure at 125°C = (398/295) × 1 atm = 1.35 atm

Specification test procedure, the lower pressures and longer times being more applicable to larger packages. The lowest allowed pressure is 30 psia for a bombing time of 23.5 hr.

After bombing, the package is placed in a container of a fluorocarbon liquid (indicator liquid) having a boiling temperature from 140°C to 200°C, and maintained at a temperature of 125°C ±5°C. The package is immersed in the indicator liquid for at least 30 sec unless a failure occurs earlier. The failure criteria states: "A definite stream of bubbles or two or more large bubbles originating from the same point shall be cause for rejection." Observing these bubbles is difficult and specific visual test conditions are required. They are a non-reflecting dull black background, a magnifier from 1.5x–30x, and a 15 thousand foot candle light source. This test is a strain on the operator's eyes when testing for hours at a time, and results in escapes as well as false failures. This test is qualitative, and most people believe that there is a decade gap in leak rates between this method and the fine leak test.

5.0 THE VAPOR DETECTION TEST

In the early 1980's, an instrument was developed to detect the fluorocarbon vapors exiting a package when conditioned in a manner similar to the bubble test. The new test is more economical than the bubble test because of its greater throughput and it does not use the expensive indicator liquid. It is more accurate than the bubble test because the total pressure difference is greater as the fluorocarbon gases are being leaked into a vacuum instead of a liquid at 1 atm. Although this test is generally a pass/fail test, it is inherently semi quantitative as the reading at test is numerical and proportional to the quantity of vapor escaping from the package.

The measurement is based on the infrared absorption of the fluorocarbon gases in the eight micron range. The gases escaping the package under test enter a cell which is irradiated by energy in the eight micron range. At the other end of the cell is an infrared detector covering the eight micron range. The amplitude of the infrared emission is decreased when the fluorocarbon gases absorbs the radiation. This decrease in amplitude is inversely amplified and converted to a digital read out which is proportional to the amount of fluorocarbons in the cell.

MIL-STD-883, Method 1014, test condition C3, sets the failure limit at 0.167 μl of detector liquid in the package. The instrument is calibrated by injecting one micro liter into the instrument and setting the gain so the instrument read 1200 counts. One sixth of a micro liter (0.167 μl) would then read 200 counts, which is the pass/fail limit. Table 4 shows the relationship between the package volume and the total fluorocarbon leak rate exiting the package. All packages contain 0.167 μl of detector liquid at room temperature.

The data in Table 8-4 is plotted in Figs. 8-3– 8-5. Figure 3 is a log-log plot of the package volume versus the total fluorocarbon pressure in the package at 125°C. Figure 8-4 is a semi-log plot of the package volume versus the total pressure in the package at 125°C. Figure 8-5 is a log-log plot of the package volume versus the total fluorocarbon leak rate including molecular as well as viscous.

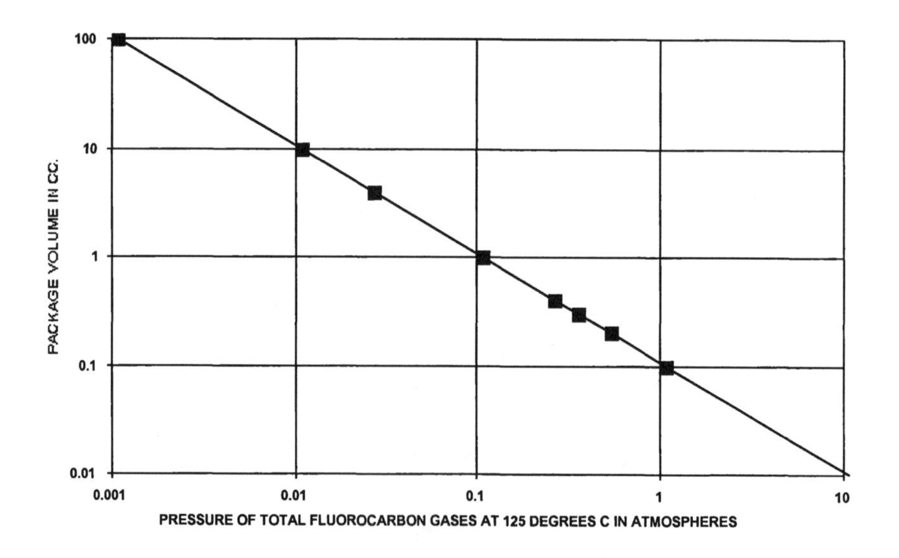

Figure 8-3. Pressure of the total fluorocarbon gases in packages as a function of package volume when the package contained 0.167 microliters of detector liquid.

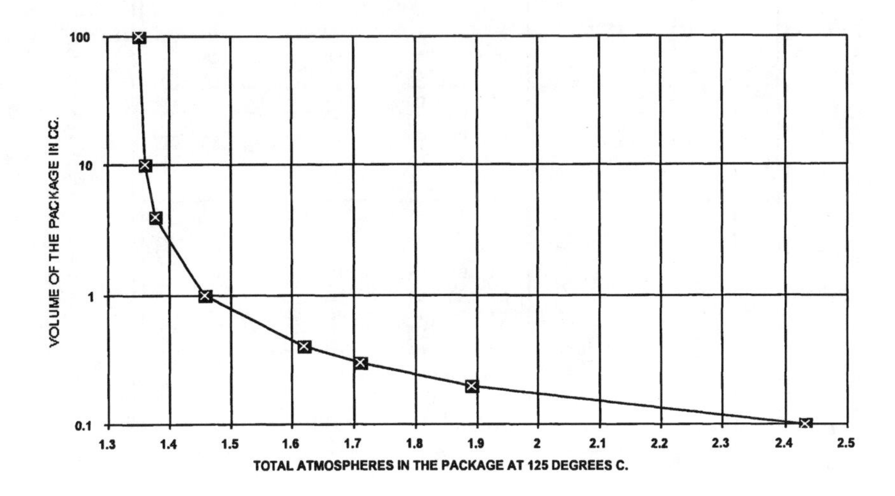

Figure 8-4. Total pressure in package at 125°C as a function of package volume, when the package contained 0.167 microliters of detector liquid.

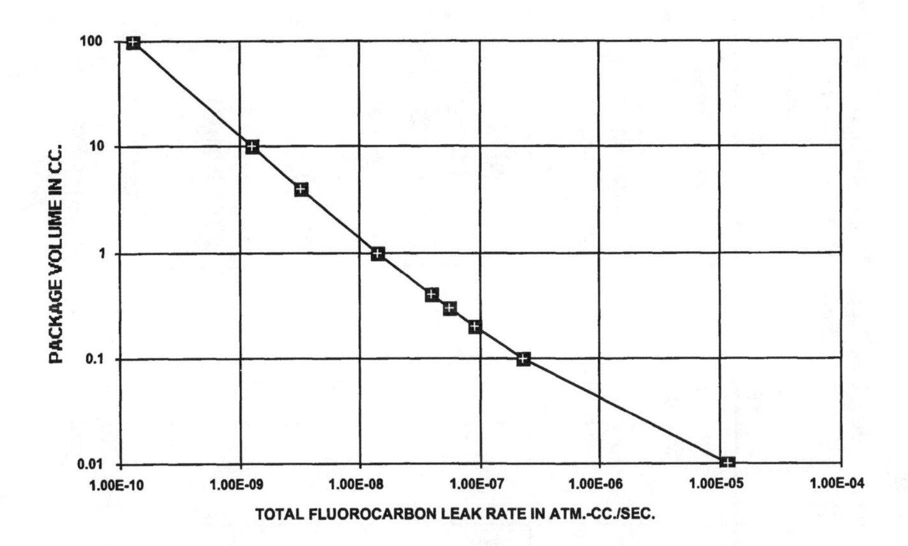

Figure 8-5. Total fluorocarbon leak rate at 125 °C as a function of package volume, when the package contained 0.167 microliters of detector liquid.

Table 8-4. Fluorocarbon Gas Leak Rates for Different Size Packages when All Packages Contain 0.167 Microliters of Detector Liquid

"A"	"B"	"C"	"D"	"E"	"F"
Package Volume at 25°C (cc)	Pressure of Sealed Gas at 125 °C = (398°K/295°K) x1 atm	Pressure of Fluorocarbon Gas at 125°C = (0.10810/"A") (atm)**	Total Pressure in the Package at 125°C = ("B" + "C") (atm)	Total Viscous Leak Rate = (Helium Viscous Leak Rate x "D"^{2}) = (8.93E-8 x "D"^{2})	Total Fluorocarbon Leak Rate = ("C" x "E"/"D")
0.01	1.349	10.81	12.159	1.32E-5	1.17E-5
0.1	1.349	1.081	2.4300	5.28E-7	2.35E-7
0.2	1.349	0.5405	1.8900	3.19E-7	9.13E-8
0.3	1.349	0.3603	1.7090	2.61E-7	5.50E-8
0.4	1.349	0.2703	1.6190	2.34E-7	3.91E-8
1.0	1.349	0.1081	1.4570	1.90E-7	1.41E-8
4.0	1.349	0.02703	1.376	1.69E-7	3.32E-9
10	1.349	0.01081	1.3600	1.65E-7	1.28E-9
100	1.349	0.001081	1.3501	1.63E-7	1.31E-10

*Leak rates are in atm-cc/sec

**0.167 μl of detector liquid yields 0.167 × 2 × 296 = 98.864 μl of fluorocarbon gas at 91°C. At 125°C, the gas volume is (398/364) × 98.864 = 108.10 μl = 0.1081 μl

6.0 THE WEIGHT GAIN TEST

This test measures the amount of detector liquid that has been forced into the package from bombing, by weighing the package before and after the bombing. The bombing pressure and time are variable and a function of the package size, similar to the conditions in the vapor detection method. Details of the test procedure is given in MIL-STD-883, Method 1014. Method 1014 allows the grouping of packages with volumes less than 0.01 cc into bins of 0.5 mg increments prior to bombing. Packages weighing between 5.0000 and 5.0005 g would be in one bin, and those between 5.0006 and 5.0010 g would be in another bin. Packages with volumes equal to or greater than 0.01 cc are allowed to be in bins of 1.0 mg increments. The use of bins is not required. The actual difference in weight can be used.

The failure criteria is a gain in weight of 1.0 mg or more for packages having a volume less than 0.01 cc. For packages greater than 0.01 cc, the failure criteria is 2.0 mg or more. If the packages have been categorized into bins, a shift from one bin to a higher bin is a failure. Weighing these differences is easily accomplished by using an automatic electronic balance and normal weighing precautions. Balances with capacities of 200–400 grams, with readings to 0.1 mg are readily available.

The density of the detector liquid is 1.7 mg per microliter. If the 0.167 μl limit is used, the failure criteria would be $1.7 \times 0.167 = 0.28$ mg. Weight gains for various leak channels are given in Tables 8-1 and 8-2.

7.0 OPTICAL LEAK TEST

Optical leak testing is based on the deflection of the package lid when the pressure outside the package differs from that inside the package. The amount of deflection is measured by an interferometer, often a laser interferometer.[2]

If a package has a leak, and it is placed in a chamber, and the *chamber is then evacuated*, the lid will exhibit the following deflections:

1. The lid will initially bow outwards (away from the package interior) as the pressure inside the package is greater than the vacuum.

2. As gas leaks out of the package the pressure inside the package will decrease and the lid will start to deflect towards the interior of the package. The magnitude and rate of this deflection is a function of the package geometry, the thickness and material of the lid, and the leak rate of the package.

If a package has a leak, and it is placed in a chamber, and the *chamber then pressurized* to two atmospheres, the lid will exhibit the following deflections:

1. The lid will initially bow inwards as the pressure outside the package is greater than that inside the package.
2. As the surrounding two atmosphere of gas leaks into the package, the lid will start to deflect outwards as the pressure inside the package increases. Here again, the magnitude and rate of this deflection is a function of the package geometry, the thickness and material of the lid, and the leak rate of the of the package.

Method 1014 of MIL-STD-883 includes a procedure and failure criteria for optical leak testing, for both gross and fine leaks. The required condition for gross leak testing is:

$$R^4/ET^3 > 1.0 \times 10^{-4}$$

where: R = the minimum inside width of the package (inside braze or cavity dimension in inches)

E = modulus of elasticity of the lid in pounds/in^2

T = the thickness of the lid in inches

The required condition for fine leak testing is:

$$R^4/ET^3 > 1.0 \times 10^{-3}$$

Table 8-5 gives the value of R^4/ET^3 for selected values of R and T.

Table 8-5. Mechanical Package Requirements for Optical Leak Testing*

R (in.)	R^4	T of Kovar Lid (in.)	T^3	R^4/ET^3 **	OK for Fine Leak	Deflection (in) for 1 Atm***
0.15	5.06E-4	0.005	1.25E-7	2.03E-4	NO	0.000088
0.15	5.06E-4	0.010	1.0E-6	2.53E-5	NO	0.000011
0.20	1.60E-3	0.005	1.25E-7	6.40E-4	NO	0.000280
0.20	1.60E-3	0.010	1.0E-6	8.00E-5	NO	0.000035
0.25	3.91E-3	0.005	1.25E-7	1.56E-3	YES	0.000683
0.25	3.91E-3	0.010	1.0E-6	1.96E-4	NO	0.000085
0.30	0.0081	0.005	1.25E-7	3.24E-3	YES	0.001420
0.30	0.0081	0.010	1.0E-6	4.05E-4	NO	0.000177
0.03	0.0081	0.015	3.375E-6	1.20E-4	NO	0.000052
0.60	0.1296	0.005	1.25E-7	5.18E-2	YES	0.022654
0.60	0.1296	0.010	1.0E-6	6.48E-3	YES	0.002832
0.60	0.1296	0.015	3.375E-6	1.92E-3	YES	0.000839
0.60	0.1296	0.020	8.0E-6	8.10E-4	NO	0.000354
0.60	0.1296	0.025	1.56E-5	4.15E-4	NO	0.000181
0.90	0.6561	0.010	1.0E-6	3.28E-2	YES	0.014330
0.90	0.6561	0.015	3.375E-6	9.72E-3	YES	0.004250
0.90	0.6561	0.020	8.0E-6	4.10E-3	YES	0.001792
0.90	0.6561	0.025	1.56E-5	2.10E-3	YES	0.001401
1.20	2.0736	0.015	3.375E-6	3.07E-2	YES	0.013425
1.20	2.0736	0.020	8.0E-6	1.30E-2	YES	0.005663
1.20	2.0736	0.025	1.56E-5	6.65E-3	YES	0.002904
1.50	5.0625	0.020	8.0E-6	3.16E-2	YES	0.013827
1.50	5.0625	0.025	1.56E-5	1.62E-2	YES	0.007091
1.80	10.4976	0.020	8.0E-6	6.56E-2	YES	0.028667
1.80	10.4976	0.025	1.56E-5	3.36E-2	YES	0.014703
2.10	19.448	0.025	1.56E-5	6.23E-2	YES	0.027240

*All examples in the table meet the gross leak requirements except $R = 0.15$, $T = 0.010$, and $R = 0.20$, $T = 0.010$.

**E for Kovar = 20×10^6

***Using [Eq. (8-9) + Eq. (8-10)]/2

The failure criteria in Method 1014 for gross leaks are:

- If the optical interferometer did not detect deflection of the lid as the chamber pressure was initially changed.
- If the interferometer detects the lid deflecting as the chamber pressure is held constant (or equivalent procedure).

The procedure for gross/fine leak testing consists of placing the package in a chamber with the optical interferometer observing the lid. The pressure in the chamber is then evacuated and the chamber held at this vacuum for a time (t_1). The package is then placed in a chamber with an interferometer and the chamber pressurized to 30 psig of helium, and held at this pressure for time (t_2). The failure criteria for the gross/fine leak test are:

- If the interferometer did not detect deflection of the lid as the chamber pressure was initially changed.
- If the interferometer detects the lid deflecting from the package leaking its entrapped internal pressure during time (t_1) as the pressure is held constant (or equivalent procedure).
- If the interferometer detects the lid deflecting from the package leaking in the pressurized helium gas during time (t_2) as the pressure is held constant (or equivalent procedure).

The quantitative deflection of the lid can be calculated using equations from a book by Raymond J. Roark, pp. 214–226.[3] Considering a square package, there are two applicable equations. The first equation (Eq. 8-10) is for the "edges supported above and below (corners held down), uniform load over the entire surface."

Eq. (8-10) $$\text{Max } y = \frac{-0.0487wa^4(m^2-1)}{m^2ET^3}$$

where y = the deflection at the center of the lid

w = pounds per square inch

m = the reciprocal of Poissons Ratio (0.25) = 4.0 for kovar

E = modulus of elasticity in pounds per square inch = 20×10^6 for kovar

T = the thickness of the lid in inches

The second equation is for "all edges fixed, uniform load over the entire surface."

Eq. (8-11) $$\text{Max } y = \frac{-0.0138wa^4}{ET^3}$$

This author has found that the average of these two equations agree well with the experimental results for square welded packages. Equations for rectangular and round shapes are also found in Roark. The applicability of these equations should be determined experimentally. The following is an example of the calculation for a square package.

Example 3. A square welded package with an inside cavity of 0.90 in. is sealed with 1 atm using a lid 0.015 in. thick. The package is put in a vacuum.

The modulus of elasticity of kovar (E) = 20×10^6 lb/in^2

m for kovar = 1/0.25 = 4.0

Using Eq. (8-10):

$$\max y = -\frac{0.0487(14.7)(0.9)^4(4^2-1)}{(4)^2(20\times 10^6)(0.015)^3}$$

$$\max y = -\frac{0.0487(14.7)(0.6561)(15)}{(16)(20\times 10^6)(3.375\times 10^{-6})} = \frac{7.045}{1080}$$

$$\max y = -0.0065 \text{ in.}$$

Using Eq. (8-11):

$$\max y = -\frac{0.0138(14.7)(0.9)^4}{(20\times 10^6)(3.375\times 10^{-6})} = \frac{0.1331}{67.5}$$

$$\max y = -0.0020 \text{ in.}$$

The average of the two calculations is (0.0065 + 0.0020)/2 = 0.00425 in.

The last column of Table 8-5 lists the deflection for various size square packages and for different lid thicknesses. Figures 8-6–8-8 are plots of deflection versus the size of square packages for different lid thicknesses.

The amount of deflection that the interferometer can detect depends upon the wavelength of the interferometer and the amount of lid deflection for a change in pressure of one atmosphere. The interferometer can measure to one half a wave length. A typical (green) laser interferometer has a wavelength = 2.5×10^{-5} in. Half this wavelength = 1.25×10^{-5} in. If a package has a lid that deflects 0.00425 in. when the pressure difference is one atmosphere, the minimum atmospheric change (ATM_{MIN}) that can be detected is easily calculated.

$$\frac{0.00425 \text{ in.}}{1 \text{ atm}} = \frac{1.25 \times 10^{-5} \text{ in.}}{ATM_{MIN}}$$

$$ATM_{MIN} = 0.00294 \text{ atm}$$

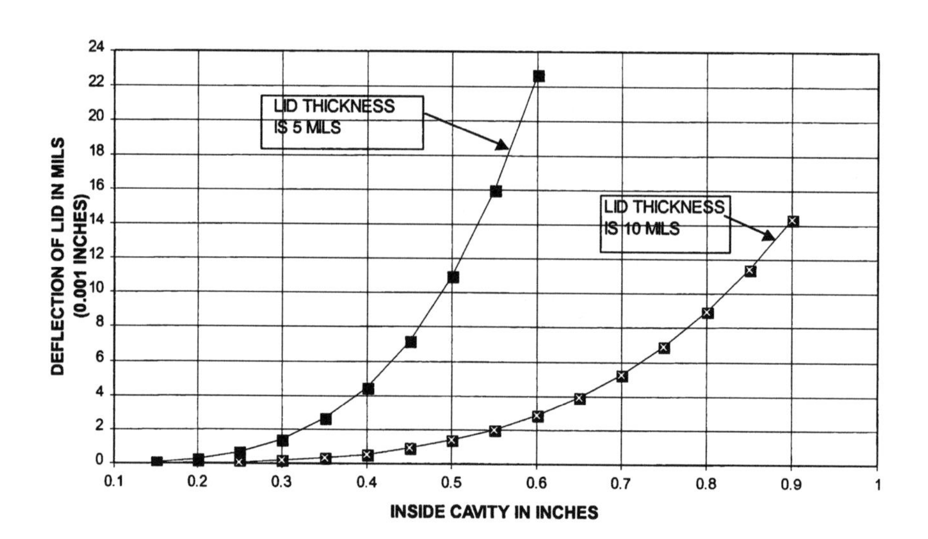

Figure 8-6. Deflection of 5 and 10 mil kovar lids for welded square packages, due to a difference in pressure of one atmosphere.

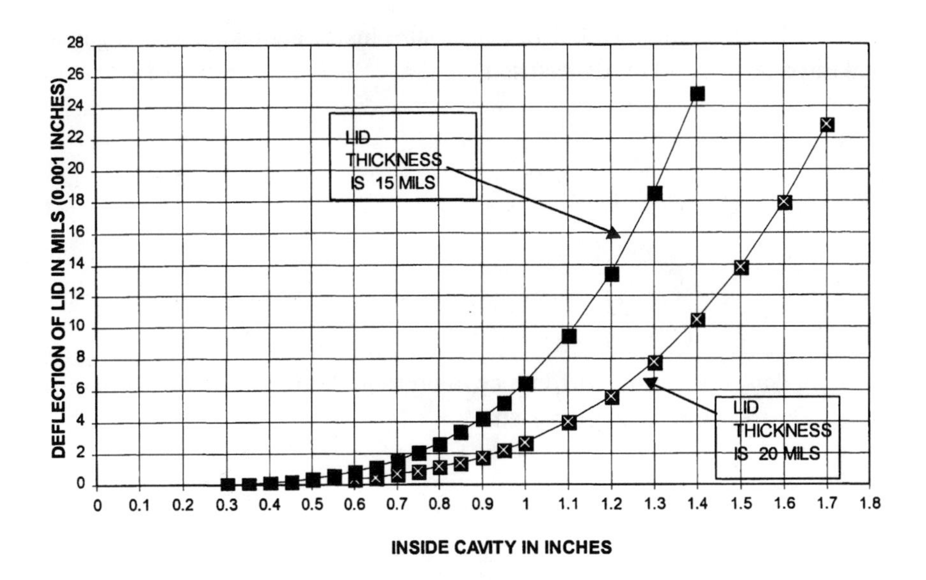

Figure 8-7. Deflection of 15 and 20 mil kovar lids for welded square packages, due to a difference in pressure of one atmosphere.

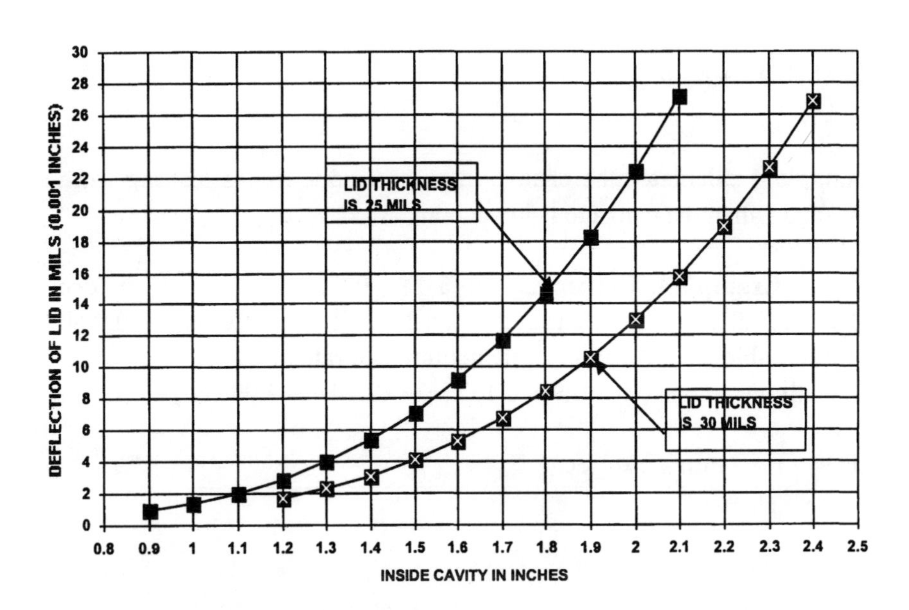

Figure 8-8. Deflection of 25 and 30 mil kovar lids for welded square packages, due to a difference in pressure of one atmosphere.

To calculate the change in pressure inside a package when gas is leaking out, Eq. (3-12) can be used:

$$\Delta p_t = \Delta p_i \left(e^{-\frac{Lt}{V}} \right)$$

where: Δp_t = the difference in pressure between the inside and outside of the package at time t

Δp_i = the initial difference in pressure between the inside of the package = 1 atm

Then $\Delta p_t = \left(e^{-\frac{Lt}{V}} \right)$, and $1 - \Delta p_t = 1 - \left(e^{-\frac{Lt}{V}} \right)$

Setting $1 - \Delta p_t = \Delta p_{change}$ then:

Eq. (8-12) $$\Delta p_{change} = 1 - \left(e^{-\frac{Lt}{V}} \right)$$

Example 4 calculates the change in pressure due to gas leaking from a package, when the package is in vacuum.

Example 4. The package has a 0.9 × 0.9 in. cavity, 0.157 in. deep. The lid is 0.015 in. thick. The deflection due to 1 atm difference = 0.00425 in. (see Table 8-5). Assume a true helium leak rate of 1×10^{-5} atm-cc/sec. The volume equals 0.9 × 0.9 × 0.157 in. × $(2.54)^3$ = 2.09 cc.

Let $t = 2$ min = 120 sec

$$\Delta p_{change} = 1 - e^{-\frac{(1\times10^{-5})(120)}{2.09}} = 1 - e^{-0.000574} = 1 - 0.999426$$

$$\Delta p_{change} = 0.000574 \text{ atm}$$

Let $t = 4 \text{ min} = 240 \text{ sec}$

$$\Delta p_{change} = 1 - e^{-0.001148} = 1 - 0.9988527$$

$$\Delta p_{change} = 0.001147 \text{ atm}$$

Let $t = 6 \text{ min} = 360 \text{ sec}$

$$\Delta p_{change} = 1 - e^{-0.001722} = 1 - 0.9982795$$

$$\Delta p_{change} = 0.001721 \text{ atm}$$

Let $t = 8 \text{ min} = 480 \text{ sec}$

$$\Delta p_{change} = 1 - e^{-0.002294} = 1 - 0.9977086$$

$$\Delta p_{change} = 0.002291 \text{ atm}$$

Let $t = 10 \text{ min} = 600 \text{ sec}$

$$\Delta p_{change} = 1 - e^{-0.00287} = 1 - 0.9917341$$

$$\Delta p_{change} = 0.002866 \text{ atm}$$

Let $t = 12 \text{ min} = 720 \text{ sec}$

$$\Delta p_{change} = 1 - \mathrm{e}^{-0.003444} = 1 - 0.996562$$

$$\Delta p_{change} = 0.003438 \text{ atm}$$

The results of these calculations are plotted in Fig. 8-9. The ATM_{MIN} value, 0.00294 atm, occurs between 10 and 12 min. It would take that long before the interferometer could detect any change in the lid deflection. The exact time for this minimum detectable deflection to take place can be calculated directly. Rearranging Eq. (8-12):

$$1 - \Delta p_{change} = +e^{-\frac{Lt}{V}}$$

$$\ln(1 - \Delta p_{change}) = -Lt/V$$

$$t = -(V/L)\ln(1 - \Delta p_{change})$$

where $\Delta P_{change} = ATM_{MIN}$

Eq. (8-13) $\qquad t = -(V/L)\ln(ATM_{MIN})$

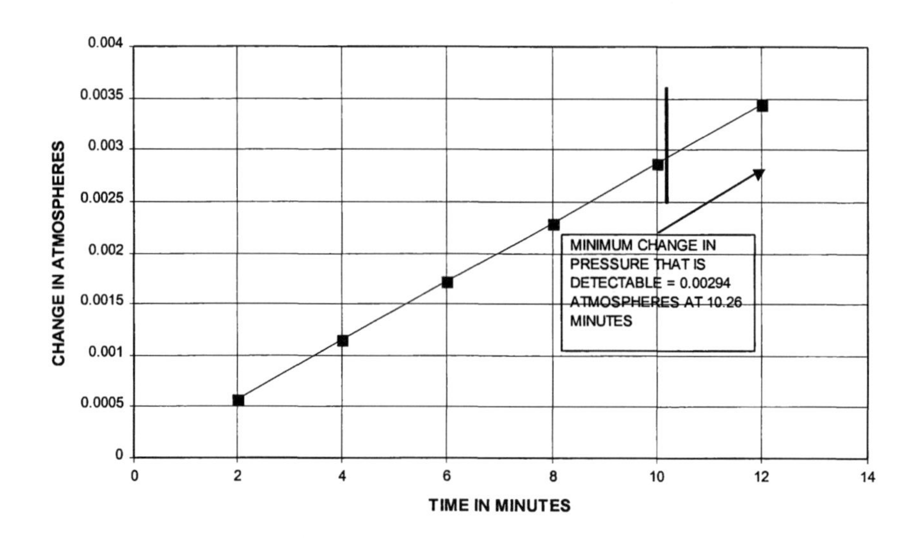

Figure 8-9. Time versus a decrease in pressure, for a package with inside dimensions of 0.9 × 0.9 × 0.16 in. with a lid 0.015 in. thick, with vacuum outside and (initially) 1 atm inside.

For the previous example:

$$t = (2.09/10^{-5}) \ln (1 - 0.00294)$$
$$= (2.09/10^{-5}) \ln 0.99706$$
$$= (2.09 / 10^{-5}) (-0.00294433)$$

$$t = 615.4 \text{ sec} = 10.26 \text{ min}$$

The time to detect the minimum change in pressure is given in Table 8-6 for selected packages, lid thickness, volumes and leak rates.

Column D is the true helium leak rate. Column E is the deflection of the lid in inches when the package is in an ambient of a vacuum or two atmospheres. Column F is the resolution of the interferometer divided by the deflection of the lid under a pressure difference of one atmosphere. This is the minimum pressure change in the package that can be detected:

$$\frac{1.25 \times 10^{-5} \text{ in.}}{\text{deflection in inches/1 atm}} = ATM_{MIN}$$

Column I is Eq. (8-12) in minutes instead of seconds to detect the change in pressure.

Looking at the first row in the table, the lid deflects outwards 0.000088 inches when the package is placed in a vacuum. As the gas in the package leaks out at a rate of 1×10^{-4} atm-cc/sec, the pressure in the package decreases. When the package has lost 0.142045 atm (Column F), the interferometer detects the lids movement. This only takes 1.48 min (Column I).

Table 8-6. Time It Takes to Detect a Lid Deflection Towards the Package Interior, after the Lid has been Deflected Outward Due to the Package Being Subjected to a Vacuum

A	B	C	D	E	F	G	H	I
L=W (in)	V0lume "A"^{2} × 0.4 × 2.54^2*	Lid Thickness (in)	Leak Rate (atm-cc/sec)	Deflection due to 1 atm**	Atm min 1.25E-05/"E" (atm)	1 – "F" (atm)	ln "G"	Time (min) "B" × "H"/ 60 × "D"
0.15	0.058064	0.005	1.00E-03	0.000088	0.142045	0.857954	–0.153204	1.482617
0.15	0.058064	0.005	1.00E-04	0.000088	0.142045	0.857954	–0.153204	14.826170
0.2	0.103225	0.005	1.00E-03	0.000280	0.044642	0.955357	–0.045670	0.785719
0.2	0.103225	0.005	1.00E-04	0.000280	0.044642	0.955357	–0.045670	7.857194
0.2	0.103225	0.010	1.00E-03	0.000035	0.357142	0.642857	–0.441832	7.601408
0.2	0.103225	0.010	1.00E-04	0.000035	0.357142	0.642857	–0.441832	76.014080
0.25	0.161290	0.005	1.00E-03	0.000683	0.018301	0.981698	–0.018471	0.496535
0.25	0.161290	0.005	1.00E-04	0.000683	0.018301	0.981698	–0.018471	4.965354
0.25	0.161290	0.010	1.00E-03	0.000085	0.147058	0.852941	–0.159064	4.275924
0.25	0.161290	0.010	1.00E-04	0.000085	0.147058	0.852941	–0.159064	42.759240
0.3	0.232257	0.005	1.00E-03	0.001420	0.008802	0.991197	–0.008841	0.342262
0.3	0.232257	0.005	1.00E-04	0.001420	0.008802	0.991197	–0.008841	3.422621
0.3	0.232257	0.010	1.00E-03	0.000177	0.070621	0.929378	–0.073239	2.835058
0.3	0.232257	0.010	1.00E-04	0.000177	0.070621	0.929378	–0.073239	28.350580
0.3	0.232257	0.015	1.00E-03	0.000052	0.240384	0.759615	–0.274943	10.642930

(Cont'd.)

Table 8- 6. (*Cont'd.*)

A	B	C	D	E	F	G	H	I
L=W (in)	V0lume "A"^{2} × 0.4 × 2.54^{2}*	Lid Thickness (in)	Leak Rate (atm-cc/sec)	Deflection due to 1 atm**	Atm min 1.25E-05/"E" (atm)	1 – "F" (atm)	ln "G"	Time (min) "B"× "H"/ 60 × "D"
0.6	0.929030	0.005	1.00E-03	0.022654	0.000551	0.999448	–0.000551	0.085460
0.6	0.929030	0.005	1.00E-04	0.022654	0.000551	0.999448	–0.000551	0.854601
0.6	0.929030	0.005	1.00E-05	0.022654	0.000551	0.999448	–0.000551	8.546014
0.6	0.929030	0.010	1.00E-03	0.002832	0.004413	0.995586	–0.004423	0.684944
0.6	0.929030	0.010	1.00E-04	0.002832	0.004413	0.995586	–0.004423	6.849449
0.6	0.929030	0.015	1.00E-03	0.000839	0.014898	0.985101	–0.015010	2.324246
0.6	0.929030	0.015	1.00E-04	0.000839	0.014898	0.985101	–0.015010	23.242460
0.6	0.929030	0.020	1.00E-03	0.000354	0.035310	0.964689	–0.035949	5.566321
0.6	0.929030	0.025	1.00E-03	0.000181	0.069060	0.930939	–0.071561	11.080430
0.9	2.090318	0.010	1.00E-03	0.014330	0.000872	0.999127	–0.000872	0.304028
0.9	2.090318	0.010	1.00E-04	0.014330	0.000872	0.999127	–0.000872	3.040286
0.9	2.090318	0.010	1.00E-05	0.014330	0.000872	0.999127	–0.000872	30.402860
0.9	2.090318	0.015	1.00E-03	0.004250	0.002941	0.997058	–0.002945	1.026175
0.9	2.090318	0.015	1.00E-04	0.004250	0.002941	0.997058	–0.002945	10.261750
0.9	2.090318	0.020	1.00E-03	0.001792	0.006975	0.993024	–0.006999	2.438665
0.9	2.090318	0.020	1.00E-04	0.001792	0.006975	0.993024	–0.006999	24.386650
0.9	2.090318	0.025	1.00E-03	0.001401	0.008922	0.991077	–0.008962	3.122322
0.9	2.090318	0.025	1.00E-04	0.001401	0.008922	0.991077	–0.008962	31.223220

(*Cont'd.*)

Table 8- 6. (*Cont'd.*)

A	B	C	D	E	F	G	H	I
L=W (in)	V0lume "A"^{2} × 0.4 × 2.54^2*	Lid Thickness (in)	Leak Rate (atm-cc/sec)	Deflection due to 1 atm**	Atm min 1.25E-05/"E" (atm)	1 – "F" (atm)	ln "G"	Time (min) "B" × "H"/ 60 × "D"
1.2	3.716121	0.015	1.00E-03	0.013425	0.000931	0.999068	–0.000931	0.576947
1.2	3.716121	0.015	1.00E-04	0.013425	0.000931	0.999068	–0.000931	5.769479
1.2	3.716121	0.020	1.00E-03	0.005663	0.002207	0.997792	–0.002209	1.368616
1.2	3.716121	0.020	1.00E-04	0.005663	0.002207	0.997792	–0.002209	13.686160
1.2	3.716121	0.025	1.00E-03	0.002904	0.004304	0.995695	–0.004313	2.671704
1.2	3.716121	0.025	1.00E-04	0.002904	0.004304	0.995695	–0.004313	26.717040
1.8	8.361273	0.020	1.00E-03	0.028667	0.000436	0.999563	–0.000436	0.607776
1.8	8.361273	0.020	1.00E-04	0.028667	0.000436	0.999563	–0.000436	6.077761
1.8	8.361273	0.025	1.00E-03	0.014703	0.000850	0.999149	–0.000850	1.185249
1.8	8.361273	0.025	1.00E-04	0.014703	0.000850	0.999149	–0.000850	11.852490
2.1	11.380620	0.025	1.00E-03	0.027240	0.000458	0.999541	–0.000458	0.870597
2.1	11.380620	0.025	1.00E-04	0.027240	0.000458	0.999541	–0.000458	8.705973

*Inside depth = 0.157 in
**From Table 8-5

8.0 PENETRANT DYE TEST

This test is destructive, but is useful for determining the location of the gross leak. The test consists of bombing the package in a penetrant type dye solution at 105 psia for three hours or at 60 psia for ten hours. After bombing, the package is washed in a solvent for the particular dye. The package is then observed under a compatible ultraviolet light source. Penetrant dyes and their corresponding light sources are:

Zyglow	3650 angstroms
Fluorescein	4935 angstroms
Rhodamine B	5560 angstroms

The test is especially useful for finding cracks in ceramic packages. To verify that the dye has penetrated the package, the package should be delidded and the inside observed under the ultraviolet light.

9.0 FLUOROCARBONS FROM A RESIDUAL GAS ANALYSIS

When fluorocarbons are present in a package, a Residual Gas Analysis (RGA) of that package will produce peaks at 31 (CF), 50 (CF_2), 69 (CF_3) and sometimes at 119 (CF_2—CF_3). The RGA is performed on the package when it is at 100°C. The vaporization equations are:

$$2(CF_3 — \underset{\displaystyle |\atop \displaystyle CF_3}{OCF} — CF_2 — O\,CF_2 — OCF_3) = 6CF_3 + 4CF_2 + 2CF + 3O_2$$

or, but not often

$$2(CF_3 — \overset{\displaystyle CF_3 \atop \displaystyle |}{OCF} — CF_2 — O\,CF_2 — OCF_3)$$
$$= 5CF_3 + 3CF_2 + 2CF + 3O_2 + 1(CF_3 — CF_2)$$

The molecular ratio of the fluorocarbon gases is: $3CF_3$, $2CF_2$, and 1CF; or occasionally $5CF_3$, $3CF_2$, 2CF and $1(CF_3 — CF_2)$. In either case,

CF_3 (mass 69) is the predominant gas. Fluorocarbons are reported as mass 69. The percentage of the gas reported at mass 69 includes the gases at the other mass peaks. A 1% at mass 69 states that the total fluorocarbon gases in the package is 1%.

The percentage of fluorocarbons reported in an RGA can be calculated from the microliters of fluorocarbon liquid in the package.

1. Calculating the microliters of fluorocarbon gases from the microliters of detector liquid at 91°C = microliters of detector liquid × 592. The 592 is the sum of 296 for the CF_3 + 2/3 of 296 for CF_2 + 1/3 of 296 for CF.
2. Convert the microliters of fluorocarbon gases to ml by dividing by 1000.
3. Calculating the fluorocarbon volume at 100°C = ml fluorocarbons gases at 91°C × (373°K/364°K). The milliliters of fluorocarbon gas for each microliter of fluorocarbon liquid = 0.592 × (373/364) = 0.607.
4. Dividing the ml of fluorocarbon gases by the volume of the package converts the volume to atmospheres of fluorocarbon gases.
5. The atmospheres of fluorocarbon gases × 100, divided by the total atmospheres in the package will yield the percentage of fluorocarbons. The total atmospheres in the package is the sum of the fluorocarbon gases + 1.26 atm. The 1.26 atm is the pressure of the gas that has been sealed in the package but is now at 100°C.

Several general equations can be written:

Eq. (8-14) $$\text{atm of fluorocarbon gas} = 0.607\,\frac{\mu\text{l of detector liquid}}{\text{package volume}}$$

Eq. (8-15) $$\%\ \text{fluorocarbons} = \frac{\text{atm fluorocarbons gas} \times 100}{1.26 + \text{atm fluorocarbons gas}}$$

Eq. (8-16) $$\text{atm fluorocarbons gas} = \frac{0.0126 \times \%\ \text{fluorcarbons}}{1 - \dfrac{\%\ \text{fluorocarbons}}{100}}$$

Eq. (8-17)

$$\mu\text{l detector liquid} = \frac{\text{atm fluorocarbons gas} \times \text{package volume}}{0.607}$$

Following are some examples using these equations.

Example 5. The microliters of detector liquid in a 4 cc package is 0.167. What is the expected percentage of fluorocarbon gases as reported in the RGA?

Using Eq. (8-14):

$$\text{atm fluorocarbon gas} = 0.607 \times 0.167/4 = 0.025$$

From Eq. (8-15):

$$\%\ \text{fluorocarbons} = \frac{0.025 \times 100}{1.26 + 0.025} = 1.95$$

Example 6. The RGA of a 4.0 cc package shows 1.0% of fluorocarbons. How much detector liquid was in the package?

Using Eq. (8-16):

$$\text{atm fluorocarbon gas} = \frac{0.0126 \times 1.0}{1 - \dfrac{1.0}{100}} = 0.0126/.99 = 0.0127$$

From Eq. (8-17):

$$\mu l\ \text{detector liquid} = \frac{0.0127 \times 4.0}{0.607} = 0.084$$

Example 7 (From Actual RGA Report).

Fluorocarbons in package = 17,000 ppm

Volume of package = 10 cc

How many microliters of detector liquid were in the package?

$$\% = \text{ppm} \times 100 = 17{,}000 \times 10^{-6} \times 100 = 1.7$$

Using Eq. (8-16):

$$\text{atm fluorocarbons} = \frac{0.0126 \times 1.7}{1 - \frac{1.7}{100}} = 0.022$$

Using Eq. (8-17):

$$\text{microliters detector liquid} = \frac{0.022 \times 10}{0.607} = 0.362$$

The RGA can detect 100 ppm of fluorocarbons. The amount of detector liquid corresponding to this 100 ppm depends upon the package volume. Using Eq. (8-16), the atmospheres corresponding to 100 ppm of fluorocarbons can be calculated.

$$100 \text{ ppm} = 100 \times 10^{-6} \times 100 = 0.01\%$$

$$\text{atm fluorocarbons} = \frac{0.0126 \times 0.01}{1 - \frac{0.01}{100}} = 0.000126$$

Note that the atmospheres of fluorocarbons are independent of package volume.

Using Eq. (8-17):

$$\text{microliters detector liquid} = \frac{0.000126 \times 10}{0.607} = 0.0021$$

If the package volume was 1.0 cc, then the microliters detector liquid = 0.00021.

These detector liquid quantities, corresponding to 100 ppm, are much less than the quantities listed in Tables 8-1–8-3, and may not represent a real physical leak.

Figure 8-10 gives the quantity of detector liquid in the package as a function of the package volume, when the RGA shows 100 ppm of fluorocarbons.

An actual RGA report shows 1910 ppm in a 10 cc package.

$$\text{atm fluorocarbons} = \frac{0.0126 \times 0.191}{1 - \dfrac{0.191}{100}} = 0.00241$$

$$\text{microliters detector liquid} = \frac{0.00241 \times 10}{0.607} = 0.0397$$

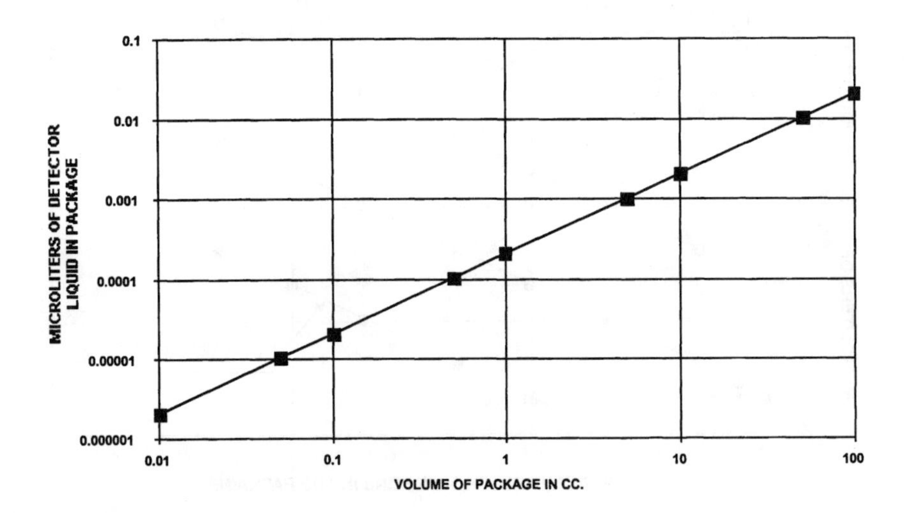

Figure 8-10. Microliters of detector liquid in the package when the RGA shows 100 ppm of fluorocarbons.

This value corresponds to a cylindrical leak channel having a true helium leak rate of 1.2×10^{-7} atm-cc/sec. For a rectangular shaped channel, the true helium leak rate would be approximately 1.5×10^{-7} atm-cc/sec.

The quantity of detector liquid in a package for a given percentage of fluorocarbons, as reported in an RGA, is a function of the package volume. Figure 8-11 shows this relationship for three volumes. Figure 8-12 shows how the percentage of fluorocarbons in a package varies with package volume when the microliters in the package = 0.167.

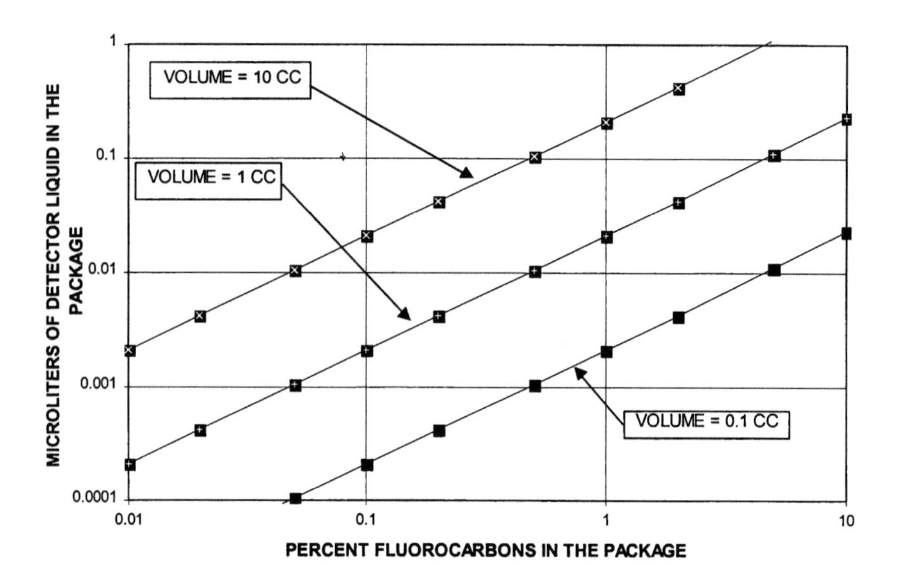

Figure 8-11. Microliters of detector liquid versus the percentage of fluorocarbons in the package for different volumes.

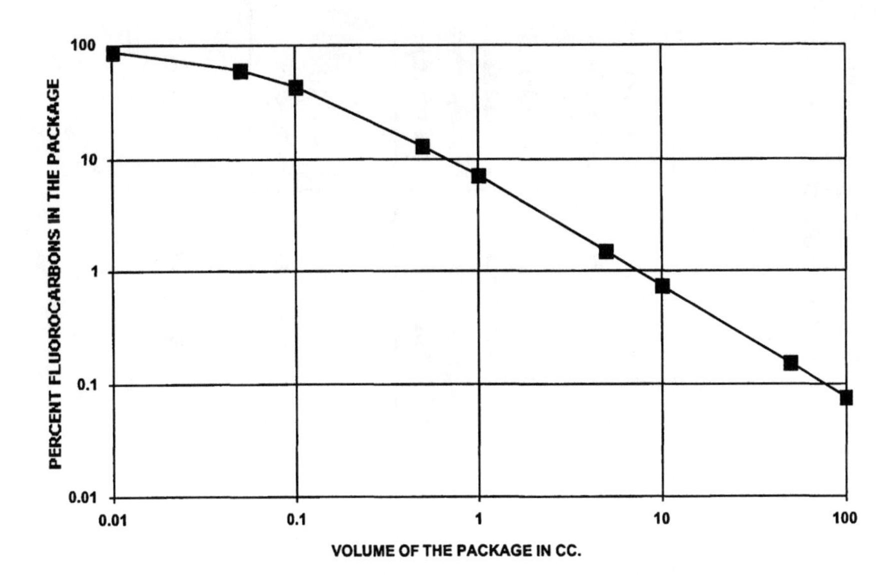

Figure 8-12. Percent fluorocarbons in the package as a function of package volume, when the microliters of detector liquid equals 0.167.

10.0 QUANTITATIVE COMPARISON OF GROSS LEAK TEST METHODS

The bubble, penetrant dye, and optical gross leak tests are qualitative. The weight gain, vapor detection, and RGA test methods are quantitative Table 8-7 compares the weight gain test with the vapor detection test.

Although the RGA test method can detect 100 ppm of fluorocarbons, this small amount will not be present in most packages, as the 100 ppm would correspond to a leak channel to small for any detector liquid to enter the package. This can be seen in Fig. 8-10.

Table 8-7. Quantitative comparison of the Weight Gain and Vapor Detection Gross Leak Test Methods and Their Limits

Test Method	Mil-Spec Limit	True Helium Leak Rate For Rectangular Channel (Mil-Spec)	True Helium Leak Rate For Cylindrical Channel (Mil-Spec)	Limit Of Equipment	True Helium Leak Rate For Rectangular Channel (Equipment Limited)	True Helium Leak Rate For Cylindrical Channel (Equipment Limited)
Vapor Detection	0.167 Micro-liters	5×10^{-7} atm-cc/sec	2.3×10^{-7} atm-cc/sec	0.1 Micro-liters	$\cong 3 \times 10^{-7}$ atm-cc/sec	2×10^{-7} atm-cc/sec
Weight Gain	2 mg, FOR $V > 0.01$ cc	$\cong 3 \times 10^{-6}$ atm-cc/sec	$\cong 2 \times 10^{-6}$ atm-cc/sec	0.1 MG	$\cong 2.3 \times 10^{-7}$ atm-cc/sec	$\cong 1.5 \times 10^{-7}$ atm-cc/sec

11.0 PROBLEMS AND THEIR SOLUTIONS

Problem 1. What is the viscous conductance of detector liquid through a cylindrical leak channel 0.2 cm long with a diameter of 5×10^{-4} cm? What is the viscous leak rate of the liquid if the bombing was at 3 atm absolute?

Solution. The viscous conductance of a cylindrical leak channel is given in Eq. (8-1):

$$F_{VC} = \frac{\pi D^4 P_a Y}{128\eta\left[\ell + \left(\frac{Y}{Z}\right)D\right]}$$

where: F_{VC} = viscous conductance in cc/sec

D = the diameter of the leak channel in cm = 5×10^{-4} cm

P_a = average pressure in dynes/cm^2 $(P_1 + P_2)/2 = (3 + 1)/2$
= 2 atm = $2 \times 1.01.\times 10^6$ dynes/ cm^2 = 2.02×10^6 dynes/ cm^2

η = the viscosity in poise [(dyne-sec)/cm^2] = 0.012 poise

ℓ = length of the leak channel in cm = 0.2 cm

Y and *Z* are calculated using Eqs. (8-2) and (8-3)

mfp of detector liquid = 1×10^{-7} cm

D/mfp = $5 \times 10^{-4}/10^{-7} = 5{,}000$

Therefore,

Y=1, Z = 1.7, Y/Z = 0.588, $(Y/Z) \times D$
= $0.588 \times 5 \times 10^{-4} = 2.94 \times 10^{-4}$ cm

$$F_{VDET} = \frac{\pi\left(5\times10^{-4}\right)^4\left(2.02\times10^6\right)(1)}{128\times(0.012)\times\left(0.2+2.94\times10^{-4}\right)}$$

$F_{VDET} = 1.29 \times 10^{-6}$ cc/sec

The viscous leak rate of the detector liquid = $R_{DET} = \Delta P \times F_{VDET}$.

$$R_{DET} = 2 \text{ atm} \times 1.29 \times 10^{-6} \text{ cc/sec} = 2.58 \times 19^{-6} \text{ atm-cc/sec}$$

Problem 2. For the leak channel in Problem 1, how much detector liquid will be forced into a 10.0 cc package when bombed for 10 hr at 3 atm absolute?

Solution. Knowing the leak rate of the detector liquid, the atmospheres of detector liquid are calculated using Eq. (8-6):

$$Q_{invP} = \Delta P_i \left(1 - e^{-\frac{R_V t}{V \Delta P_i}}\right)$$

$$Q_{DET} = 2\left[1 - e^{-\frac{(2.58 \times 10^{-6})(36{,}000)}{10 \times 2}}\right]$$

$$Q_{DET} = 2(1 - e^{-0.004644}) = 2(1 - 0.99537) = 2(0.004633)$$

$$Q_{DET} = 0.00927 \text{ atm}$$

These are the atmospheres of detector liquid.

Using the ratio:

$$\frac{\text{Detector Liquid atm}}{\text{Total atm in Package}} = \frac{\text{Volume of Detector Liquid}}{\text{Package Volume}}$$

$$\text{Volume of Detector Liquid} = \frac{0.00927 \times 10}{1.00927} = 0.0918 = 91.8 \ \mu\text{l}$$

Problem 3. A 1 cc package having a cylindrical leak channel 0.1cm. long, has a true helium leak rate of 1×10^{-7} atm-cc/sec. What volume of liquid has been injected into the package after bombing it for 4 hr at 3 atm absolute of detector liquid?

Solution. The volume of liquid injected into a package = the atmospheres of liquid times the package volume divided by the total pressure in the package. This total pressure is usually almost equal to one, as the volume of the liquid is very small compared to the package volume. If this is not true, a correction can be made.

The equation for the atmospheres of detector liquid is:

$$Q_{DET} = \Delta P_i \left(1 - e^{-\frac{R_{DET} t}{V \Delta P}} \right)$$

where:

$t = 4 \text{ hr} = 4 \times 3600 = 14{,}400 \text{ sec}$

$V = 1 \text{ cc}$

$\Delta P = 2 \text{ atm}$

$R_{DET} = \Delta P \times F_{DET}$

F_{DET} is calculated using Eq. (8-7):

$$F_{DET} = F_{VCHE} \frac{(\eta_{HE})(P_{AVEDET})(Y_{DET})}{(\eta_{DET})(P_{AVEHE})(Y_{HE})}$$

The viscous part of the true helium leak rate is F_{VCHE}, and can be obtained from Fig. 3-3. This figure shows that when the true helium leak rate is 1×10^{-7}, the viscous part is 45% or 4.5×10^{-8} atm-cc/sec.

$\eta_{HE} = 194 \times 10^{-6}$ poise

$\eta_{DET} = 0.012$ poise

$P_{AVEDET} = (3 + 1)/2 = 2$ atm

$P_{AVEHE} = (1 + 0)/2 = 0.5$ atm

$Y_{DET} = 1$ because $D/\text{mfp} \cong 1{,}000$

$Y_{HE} = 2.40$, (from Table 8-1)

$$F_{DET} = 4.5 \times 10^{-8} \times \frac{(194 \times 10^{-6})(2)(1)}{(0.012)(0.5)(2.40)} = 1.213 \times 10^{-9} \text{ cc/sec}$$

$$R_{DET} = \Delta P \times F_{DET} = 2 \times 1.213 \times 10^{-9} = 2.426 \times 10^{-9} \text{ atm-cc/sec}$$

$$Q_{DET} = 2\left[1 - e^{-\frac{(2.426\times10^{-9})(14,400)}{1\times2}}\right] = 2\left[1 - e^{-(1.747\times10^{-5})}\right]$$

$$Q_{DET} = 2(1 - 0.9999825) = 2(1.747 \times 10^{-5}) = 3.493 \times 10^{-5} \text{ atm}$$

$$\text{Volume} = Q_{DET} \times \text{Package Volume} = 3.493 \text{ atm} \times 10^{-5} \times 1 \text{ cc} = 3.49 \times 10^{-5} \text{ ml}$$

$$\text{Volume} = 0.035\ \mu\text{l of detector liquid}$$

Problem 4. A 0.1 cc package is bombed at 2 atm absolute of helium for 16 hr, and then fine leak screened to $R_{HE} = 1 \times 10^{-7}$ atm-cc/sec. The package is then bombed in detector liquid at 3 atm absolute for 4 hr, and gross leak screened to 0.167 μl. The package passes the fine leak screen but fails the gross leak screen. Assuming that the leak channel is rectangular, what is the true helium leak rate?

Solution. Figure 4-4 shows that if $R_{HE} = 1 \times 10^{-7}$ atm-cc/sec, $L_{HE}/V = 1 \times 10^{-5}$ atm-cc/sec. Since $V = 0.1$ cc, $L_{HE} = 1 \times 10^{-6}$ atm-cc/sec. This is the true helium fine leak limit to which the package was screened. The true helium leak rate, L_{HE}, is equal to or less than 1×10^{-6} atm-cc/sec. The leak rate of the detector liquid R_{DET}, is independent of the package volume when the volume of the liquid is small. From Table 8-2 or Fig. 8-2, 0.167 μl of detector liquid corresponds to a true helium leak rate of 5×10^{-7} atm-cc/sec. Failing the gross leak screen indicates a true helium leak rate greater than 5×10^{-7} atm-cc/sec. The true helium leak rate of the package is between 5×10^{-7} and 1×10^{-6} atm-cc/sec.

Problem 5. A package gains 3 mg after bombing for 4 hr at 3 atm absolute of detector liquid. Assuming a cylindrical leak channel 0.1 cm long, what is the true helium leak rate?

Solution. From Table 8-1, $L_{HE} \cong 2 \times 10^{-6}$ atm-cc/sec. Although the package volume is not stated, small amounts of liquid entering a package are independent of the package volume.

Problem 6. A 4.0 cc package sealed with 90% nitrogen and 10% helium has a cylindrical leak channel. The package measures 5×10^{-7} atm-cc/sec without bombing. The package is bombed in detector liquid for 4 hr at 3 atm absolute. What is the expected weight gain?

Solution. The true helium leak rate,

$$L_{HE} = \frac{\text{Measured leak rate}}{\text{Atmospheres of helium in the package}}$$

$$L_{HE} = \frac{5 \times 10^{-7}}{0.1} = 5 \times 10^{-6} \text{ atm-cc/sec}$$

From Fig. 8-1, 5×10^{-6} corresponds to approximately 5 µl. The density is 1.7 mg/µl. The expected weight gain = 5 × 1.7 = 8.5 mg.

Problem 7. A 0.5 mm diameter pin, set in a glass bead 0.1 cm thick, has a crack completely around it. The width of the crack measures 1×10^{-4} cm. The internal package volume is 2.0 cc. What is the weight gain after bombing 4 hr in detector liquid at an absolute pressure of 3 atm?

Solution. The length of the crack is

$$\pi D = 1.571 \text{ mm} = 0.1571 \text{ cm}$$

The viscous conductance of helium is F_{VRHE}, and for a rectangular channel the equation used is Eq. (8-8):

$$F_{VRHE} = \frac{(0.24)(a)^2(b)^2(P_{AVE})(Y)}{\ell} \text{ cc/sec}$$

where:
$a = 1 \times 10^{-4}$ cm
$b = 0.157$ cm
$a/b = 0.00064$
$Y = 0.005$, from Table 2-1a
$P_{AVE} = (1 + 0)/2 = 0.5 \text{ atm} = 380 \times 10^6$ microns
$\ell = 0.1$ cm

$$F_{VRHE} = \frac{(0.24)(10^{4})^{2}(0.157)^{2}(380\times10^{3})(0.005)}{0.1}$$

$$F_{VRHE} = 1.124 \times 10^{-6} \text{ cc/sec}$$

Correcting for the viscosity and the average pressure,

$$F_{DET} = F_{VRHE} \times \frac{\eta_{HE}}{\eta_{DET}} \times \frac{P_{AVEDET}}{P_{AVEHE}}$$

$$F_{DET} = 1.124\times10^{-6} \times \frac{194\times10^{-6}}{0.012} \times \frac{2}{0.5}$$

$$F_{DET} = 7.269 \times 10^{-8} \text{ cc/sec}$$

$R_{DET} = F_{DET} \times \Delta P = 7.269 \times 10^{-8}$ cc/sec × 2 atm = 1.454×10^{-7} atm-cc/sec

Using Eq. (8-6):

$$Q = \Delta P\left(1 - e^{-\frac{Rt}{V\Delta P}}\right)$$

$$Q = 2\left[1 - e^{\frac{(1.454\times10^{-7})(14{,}400)}{2\times2}}\right]$$

$$Q = 2\left(1 - e^{-5.234\times10^{-4}}\right)$$

$$Q = 2(1 - 0.999477) = 2(5.23 \times 10^{-4})$$

$Q = 1.046 \times 10^{-3}$ atm of detector liquid

Volume of liquid = atm of liquid × package volume

Volume of liquid = 1.046×10^{-3} atm × 2 ml = 2.092×10^{-3} ml = 2.092 μl

Weight of the liquid = μl × mg/μl = 2.092 × 1.7 = 3.6 mg

Problem 8. An MCM has been solder mounted onto a printed circuit board. The MCM had a true helium leak rate of 5×10^{-8} atm-cc/sec prior to mounting. The MCM failed the electrical test after mounting, and there is a question as to the MCM's hermeticity. The internal dimensions of the MCM is 1 in. by 1 in. by 0.16 in.. The lid to the package is 0.020 in. thick. The MCM, still mounted on its board, is sent to a laboratory for optical leak testing. The MCM and its board was placed in a vacuum, and six minutes after the lid had stopped deflecting outwards, the green laser interferometer starts to see the lid deflect inwards. What is the leak rate of the MCM?

Solution. Equation (8-12) is the equation relating to optical leak testing:

$$t = -\frac{V}{L}\ln\left(1 - ATM_{MIN}\right)$$

where: t = the time in seconds = 6 × 60 sec = 360 sec

V = the internal package volume in cc

= 2.54 cm × 2.54 cm × 0.4 cm = 2.58 cc

ATM_{MIN} = the minimum *atmospheres* the interferometer can detect

= the minimum *distance* the interferometer can detect divided by the deflection of the lid per atm

The deflection per atmosphere for this size package and a 0.020 in. thick lid is obtained from Fig. 8-7, and is 0.0027 in./atm. The minimum distance that the interferometer can detect is 1.25×10^{-5} in. (half the wavelength).

$$ATM_{MIN} = \frac{1.25 \times 10^{-5}\text{ in.}}{\dfrac{0.0027\text{ in.}}{1\text{ atm.}}} = 0.004630\text{ atm.}$$

Rearranging the equation to solve for L,

$$L = -\frac{V}{t}\ln(1 - ATM_{MIN})$$

$$L = -\frac{2.58}{360}\ln(1 - 0.004630)$$

$$L = -0.007167 \ln(0.99537) = -0.007167(-0.00464)$$

$$L = 3.33 \times 10^{-5} \text{ atm-cc/sec}$$

Problem 9. A package with inside dimensions of 0.5 in. by 0.5 in. by 0.16 in. is sealed with a 0.005 in. thick lid. The lid develops a circular hole of diameter 3×10^{-4} cm. The package is bombed for 10 hr at 2 atm absolute of detector liquid. How much liquid is forced into the package?

Solution.

$$Volume = 0.5 \text{ in.} \times 0.5 \text{ in.} \times 0.16 \text{ in.} = 0.04 \times (2.54)^3 = 0.655 \text{ cc}$$

The depth of the hole = the thickness of the lid = 0.005 in. = 0.0127 cm

Equation (8-1) is for viscous conductance:

$$F_{VC} = \frac{\pi D^4 P_a Y}{128\eta\left[\ell + \left(\frac{Y}{Z}\right)D\right]}$$

where: $D = 3 \times 10^{-4}$ cm

$P_a = (2 + 1)/2 = 1.5$ atm $= 1.5 \times 1.01 \times 10^6$ (dynes/cm^2)/atm

$Y = 1$ because D/m.f.p. $= 3 \times 10^{-4}/10^{-7} = 3{,}000$

η = the viscosity of the detector liquid = 0.012 poise

$\ell = 0.0127$ cm

$(Y/Z)D$ is very small and can be neglected

$$F_{DET} = \frac{\pi(3\times10^{-4})(1.5\times1.01\times10^{6})}{(128)(0.012)(0.0127)} = \frac{3.855\times10^{-8}}{0.0195}$$

$$F_{DET} = 1.976 \times 10^{-6} \text{ cc/sec}$$

$$R_{DET} = F_{DET} \times \Delta P = 1.976 \times 10^{-6} \times (2-1) = 1.976 \times 10^{-6} \text{ atm-cc/sec}$$

Using Eq. (8-6) to solve for the atmospheres of liquid in the package:

$$Q = \Delta P\left(1 - e^{-\frac{Rt}{V\Delta P}}\right)$$

$$Q = 1\left[1 - e^{-\frac{(1.976\times110^{-6})(36{,}000)}{(0.655)(1)}}\right]$$

$$Q = 1 - e^{-0.1086} = 1 - 0.89708 = 0.1029 \text{ atm}$$

Volume = Q × package volume = 0.1029 × 0.655 = 0.0674 ml = 67.4 μl

Problem 10. A rectangular package with inside dimensions of 0.5 in. by 1.0 in. by 0.16 in. has a 0.010 in. thick kovar lid. How long will it take to detect the lid moving inwards after the lid has stopped moving outwards when the package is in a vacuum? The true helium leak rate is 5 × 10^{-5} atm-cc/sec. Assume the deflection equation is:

$$\max y = -0.111\frac{wb^4}{ET^3}$$

Solution. In the above equation:

b = 0.5 in.

E = 20 × 10^6 lb/in.^{2}

T = 0.010 in.

w = 14.7 lb

$$\max y = {}^{-}0.111 \frac{14.7(0.5)^4}{(20\times 10^6)(0.010)^3} = -\frac{0.1020}{20} = -0.0051 \text{ inches}$$

$$\text{Volume} = 0.5 \times 1 \times (2.54)^2 \times 0.4 = 1.29 \text{ cc}$$

$$ATM_{MIN} = \frac{1.25\times 10^{-5}}{-0.0051} = -0.002451 \text{ atm}$$

From Eq. (8-12):

$$t = -\frac{1.29}{5\times 10^{-5}} \ln(1-0.002451) = -25{,}800 \ln(0.997549)$$

$$t = -25{,}800 \times -0.002454$$

$$t = 63.3 \text{ sec}$$

Problem 11. A package with a 1.0 cc volume is sealed in a 100% nitrogen atmosphere. After bombing in helium for 10 hr in 3 atm absolute, the measured leak rate is 5×10^{-7} atm-cc/sec. What is the expected weight gain after bombing for 4 hr at 3 atm absolute of detector liquid? Assume the leak channel has a uniform rectangular cross-section with a ratio of 100:1 and is 0.1 cm deep.

Solution. If we knew the helium viscous leak rate, we could calculate the viscous leak rate of the detector liquid. Knowing that, the volume and weight of the liquid could be calculated. To get the helium viscous leak rate, we must first know the true helium leak rate. Equation (4-22) calculates the true helium leak rate from the measured rate and bombing conditions.

$$L_{\text{HE}} = \frac{R_{\text{HE}}}{\Delta P_{\text{HE}}\left(1 - e^{-\frac{L_{\text{HE}}T}{V}}\right)}$$

where: L_{HE} is the true helium leak rate

R_{HE} is the measured helium leak rate = 5×10^{-7} atm-cc/sec

$\Delta P_{HE}\left(1-e^{-\frac{L_{HE}T}{V}}\right)$ = the atmospheres of helium in the package

DP_{HE} = difference in the helium partial pressure = (3 – 0) = 3 atm

V = 1 cc

T = time in sec = 10 hr × 3600 sec/hour = 36,000 sec

Transposing Eq. (8-22):

$$R_{HE} = L_{HE} \times DP_{HE}\left(1-e^{-\frac{L_{HE}T}{V}}\right)$$

L_{HE} is greater than R_{HE} because L_{HE} corresponds to 100% helium in the package. Plotting L_{HE} versus R_{HE} on log-log paper will result in a straight line.

Let $L_{HE} = 1 \times 10^{-5}$ atm-cc/sec.

$$R_{HE} = 10^{-5}\,(3)\left[1-e^{-\frac{(10^{-5})(36{,}000)}{1}}\right] = 3 \times 10^{-5}\,(1 - e^{-0.36})$$

$$R_{HE} = 3 \times 10^{-5}(1 - 0.697676) = 3 \times 10^{-5}(0.3023) = 9.07 \times 10^{-6} \text{ atm-cc/sec}$$

Let $L_{HE} = 1 \times 10^{-6}$ atm-cc/sec.

$$R_{HE} = 3 \times 10^{-6}(1 - e^{-0.036}] = 3 \times 10^{-6}(1 - 0.96464) = 3 \times 10^{-6}(0.035359)$$

$$R_{HE} = 1.061 \times 10^{-7} \text{ atm-cc/sec}$$

The log-log plot shows that when $R_{HE} = 5 \times 10^{-7}$, $L_{HE} = 2.256 \times 10^{-6}$ atm-cc/sec (also see Fig. 6-3 for the value of L_{HE}).

Figure 3-4 gives the percent viscous as a function of L_{HE} for different length to width ratios. For $L_{HE} = 2.25 \times 10^{-6}$ and for an aspect ratio of 100:1, the percent viscous is approximately 15%. Therefore the viscous part of L_{HE} equals $0.15 \times 2.25 \times 10^{-6} = 3.38 \times 10^{-7}$ atm-cc/sec.

The viscous conductance of the detector liquid = F_{DET}

Correcting for the difference in viscosities and average pressure:

$$F_{CET} = F_{VCHE} \frac{\eta_{HE}}{\eta_{DET}} x \frac{P_{AVEDET}}{P_{AVEHE}}$$

$$= 3.38 \times 10^{-7} \times \frac{194 \times 10^{-6}}{0.012} \times \frac{2}{0.5}$$

$$= 2.19 \times 10^{-8} \text{ cc/sec}$$

The detector leak rate = $\Delta P \times F_{DET} = 2 \times 2.19 \times 10^{-8} = 4.37 \times 10^{-8}$ atm-cc/sec.

$$Q = \Delta P \left(1 - e^{-\frac{R_{DET} t}{V \Delta P}}\right) = 2\left[1 - e^{-\frac{(4.37 \times 10^{-8})(14,400)}{1 \times 2}}\right]$$

$$= 2\left(1 - e^{-3.146 \times 10^{-8}}\right)$$

$$= 2(1 - 0.999685) = 2(3.1459 \times 10^{-4})$$

$$= 6.29 \times 10^{-4} \text{ atm}$$

Volume = $Q \times$ package volume
= $6.29 \times 10^{-4} \times 1 = 6.29 \times 10^{-4}$ ml = 0.629 μl

Weight gain = μl × mg/μl = 0.629 × 1.7 = 1.07 mg

Problem 12. Assuming the same conditions as Problem 11, what is the expected reading when performing the vapor detection test?

Solution. The volume of detector liquid is 0.629 μl. The instrument reads 200 when the volume is 0.167 μl. The instrument will read (0.629/0.167) × 200 counts = 753 counts for the 0.629 μl.

Problem 13. An RGA shows two percent fluorocarbons in a 2.0 cc package. How much detector liquid is in the package?

Solution. Equation (8-16) calculates the atmospheres of fluorocarbons from the percentage of fluorocarbons in the package.

$$\text{atm of fluorocarbons} = \frac{0.0126 \times \%\ \text{fluorocarbons}}{1 - \dfrac{\%\ \text{fluorocarbons}}{100}}$$

$$\text{atm. of fluorocarbons} = \frac{0.0126 \times 2}{1 - \dfrac{2}{100}} = \frac{0.0252}{0.98} = 0.02571$$

Equation (8-17) calculates the microliters of fluorocarbons from the atmospheres of fluorocarbons in the package.

$$\mu\text{l of fluorocarbons} = \frac{\text{atm of fluorocarbons} \times \text{package volume}}{0.607}$$

$$\mu\text{l of fluorocarbons} = \frac{0.02571 \times 2}{0.607}$$

$$\mu\text{l of fluorocarbons} = 0.0847$$

Problem 14. A 3.0 cc package has gained 3 mg when bombed for 4 hr at 3 atm absolute of detector liquid. The package is sent to a laboratory for a RGA. What is the expected percent fluorocarbons?

Solution. The bombing conditions are superfluous, as the weight gain defines the liquid in the package. First, the 3 mg is converted to μl:

$$\frac{3\,\text{mg}}{1.7\,\text{mg}/\mu\text{l}} = 1.76\,\mu\text{l of detector liquid}$$

From Eq. (8-14):

$$\text{atm fluorocarbon gas} = 0.607\,\frac{\mu\text{l detector liquid}}{\text{package volume}}$$

$$\text{atm fluorocarbon gas} = 0.607 \times \frac{1.76}{3} = 0.356$$

Using Eq. (8-15):

$$\%\ \text{fluorocarbons} = \frac{\text{atm fluorocarbons gas} \times 100}{1.26 + \text{atm fluorocarbons gas}}$$

$$\%\ \text{fluorocarbons} = \frac{0.356 \times 100}{1.26 + 0.356} = 22\%$$

Problem 15. A 5.0 cc package has a true helium leak rate of 1×10^{-7} atm-cc/sec. What is the expected fluorocarbon percentage in the package after bombing in 5 atm absolute of detector liquid for 2 hr? Assume a cylindrical leak channel 0.1 cm deep.

Solution. Figure 3-3 shows the percent viscous versus the true helium leak rate for a cylindrical leak channel. For a true helium leak rate of 1×10^{-7} atm-cc/sec, the viscous rate is about 50% or 5×10^{-8} atm-cc/sec. This is also the value of the viscous conductance (F_{VCHE}) as the total pressure difference is 1 atm. Using Eq. (8-7):

$$F_{CDET} = F_{VCHE} \times \frac{\eta_{HE}}{\eta_{DET}} \times \frac{P_{AVEDET}}{P_{AVEHE}} \times \frac{Y_{DET}}{Y_{HE}}$$

where: F_{CDET} = viscous conductance of the detector liquid through the cylindrical channel

F_{VCHE} = viscous conductance of helium through the cylindrical channel

η_{HE} = the viscosity of helium 194×10^{-6}

η_{DET} = the viscosity of the detector liquid 0.012

P_{AVEDET} = the average pressure of the detector liquid = (5 + 1)/2 =3

P_{AVEHE} = the average pressure of the helium = (1 + 0)/2 = 0.5

Y_{DET} = 1 because the mfp is so small (1×10^{-7} cm)

Y_{HE} = 2.4, and is calculated using Eq. 8-2 and the diameter in Table 8-1 for $H_{VCHE} = 5 \times 10^{-8}$

$$F_{CDET} = V5 \times 10^{-8} \times \frac{194 \times 10^{-6}}{0.012} \times \frac{3}{0.5} \times \frac{1}{2.4}$$

$$F_{CDET} = 2.021 \times 10^{-9} \text{ cc/sec}$$

$$R_{DET} = \Delta P \times F_{CDET} = (5-1) \times 2.021 = 8.08 \times 10^{-9} \text{ atm-cc/sec}$$

The atmospheres of detector liquid = Q_{DET}, and

$$Q_{DET} = \Delta P \left(1 - e^{-\frac{R_{CDET} x t}{V \Delta P}}\right) = 4\left[1 - e^{-\frac{\left(8.08 \times 10^{-9}\right)(7{,}200)}{5 \times 4}}\right]$$

$$= 4\left(1 - e^{-2.909 \times 10^{-6}}\right) = 4(1 - 0.9999709)$$

$$= 4(2.909 \times 10^{-5}) = 1.164 \times 10^{-4} \text{ atm}$$

The volume of liquid = atm of liquid × package volume
= $2.91 \times 1.164 \times 10^{-4}$

The volume of liquid = 5.82×10^{-4} ml = 0.582 μl

Applying Eq. (8-14):

$$\text{atm of fluorocarbon gas}\ \frac{\text{l liquid}}{\text{package volume}} = 0.607 \times \frac{0.582}{5} = 0.0706$$

Using Eq. (8-15) to solve for the percentage of fluorocarbons:

$$\%\ \text{fluorocarbons} = \frac{\text{atm fluorocarbon gases} \times 100}{1.26 + \text{atm fluorocarbon gases}}$$

$$\%\ \text{fluorocarbons} = \frac{0.0706 \times 100}{1.26 + 0.0706} = 5.31$$

REFERENCES

1. Hedley, W. H., et al., "Flow Through Micro-openings," Final Comprehensive Report, Monsanto Research Corporation under contract with U.S. Army Biological Center, Fort Detrick [Contract DA 18064-AMC-563(A)] (1968)
2. Tyson, J., II, *ISHM 1992 Proceedings*, pp. 348–351 (1992)
3. Roark, R. J., *Formulas for Stress and Strain*, pp. 194–226, McGraw-Hill (1988)

9

The Permeation of Gases Through Solids

1.0 DESCRIPTION OF THE PERMEATION PROCESS

The passage of a gas through a solid is called permeation. Permeation is a two phase process. The first phase is the adsorption of the gas on the surface of the solid. The second phase is the diffusion of the gas through the solid. The diffusion rate through the solid depends on:

1. The non-perfect density of the solid due to:
   a. An internal porous structure that has access to the outside
   b. The space between grain boundaries and the number and size of the grains
   c. Slip planes in the solid
   d. Other defects in the solid
2. The chemical affinity of the gas for the solid. Active gases such as hydrogen or oxygen have an affinity to some solids that an inert gas would not. Hydrogen passing through a glass will have some chemical affinity to the oxygen in the glass thereby slowing down the diffusion. In a similar way, oxygen will be slowed down when passing through a material that could be oxidized.

It is difficult to measure the gas solubility on the surface and the diffusion through the solid separately. Solubilities for several gas-solid combinations can be found in Redhead, Hobson and Kornrlson, pp. 95–98.[1] Permeation, the two phases together, has been measured for many gases through solids.

The property of the solid that characterizes the amount of gas that can pass through the solid is called permeability. There is a different permeability for each gas-solid combination. For example, there are permeabilities for:

helium through quartz

hydrogen through quartz

nitrogen through quartz

oxygen through quartz

argon through quartz

helium through soda lime glass

helium through Pyrex

helium through aluminum oxide

helium through kovar

helium through neoprene

The quantity of gas that permeates through a solid in one second can be measured and represented as Eq. (1):

Eq. (9-1) $$R = \frac{SDA\Delta P}{d}$$

where: A = the area in cm^2

S = the solubility of the gas in the solid

D = the diffusion of the gas through the solid per sec

ΔP = the partial pressure difference of the gas across the solid

d = the thickness of the solid

Combining the solubility and the diffusion, and calling it permeability, we have Eq. (9-2):

Eq. (9-2) $$R = \frac{KA\Delta P}{d}$$

where: A = the area in cm^2

ΔP = the partial pressure difference of the gas across the solid

d = the thickness of the solid

K = the permeability of the gas-solid

The same units are not always used for P and d, so that K is also not always in the same units. When P is in atmospheres and d is in centimeters, R is in atm-cc/sec K is then in cm^2/sec. When P is in cm of mercury, and d is in centimeters, R is in cm-cc/sec. The thickness d is often given in millimeters, so that a factor of ten is introduced. The leak rate, R, is often measured or calculated in micron-liters/hour. The following permeation equations can be written:

Eq. (9-3) $$R = \frac{KA\Delta P_{ATM}}{d_{CM}} \text{ in atm-cc/sec}$$

where K is in cm^2/sec, and it is for the condition of an area of 1 cm^2, a thickness of 1 cm, and a partial pressure difference of 1 atm.

Eq. (9-4) $$R = \frac{KA\Delta P_{ATM}}{d_{mm}} \text{ in atm-cc/sec}$$

where K is in $0.1 \times cm^2$/sec, and it is for the condition of an area of 1 cm^2, a thickness of 1 mm, and a partial pressure difference of 1 atm.

Eq. (9-5) $$R = \frac{KA\Delta P_{cm}}{d_{mm}} \text{ in cm of pressure-cc/sec}$$

where K is in $0.1 \times cm^2$/sec, and it is for the condition of an area of 1 cm^2, a thickness of 1 mm and a partial pressure difference of 1 cm of mercury.

The value of R, in micron-liters/hour is a much larger number than the equivalent permeation in atm-cc/sec, and is therefore also used. The value of K in Eq. (9-3) is $7.6 \times$ the value of K in Eq. (9-5). The permeabilities for several gas-solid combinations are given in Table 9-1. The use of this table is seen in Example 1.

Table 9-1. Permeabilities of Selected Gas-Solid Combinations*

Gas-Solid Combination	*K* at 25°C	*K* at 80°C	*K* at 100°C
Helium-SiO_2 (Quartz)	1.E-9		4.2E-9
Helium-Vycor	3.E-10		1.5E-9
Helium-7740 Glass	8.5E-11		7.E-10
Helium 1720 Glass	1.5E-15		1.E-13
Helium-96% Alumina	3.E-16		2.E-14
Helium-Any Metal	0		0
Nitrogen or Oxygen-7740 Glass	5.E-17		2.E-16
Neon-Vycor	2.E-16		1.5E-14
Helium-Silicone	2.4E-6	5.6E-6	
Helium-Fluorocarbon	1.3E-7	1.3E-6	
Helium-Neoprene	6.5E-8	6.0E-7	
Helium-Butyl	6.7E-8	5.4E-7	
Helium-Buna N	8.E-8	4.3E-7	

*Values of *K* are taken from, or are averages of, Refs. 2–6. Units of *K* are in cm^2/sec and have values that will yield *Q* in atm-cc/sec, when *A* is in cm^2, thickness in cm, and pressure difference in atm.

Example 1. An enclosure having a volume of 20 cc has been sealed under vacuum and has proven to be hermetic with respect to fine and gross leaks. The glass part of the enclosure is Corning 7740 and the surface area of the glass is 40 cm^2. The thickness of the glass is 0.05 cm. The enclosure is placed in a 100% nitrogen atmosphere oven at 100°C for 30 days. How much nitrogen permeated into the enclosure after the 30 days?

Using Eq. (9-3) and the value of *K* from Table 9-1:

$$R \text{ in atm-cc/sec} = \frac{KA\Delta P}{d} = \frac{(2\times10^{-16})(40)(1)}{0.05}\ 1.6\times10^{-13}$$

This is the quantity of nitrogen entering the enclosure per second. To get the quantity of nitrogen in the enclosure after 30 days:

$$\text{Quantity of nitrogen} = 1.6 \times 10^{-13} \times 30 \times 24 \times 3600 = 4.16 \times 10^{-7} \text{ atm-cc}$$

To determine the atmospheres in the package, the above value is divided by the package volume.

$$Q = 4.16 \times 10^{-7} \text{ atm-cc/20 cc} = 2.08 \times 10^{-8} \text{ atm}$$

Therefore, the equation for the atmospheres in the package is Eq. (3) times the time and divided by the volume:

Eq. (9-6) $$Q = \frac{KA\Delta Pt}{Vd}$$

2.0 EFFECT OF TEMPERATURE ON PERMEATION

As the temperature of a solid is increased, the pores and defects in the solid become more open and gases can pass through the solid more easily. Therefore the permeation increases with an increase in temperature. Figures 9-1–9-3 show the relationship between the permeation and the reciprocal of the temperature in degrees Kelvin. The units of the permeation are in square centimeters per second and have the value in Eq. (9-3) to yield Q in atm-cc/sec.

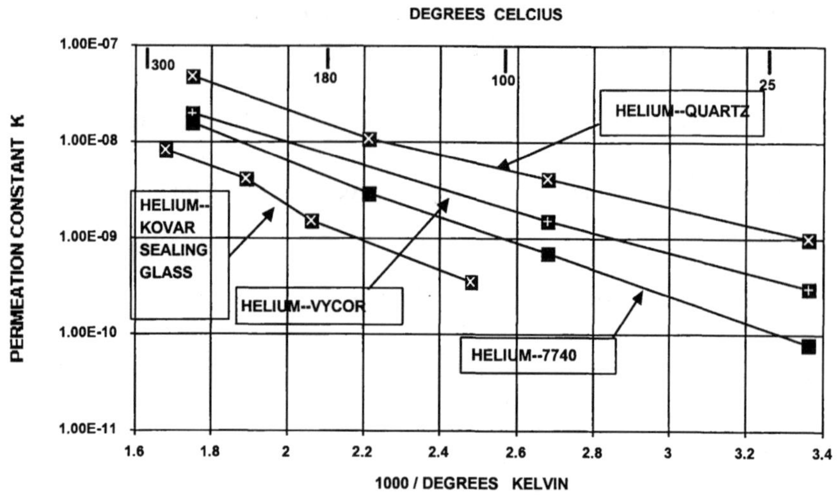

*The units of the permeation constant are in square centimeters per second and have the value in Eq. (9-3) to yield R in atm-cc/sec.

Figure 9-1. Effect of temperature on the permeation constant for helium-quartz, and selected glasses.

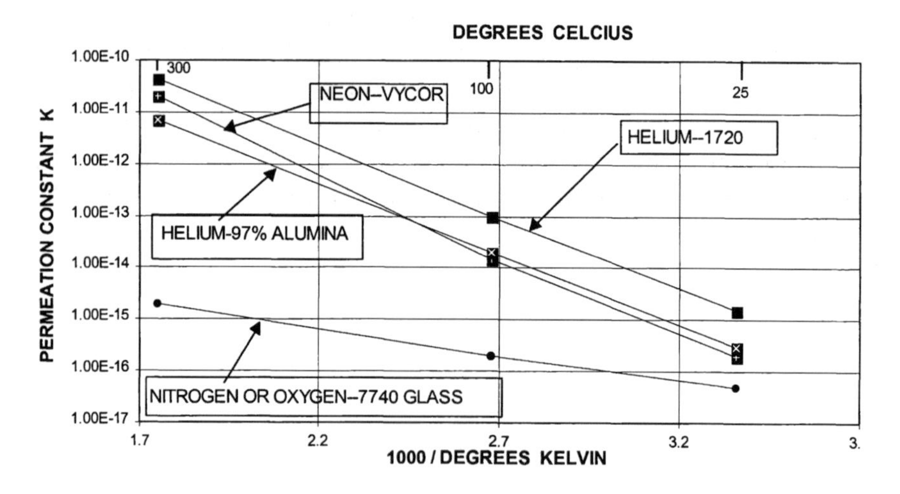

*The units of the permeation constant are in square centimeters per second and have the value in Eq. (9-3) to yield R in atm-cc/sec.

Figure 9-2. Effect of temperature on the permeation constant for helium-1720 glass, helium-97% alumina, neon-vycor, and nitrogen or oxygen-7740 glass.

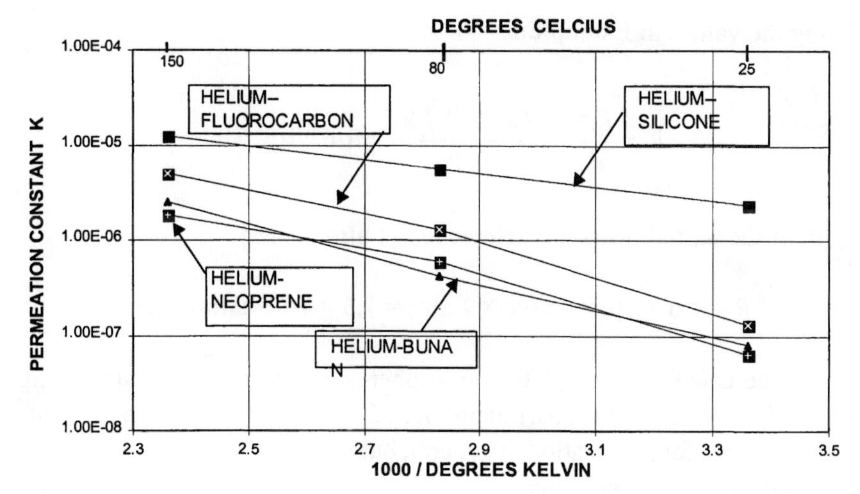

*The units of the permeation constant are in square centimeters per second and have the value in Eq. (9-3) to yield R in atm-cc/sec.

Figure 9-3. Effect of temperature on the permeation constant of helium through selected elastomers.

3.0 TREATING PERMEATION AS A LEAK RATE

The units of permeation can be atm-cc/sec. These are the usual units for leak rates, although the values for permeation are usually much smaller. The general equation for leak rates is:

Eq. (9-7) $$R = F\Delta P$$

where R = the leak rate

F = the conductance and = KA/d

ΔP = the difference in partial pressure

And therefore:

Eq. (9-8) $$R = \frac{KA\Delta P}{d}$$

Using the values in Example 1:

$$F = \frac{KA}{d} = \frac{(2 \times 10^{-16})(40)}{0.05} = 1.6 \times 10^{-13} \text{ cc/sec}$$

If the partial pressure difference is 1 atm, then:

$$R = 1.6 \times 10^{-13} \text{ cc/sec} \times 1 \text{ atm} = 1.6 \times 10^{-13} \text{ atm-cc/sec}$$

The calculation of Q, the atmospheres entering the package, using Eq. (9-6) assumes that the partial pressure difference does not change. This is usually true for permeation in general, but may not true when elastomers are used for sealing packages.

Example 2a. A 1.0 cc package is sealed in vacuum with a silicone o-ring having a permeation K value of 2.4×10^{-6} cm^2/sec. The surface area of the o-ring that is exposed to the helium is 0.3 cm^2. The thickness of the o-ring is 0.03 cm. The package is bombed at 3 atm absolute of helium for 4 hr. How much helium is in the package at the end of the 4 hr?

Using Eq. (9-3):

$$R = \frac{\left[2.4 \times 10^{-6} \frac{\text{cm}^2}{\text{sec}} (0.3 \text{ cm}^2)(3 \text{ atm})\right]}{0.03 \text{ cm}} = 7.2 \times 10^{-5} \text{ atm-cc/sec}$$

Using Eq. (9-6) to calculate the atmospheres in the package:

$$Q = \frac{KA\Delta Pt}{Vd} = \frac{(2.4 \times 10^{-6})(0.3)(3)(14400)}{1 \times 0.03} = 1.037 \text{ atm}$$

The atmospheres in the package can also be calculated using the leak rate formula, allowing for the change in pressure inside the package. This is Eq. (9-9):

Eq. (9-9) $$Q = \Delta P\left(1 - e^{-\frac{Rt}{\Delta Pv}}\right)$$

Example 2b.

$$Q = 3\left[1 - e^{\frac{(7.2\times10^{-5})(14400)}{3\times1}}\right] = 3\left(1 - e^{-0.03456}\right)$$

$$Q = 3(1 - 0.70780) = 3(0.2922) = 0.8766 \text{ atm}$$

This value is appreciably less than that using Eq. (9-6), that equation yielding an 18.3% error. It is not obvious when to use Eq. (9-9) instead of Eq. (9-6). The permeation literature uses Eq. (9-6), but most of the literature does not address electronic packages. Equation 9-9 is always correct. The value of *KAt/Vd* is the key to the usage of Eq. (9-6). In Figs.9-4 and 9-5, the value of *KAt/Vd* is plotted against the percent error when using Eq. (9-6).

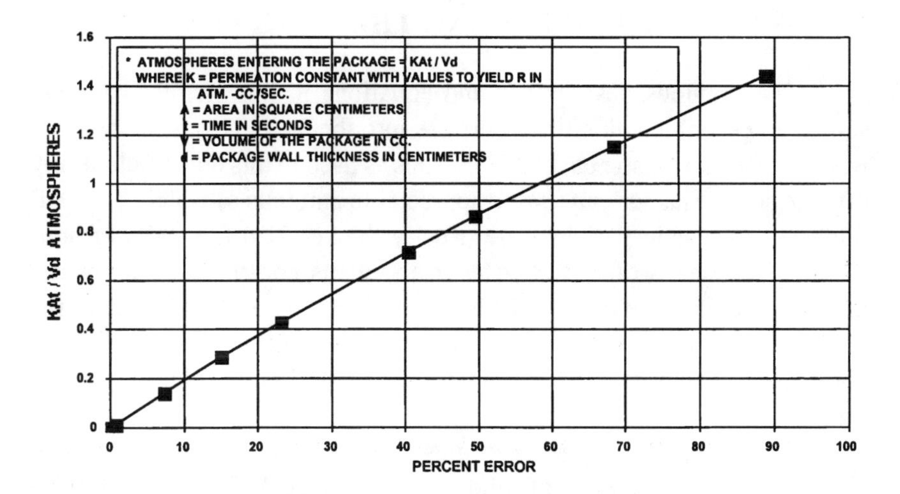

Figure 9-4. The percent error when calculating the gas entering a package due to permeation, when neglecting the pressure buildup in the package.

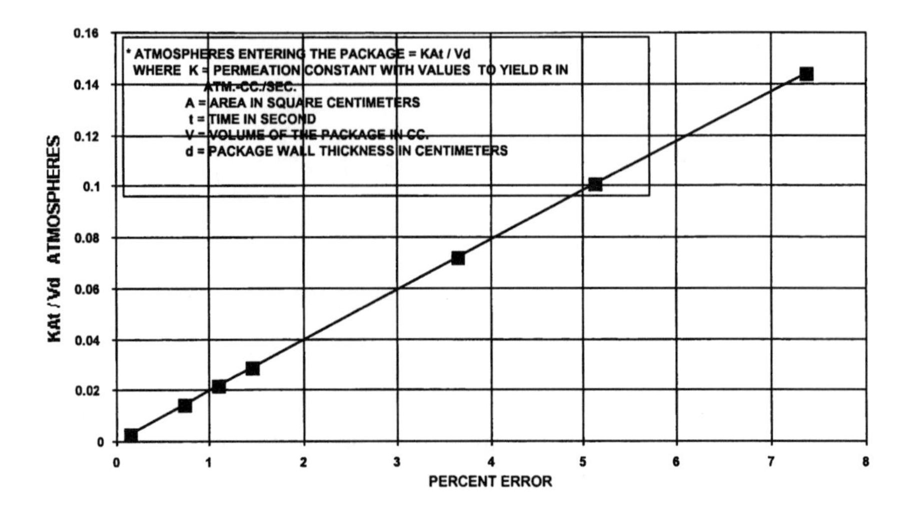

Figure 9-5. The percent error when calculating the gas entering a package due to permeation, when neglecting the pressure buildup in the package (enlargement of Fig. 9-4 for smaller percentages).

4.0 WATER VAPOR PASSING THROUGH PLASTICS

The previous descriptions and equations in this chapter have assumed a negligible solubility of the gas into the wall. This is not true for water vapor into plastic packages. Plastics as used in this book include any organic solid. Plastics can absorb appreciable amounts of water. The wall thickness of plastic packages are usually thicker than the walls of non-plastic packages, so that the water in the walls contribute to the water entering the package cavity. If the plastic contains more water than the ambient, water will leave the plastic, entering the ambient. This condition is common as the humidity of the ambient can change daily and from place to place.

The absorption of gases in glass and ceramics is in the range of 0.001–0.01 cc of gas per cc of solid. Solubilities are dimensionless but are given in cc/cc at a pressure of one atmosphere. The solubility of gases in metals is zero, so that the permeation of gases in metals is also zero. The solubility of water vapor in plastics is in the range of 1–400.

The solubility of water vapor in plastics can be calculated from the weight gain of the plastic when exposed to water vapor. The range of water absorption in plastics at room temperature and 98% relative humidity is from 0.2% to 2.0%.

Example 3. Assume that a plastic strip 1 cm thick is subjected to 98% relative humidity for 300 hr at 25°C. The plastic gains 0.5% in weight, from 10.00 g to 10.05 g. This 0.05 g (50 mg) has a corresponding vapor volume.

One molecular weight (18 g) occupies 22.4 l

18 mg occupies 22.4 cc

Volume for 50 mg = (50 mg × 22.4 cc)/18 mg = 62.2 cc

If the plastic had a specific gravity of 1.5, then:

Volume of plastic = 10 g./(1.5 g/cc) = 6.67 cc

$$\text{Solubility} = \frac{62.2\ \text{cc}}{6.67\ \text{cc}} = 9.32\ \text{cc/cc}$$

Solubilities increase with the temperature and water vapor concentration (partial pressure). Changing the temperature from 25°C to 85°C will increase the solubility by approximately a factor of ten. Increasing the relative humidity from 25% to 85% increases the solubility by a factor between two and three.

When a plastic package with no water in the cavity or the wall, is placed in a water vapor environment, the following takes place.

1. The water vapor pressure forces water into the plastic walls.
2. The water diffuses through the wall into the package.
3. The amount of water in the walls depends on the solubility of the plastic and the partial pressure of the water in the ambient.

Tencer[7] has developed a model for plastic packages having thick walls. This model required a modification to the basic permeation equa tions. The modification adds the diffusion constant, as this is the predominant mechanism for water passing through thick plastic walls.

Equation (9-9) has the general form of:

Eq. (9-10) $$Q = \Delta P\left(1 - e^{-\frac{t}{\tau}}\right)$$

where τ = the time constant, and in Eq. (9-9) = $\dfrac{\Delta PV}{R}$

Substituting the value of R from Eq. (9-8):

$$\tau = \frac{\Delta PV}{KA\frac{\Delta P}{d}} = \frac{Vd}{KA}$$

Instead of this time constant, the term $d^2/2D$ is added for diffusion, so that the time constant for thick plastic walls is:

$$\frac{Vd}{KA} + \frac{d^2}{2D}$$

One over the time constant = $\dfrac{1}{\tau} = \dfrac{1}{\dfrac{Vd}{KA} + \dfrac{d^2}{2D}} = \dfrac{KA2D}{2DVd + KAd^2}$

The new equation for the amount of water entering a package with thick walls in time t is:

Eq. (9-11) $$Q = \Delta P\left(1 - e^{-\frac{KA2Dt}{2DVd + Kd^2}}\right)$$

where: Q = the amount of water in atmospheres entering the package cavity in time t

ΔP = the difference in partial pressure between the outside and inside of the package

K = the permeation constant

A = the surface area in cm^2

D = the diffusion constant

V = the volume of the cavity in the package in cc

d = the thickness of the wall in cm

Equations (9-9) and (9-11) can be compared for various type packages.

Example 4. Assume a 20 cc package that has been sealed with 100% nitrogen. The package contains metal to glass feedthroughs, the glass being of type 7740, with a surface area of 2 cm^2. The glass is 0.1 cm thick and has a permeation constant at 25°C of 8×10^{-11}. The diffusion constant at 25°C is 1.7×10^{-8} . The package is bombed in helium at 2 atm absolute for 16 hr. How much helium is forced into the package?

Using Eq. (9-9):

$$Q = \Delta P\left(1 - e^{-\frac{KAt}{Vd}}\right) = 2\left[1 - e^{-\frac{(8.5\times10^{-11})(2)(16\times3600)}{(20)(0.1)}}\right]$$

$$Q = 2\left(1 - e^{-4.896\times10^{-6}}\right) = 2(1 - 0.9999951) = 2 \times 4.896 \times 10^{-6}$$

$$Q = 9.79 \times 10^{-6} \text{ atm}$$

Using Eq. (9-11):

$$Q = \Delta P\left(1 - e^{-\frac{KA2Dt}{2DVd + Kd^2}}\right) = 2\left[1 - e^{-\frac{(2)(8.5\times10^{-11})(2)(1.7\times10^{-8})(16\times3600)}{(2)(1.7\times10^{-8})(20)(0.1)+(2)(8.5\times10^{-11})(0.1)^2}}\right]$$

$$Q = 2\left(1 - e^{-\frac{3.329\times10^{-13}}{6.8\times10^{-8}+1.7\times10^{-12}}}\right) = 2\left(1 - e^{-4.896\times10^{-6}}\right)$$

$$Q = 2(1 - 0.9999951) = 2(4.896 \times 10^{-6})$$

$$Q = 9.79 \times 10^{-6} \text{ atm}$$

In this example, Eq. (9-11) gave the same answer as Eq. (9-9).

Example 5. Consider a package made out of a polycarbonate plastic having the dimension 2 cm by 2 cm by 1 cm with a wall thickness of 0.4 cm:

Surface area = 16 cm^2

Volume of the cavity = 0.288 cc

$K = 1.12 \times 10^{-5}$

$D = 8.06 \times 10^{-8}$

The package cavity and the walls initially contain no water. The package is now placed in 100% relative humidity at 25°C for 10 hr. How much water is in the package after the 10 hr?

The absolute water vapor pressure at 100% relative humidity at 25°C is 0.0313 atm (see Table 5-3). Using Eq. (9-9):

$$Q = 0.0313\left[1 - e^{-\frac{(1.12\times10^{-5})(16)(36000)}{(0.288)(0.4)}}\right]$$

$$Q = 0.0313(1 - e^{-56}) = 0.0313(1 - 0)$$

$$Q = 0.0313 \text{ atm}$$

Using Eq. (9-11):

$$Q = 0.0313\left[1 - e^{-\frac{(1.12\times10^{-5})(16)(2)(8.06\times10^{-8})(36000)}{(2)(8.06\times10^{-8})(0.288)(0.4)+(1.12\times10^{-5})(16)(0.4)^2}}\right]$$

$$Q = 0.0313\left(1 - e^{-\frac{1.0399\times10^{-6}}{1.85\times10^{-8}+2.867\times10^{-5}}}\right) = 0.0313\left(1 - e^{-\frac{1.0399\times10^{-6}}{2.869\times10^{-5}}}\right)$$

$$Q = 0.0313(1 - e^{-0.0362}) = 0.0313(1 - 0.9644) = 0.0313(0.0356)$$

$$Q = 0.00111 \text{ atm in the package}$$

Equation (9-9) did not allow for water in the package walls. Equation (9-11) assumes water in the wall so that there is less water in the cavity. The flow of water vapor into the cavity is slowed down by the plastic walls.

5.0 PROBLEMS AND THEIR SOLUTIONS

Problem 1. A Pyrex glass (type 7740) cylinder having outside dimensions of 4 cm high, 2 cm in diameter and with a wall thickness of 0.1 cm, is bombed at 2 atm absolute of helium for 10 hr. Assuming the fine and gross leak rates are zero, how much helium has permeated into the cylinder during the 10 hr?

Solution.

Volume of the cylinder = $(\pi/4)(2.0 - 0.2)^2 = 9.67$ cc

\External surface area = $(2 \times \pi \times 4) + 2[(\pi/4) \times 2^2] = 31.46$ cm^2

Using Eq. (9-10):

$$Q = \Delta P\left(1 - e^{-\frac{KAt}{Vd}}\right)$$

where: Q = atmospheres of helium permeated into the cylinder in time t

ΔP = 2 atm of helium

t = time in sec = 10 × 3600 = 36,000

V = 9.67 cc

K = 8.5 × 10^{-11} cm^2/sec (see Table 9-1)

A = 31.46 cm^2

d = 0.1 cm

$$Q = 2\left[1 - e^{-\frac{(8.5\times10^{-11})(31.46)(36{,}000)}{(9{,}67)(0.1)}}\right] = 2\left(1 - e^{-9.95\times10^{-5}}\right)$$

$$Q = 2(1 - 0.99990042) = 2(9.95 \times 10^{-5})$$

$$Q = 1.99 \times 10^{-4} \text{ atm of helium}$$

Problem 2. A quartz crystal resonator is in a Pyrex glass (type 7740) enclosure which has been sealed in a vacuum of 1 × 10^{-8} torr. The Q of the crystal will be greater than 1 million, as long as the pressure in the enclosure is less than 1 × 10^{-3} torr. The external surface area of the enclosure is 12 cm^2, the volume is 4 cc and the wall thickness is 0.05 cm. How long will the Q be greater than 1 million if the enclosure is in air at 25°C?

Solution. Equation (9-11) can used to calculate, this time. The permeation constant for nitrogen or oxygen through glass 7740 is 5 × 10^{-17} (see Table 9-1). The permeation constant for helium through this glass is 8.5 × 10^{-11}. The problem can be treated as two processes in parallel, the air less the helium, and the helium by itself.

$$K_{AIR} = 5 \times 10^{-17} \text{ for atm-cc/sec}$$
$$= 5 \times 10^{-17} \times 760 \text{ torr/atm}$$
$$= 3.8 \times 10^{-14} \text{ for torr-cc/sec}$$

$$K_{HE} = 8.5 \times 10^{-11} \text{ for atm-cc/sec}$$
$$= 8.5 \times 10^{-11} \times 760 \text{ torr/atm}$$
$$= 6.48 \times 10^{-8} \text{ for torr-cc/sec}$$

one atmosphere = 760 torr

$$t_{\text{AIR}} = -\frac{(4)(0.05)}{(3.8\times10^{-14})(12)}\ln\left(1-\frac{10^{-3}}{760}\right)$$

$t_{\text{AIR}} = -4.386 \times 10^{11} \ln[1 - (10^{-3}/760)]$

$t_{\text{AIR}} = -4.386 \times 10^{11} \ln(1 - 1.316 \times 10^{-6})$

$t_{\text{AIR}} = -4.386 \times 10^{11} \ln(0.99999868)$

$t_{\text{AIR}} = -4.386 \times 10^{11}(-1.316 \times 10^{-6})$

$t_{\text{AIR}} = 5.771 \times 10^{5}$ sec = 160.3 hr

The pressure of helium in the air is 5.24×10^{-6} atm = $5.24 \times 10^{-6} \times 760$ torr/atm = 3.98×10^{-3} torr.

$$t_{\text{HE}} = -\frac{(4)(0.05)}{(6.48\times10^{-8})(12)}\ln\left(1-e^{-\frac{10^{-3}}{3.98\times10^{-3}}}\right)$$

$t_{\text{HE}} = -2.572 \times 10^{5} \ln(1 - 0..25126)$

$t_{\text{HE}} = -2.572 \times 10^{5} \ln(0.74874)$

$t_{\text{HE}} = -2.572 \times 10^{5}(-0.28936)$

$t_{\text{HE}} = 7.44 \times 10^{4}$ sec = 20.67 hr

The two processes in parallel = $t_{\text{AIR+HE}}$

$$t_{\text{AIR+HE}} = \frac{(160.3)(20.67)}{160.3 + 20.67} = 18.31\ \text{hr}$$

Problem 3. A 10 cc enclosure is sealed in a 1×10^{-6} torr vacuum using a neoprene o-ring. The surface of the o-ring that makes contact with the ambient air is 12 cm^2 with a thickness of 0.4 cm. The enclosure contains 24 glass to kovar feedthroughs 0.1 cm thick,0.1 cm inside diameter and 0.2 cm outside diameter. How much helium will permeate into the enclosure when the enclosure is in air at 25°C for one year?

Solution. Equation (9-6) can be used if the value of KAt/Vd is small enough. For glass:

$$K_{GLASS} = 1.1 \times 10^{-11} \text{ (see Table 9-1)}$$

$$A_{GLASS} = 24 \times [(\pi/4)(0.2^2 - 0.1^2)] = 24 \times 0.02356 = 0.5655 \text{ cm}^2$$

$$t = 1 \text{ yr} = 3.1536 \times 10^7 \text{ sec}$$

$$V = 10 \text{ cc}$$

$$d_{GLASS} = 0.1 \text{ cm}$$

For glass,

$$\frac{KAt}{Vd} = \frac{(1.1 \times 10^{-11})(0.5655)(3.1536 \times 10^7)}{(10)(0.01)} = 1.96 \times 10^{-4}$$

Figure 9-5 shows that the error in using Eq. (9-6) is a fraction of 1%. Using Eq. (9-6):

$$Q_{GLASS} = \frac{KAt}{Vd} = 1.96 \times 10^{-4} \times \Delta P = 1.96 \times 10^{-4} \times 5.2 \times 10^{-6}$$

$$Q_{GLASS} = 1.02 \times 10^{-9} \text{ atm of helium}$$

For neoprene,

$$d_{NEOPRENE} = 0.4 \text{ cm}$$

$$A_{NEOPRENE} = 12 \text{ cm}^2$$

$$K_{NEOPRENE} = 6.5 \times 10^{-8}$$

$$\frac{KAt}{Vd}=\frac{(6.5\times10^{-8})(12)(3.1536\times10^{7})}{(10)(0.4)}=6.15$$

Figure 9-4 shows that an error of greater than 90% would occur if Eq. (9-6) were used. Therefore, Eq. (9-10) will be used.

$$Q=\Delta P\left(1-e^{-\frac{KAt}{Vd}}\right)$$

$$Q=5.2\times10^{-6}(1-e^{-6.15})=5.2\times10^{-6}(1-2.133\times10^{-3})$$

$$Q=5.2\times10^{-6}(0.9978665)$$

$$Q=5.19\times10^{-6}\text{ atm of helium}$$

Total helium in the enclosure $= 1.02\times10^{-9}+5.19\times10^{-6}=5.19\times10^{-6}$ atm.

Problem 4. A lot of polycarbonate plastic packages have zero fine and gross leaks and contain no water. The volume of the internal cavity is 0.288 cc. The external surface area is 16 cm^2 and the wall thickness is 0.4 cm. The permeation constant for water vapor is 1.12×10^{-5} cm^2/sec. The diffusion constant is 8.06×10^{-8} cm^2/sec. The packages are exposed to a water vapor pressure of 0.02 atm (relative humidity = 67%). How long will it take until the water pressure in the package is 0.018 atm?

Solution. Solving Eq. (9-11) for *t:*

$$Q=\Delta P\left(1-e^{-\frac{KA2Dt}{2DVd+Kd^2}}\right)$$

$$1-\frac{Q}{\Delta P}=e^{-\frac{KA2Dt}{2DVd+Kd^2}}$$

$$\ln\left(1-\frac{Q}{\Delta P}\right)=-\frac{KA2Dt}{2DVd+Kd^2}$$

$$t=-\frac{\left(2DVd+Kd^2\right)\ln\left(1-\frac{Q}{\Delta P}\right)}{KA2D}$$

$$t=-\frac{\left[(2)\left(8.06\times10^{-8}\right)(0.288)(0.4)+\left(1.12\times10^{-5}\right)(0.4)^2\right]\ln\left(1-\frac{0.018}{0.02}\right)}{\left(1.12\times10^{-5}\right)(16)(2)\left(8.06\times10^{-8}\right)}$$

$$t=-\frac{\left[\left(1.86\times10^{-8}\right)+\left(1.79\times10^{-6}\right)\right]\ln(0.1)}{2.89\times10^{-11}}=-\frac{\left(1.81\times10^{-6}\right)(-2.303)}{2.89\times10^{-11}}$$

$$t=1.442\times10^5\ \text{sec}=40.1\ \text{hr}$$

REFERENCES

1. Redhead, P. A., Hobson, J. P., and Kornelson, E. V., *The Physical Basis Of Ultra High Vacuum*, Chapman and Hall (1968)
2. Altemose, V. O., *J. Appl. Physics*, 32:1309–1316 (1961)
3. Barrer, R. M., *Diffusion In And Through Solids*, Cambridge University Press (1941)
4. Leiby, C. C., and Chen, C. L., *J. Appl. Physics*, 31:268 (1960)
5. Norton, F. J., *J. Amer. Ceramic Soc.*, 36:90 (1953)
6. Miller, C. F., and Shepard, R. W., *Vacuum*, 11:58 (1961)
7. Tenser, M., *IEEE Electronic Components Conference*, pp. 196–208 (1994)

10

Residual Gas Analysis (RGA)

1.0 DESCRIPTION OF THE TEST

Residual gas analysis (RGA) is performed on sealed electronic packages by means of a mass spectrometer. A mass spectrometer is an instrument that converts a gas sample into ions, separates the ions in accordance with their mass to charge ratio, and then collects the ions so as to report their relative quantity. All mass spectrometers have three common features:

1. A source of electrons to impact the sample gas, thereby creating ions.
2. A mechanism to sort the produced ions according to their mass to charge ratio.
3. A method of deflecting/collecting these ions.

There are several types of mass spectrometers, but the type used for RGA is the quadrupole mass spectrometer. A description of this type of instrument can be found in Refs. 1and 2.

The test is performed by first preconditioning the sample a 100°C for 12–24 hr. The package is then mounted in the inlet port of the mass spectrometer. An O-ring seal is made between the package cover and the mass spectrometer. The mounted sample is then enclosed in an ovenwhich maintains the sample at 100°C. The inlet port of the spectrometer contains a mechanism which can pierce the cover of the package. The sample and spectrometer are pumped down to a vacuum level to insure no interference from the residual gas of the spectrometer. The cover is then punctured and the gas in the package escapes into the spectrometer, where it is analyzed. The mechanical details, requirements of this test, and the calibration for water vapor are found in Method 1018 of MIL-STD-883, "Internal Water Vapor Content." The calibration of the other gases is left to the individual laboratories.

2.0 WHAT THE TEST MEASURES

The original purpose of this test was to measure the moisture in packages. This is still the primary purpose, as Method 1018 has a maximum moisture requirement of 5,000 ppm. However, the test also measures the amounts of the other gases in the package. The following gases are measured, if present:

ammonia
argon
carbon dioxide
fluorocarbons (see Ch. 8)
helium
hydrocarbons
hydrogen
isopropyl alcohol
methane
methanol
methyl ethyl ketone (MEK)
nitrogen
oxygen
toluene
water (moisture)

Tables 10-1–10-3 are examples of RGA reports.

Table 10-1. Verification of the Sealing Chamber Atmosphere by Residual Gas Analysis

Gas	Percent/PPM	Sample 1	Sample 2
Argon	ppm	ND	ND
Carbon dioxide	ppm	<100	<100
Fluorocarbons	ppm	ND	ND
Helium	percent	12.8	12.4
Hydrogen	ppm	ND	ND
Moisture	ppm	<100	<100
Nitrogen	percent	87.2	87.6
Oxygen	ppm	ND	ND

Table 10-2. Residual Gas Analysis of a Hybrid Subjected to Military Screening

Gas	Percent/PPM	Hybrid
Ammonia	percent	2.96
Argon	ppm	100
Carbon dioxide	ppm	5,102
Fluorocarbons	ppm	ND
Helium	percent	11.4
Hydrocarbons	ppm	<100
Hydrogen	ppm	131
Moisture	ppm	1,073
Nitrogen	percent	85.0
Oxygen	ppm	ND

Table 10-3. Residual Gas Analysis of a Hybrid Subjected to Military Screening and a 1000 Hr Life Test

Gas	Percent/PPM	Hybrid
Ammonia	percent	5.14
Argon	ppm	437
Carbon dioxide	percent	2.07
Fluorocarbons	ppm	ND
Helium	percent	9.86
Hydrocarbons	ppm	800
Hydrogen	ppm	447
Methanol	ppm	125
Moisture	ppm	956
Nitrogen	percent	82.7
Oxygen	ppm	ND

Table 10-1 shows the RGA for two pinless packages sealed with 90% nitrogen and 10% helium. This is a test to verify the integrity of the sealing chamber. The helium is greater than 10%, as 10% is the minimum requirement. ND means that the gas was not detected; <100 means that there was probably some present but less than 100 ppm.

Table 10-2 shows the RGA for a hybrid which had gone through the military screening including burn-in for 160 hr at 125°C. The ammonia is from an amine cured epoxy. The moisture and hydrogen are from the package walls and from parts within the package.

Table 10-3 is an RGA for the same type hybrid as in Table 10-2, but this hybrid has been exposed to a 1000 life test at 125°C in addition to the screening.

3.0 CALCULATION OF LEAK RATES FROM RGA DATA

One of the uses of the RGA data is the calculation of the leak rate of the package. This is particularly useful in determining the source of the water in the package. Knowing the leak rate can help determine if the water leaked in or if it was generated within the package due to outgassing or a result of a chemical reaction. The leak rate calculation can be made from the amount of argon or oxygen in a package when the package has been sealed primarily with nitrogen. For packages sealed in 100% helium or argon, the leak rate can be calculated based on the amount of nitrogen or oxygen in the package. The origin of the argon in a package may not always be from the air. Nitrogen gas derived from its liquid often contains from 100–600 ppm of argon as an impurity, but some liquid nitrogen lots contain no argon. The most reliable leak rate calculation for packages sealed with primarily nitrogen is based on argon and oxygen. Calculations based on the amount of water in the package can also be made, but are least reliable as some water is usually outgassed in the package.

The oxygen to argon ratio in the atmosphere is 0.20948/0.00934 = 22.428. This is not the oxygen to argon ratio when air has leaked into a package because the true argon leak rate is less than the true oxygen leak rate. The true argon leak rate = 0.894 of the true oxygen leak rate (see Table 3-1). Assuming that all the oxygen and argon comes from outside the package, the oxygen to argon ratio will vary with leak rate and time. Figure 10-1 shows the oxygen to argon ratio for different true oxygen leak rates divided by the package volume, as a function of time. These curves came from solving Eq. (4-1) for both oxygen and argon. This equation is repeated here as Eq. (10-1):

Eq. (10-1) $$Q_{INp} = \Delta p_i \left(1 - e^{-\frac{Lt}{V}}\right)$$

where: Q_{inP} = the quantity of gas in atmospheres that has leaked into the package in time t

Δp_i = the initial difference in partial pressure of the leaking gas

t = the time in seconds

V = the internal free volume of the package in cc

L = the true leak rate (for Δp = 1 atm) of the leaking gas

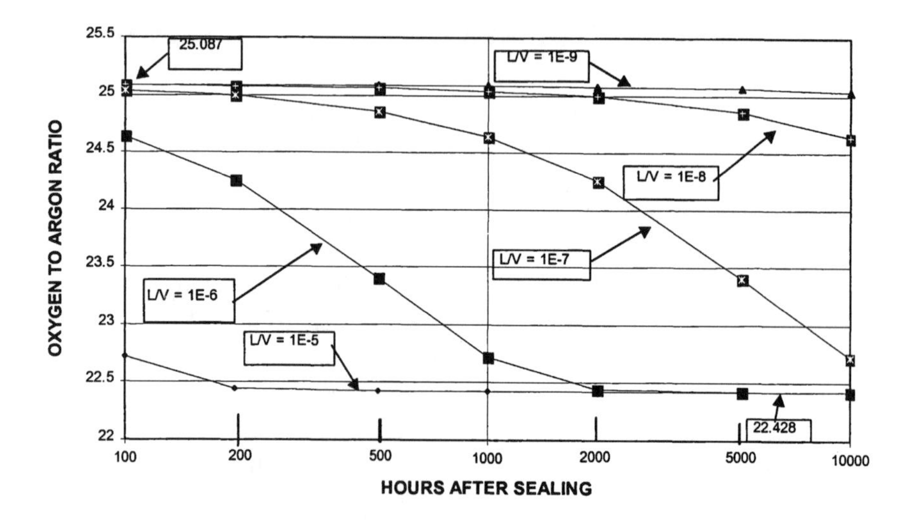

Figure 10-1. Oxygen to argon ratio for air leaking into a package as a function of the true oxygen leak rate divided by the package volume.

When the oxygen to argon ratio is greater than 25.087, some of the oxygen has come from within the package. If the leak rate is calculated based on this amount of oxygen, the calculated leak rate will be greater than the correct value. When the oxygen to argon ratio is less than 22.428, the following possibilities exist:

a. Some oxygen has oxidized something within the package.

b. There was argon in the sealing chamber.

c. Both a and b.

When the ratio is between 22.428 and 25.087, all the oxygen and argon might have come from the air. This can be verified by using Fig. 10-1 to check if the expected value of the ratio agrees with the curve for the *L/V* and *t* values.

Leak rate calculations must be based on the partial pressure of the leaking gas. The RGA report is given in percentages. The partial pressure of the gases in the package only agrees with the percentages when there is one atmosphere in the package at room temperature. The RGA being performed at 100°C increases the pressure in the package to a value

$$\frac{273+100}{273+22} \times \text{pressure at } 22°\text{C}$$

(assuming the packages are sealed at 22°C and at one atmosphere). The RGA percentages at 100°C. are equal to the percentage of the gases at the sealing temperature. This is true because the partial pressures of the gases that were sealed in the package increase in pressure the same percentage as the increase in total pressure.

Gases are leaking in and out of the package at different rates so that the total atmosphere in the package is not always equal to one. Figure 10-2 shows the total atmospheres in a package as a function of time, when the package is in ambient air. This package has been sealed with 90% nitrogen and 10% helium. There are four curves in this figure, each for a different value of the true helium leak rate divided by the package volume. The total atmospheres in the package for any time up to 1,000 days after sealing is between 0.956 and 1.000 atm. For packages sealed with 100% nitrogen, the minimum atmospheres is greater than 0.956 atm. These small variations resulting from the variation in leak rates can be neglected.

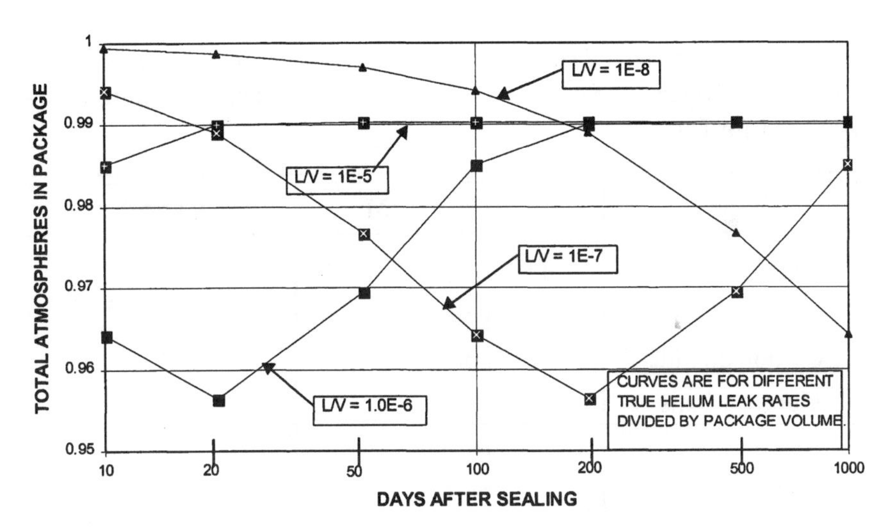

Figure 10-2. Total atmospheres in a package when sealed with 90% nitrogen and 10% helium, when in ambient air.

Gases outgassing within the package increases the total pressure. Table 10-3 shows 5.14% of ammonia, which could only come from within the package. The ammonia raises the total pressure about 5%. This small change can also be neglected. However, there may be packages that have large percentages of internally generated gases which would have to be considered. To calculate the leak rates from the RGA data, Eq. (4-10) is used and the equation is repeated here as Eq. (10-2):

Eq. (10-2) $$L = -\frac{V}{t}\left[\ln\left(1 - \frac{Q_{inP}}{\Delta P_i}\right)\right]$$

Table 10-4 is a RGA report containing an appreciable amount of oxygen. The oxygen to argon ratio is 18.7/0.8496 = 22.01. This is less than 22.428, so that some of the oxygen has reacted with something within the package or argon was an impurity in the nitrogen, or both. In any case, the calculation of the true helium leak rate using the amount of oxygen in the package will give the minimum possible leak rate. The calculation using the amount of argon in the package represented by Table 10-4, will yield the maximum or true leak rate.

Table 10-4. Residual Gas Analysis of a Hybrid Containing a Large Amount of Oxygen After Being Subjected to Military Screening and a 1000 hr Life Test

Gas	Percent/PPM	Hybrid
Ammonia	ppm	ND
Argon	ppm	8496
Carbon dioxide	ppm	714
Fluorocarbons	ppm	ND
Helium	ppm	ND
Hydrocarbons	ppm	ND
Hydrogen	ppm	ND
Methanol	ppm	ND
Moisture	ppm	7494
Nitrogen	percent	79.6
Oxygen	percent	18.7

The internal volume of the package was 1.86 cc. The RGA test was performed on December 15, and the seal date was August 31.

$$t = 106 \text{ days} \times 24 \text{ hr} \times 3600 \text{ sec} = 9{,}158{,}400 \text{ sec}$$

Using Eq. (10-2) for the oxygen:

$$L_{O_2} = -\frac{1.86}{9158400}\left[\ln\left(1 - \frac{18.7}{20.95}\right)\right]$$

$$L_{O_2} = -2.03 \times 10^{-7}[\ln(1 - 0.8926)] = -2.03 \times 10^{-7}(\ln 0.107398)$$

$$L_{O_2} = -2.03 \times 10^{-7}(-2.23) = 4.53 \times 10^{-7} \text{ atm-cc/sec of oxygen}$$

$$L_{HE} = \sqrt{\frac{32}{4}}\, L_{O_2} = 2.83 \times 4.53 \times 10^{-7} = 1.28 \times 10^{-6} \text{ atm-cc/sec of helium}$$

This is the minimum true helium leak rate.

Using Eq. (10-2) for the argon:

$$L_{AR} = -2.03 \times 10^{-7}\left[\ln\left(1 - \frac{8496}{9430}\right)\right]$$

$$L_{AR} = -2.03 \times 10^{-7}[\ln(1 - 0.900954)] = -2.03 \times 10^{-7}(n\, 0.099045)$$

$$L_{AR} = -2.03 \times 10^{-7}(-2.312) = 4.69 \times 10^{-7} \text{ atm-cc/sec of argon}$$

$$L_{HE} = \sqrt{\frac{40}{4}}\, L_{O_2} = 3.16 \times 4.69 \times 10^{-7} = 1.48 \times 10^{-6} \text{ atm-cc/sec of helium}$$

This is the maximum true helium leak rate. Based on these two calculations, the true helium leak rate is between 1.24 and 1.48×10^{-6} atm-cc/sec.

Table 10-5 is another example of a RGA containing oxygen. This package had an internal volume of 0.5 cc, and was sealed on October 7. The RGA test was performed on December 14.

$$t = 69 \text{ days} = 5{,}961{,}600 \text{ sec}$$

Table 10-5. Residual Gas Analysis of a Hybrid Containing Oxygen After Being Subjected to Military Screening

Gas	Percent/PPM	Hybrid
Ammonia	ppm	ND
Argon	ppm	2569
Carbon dioxide	ppm	719
Fluorocarbons	ppm	ND
Helium	percent	7.29
Hydrocarbons	ppm	ND
Hydrogen	ppm	ND
Methanol	ppm	ND
Moisture	ppm	1550
Nitrogen	percent	88.1
Oxygen	percent	4.15

The oxygen to argon ratio is 4.15/0.2569 = 16.15. Using Eq. (10-2), the true oxygen leak rate is:

$$L_{O_2} = -\frac{0.5}{5{,}961{,}600}\left[\ln\left(1-\frac{4.15}{20.95}\right)\right]$$

$$L_{O_2} = -8.387 \times 10^{-8}[\ln(1-0.1981)] = -8.387 \times 10^{-8}(\ln 0.801909)$$

$$L_{O_2} = -8.387 \times 10^{-8}(-0.22076)$$

$$L_{O_2} = 1.85 \times 10^{-8} \text{ atm-cc/sec of oxygen}$$

$$L_{HE} = \sqrt{\frac{32}{4}}\, L_{O_2} = 2.83 \times 1.85 \times 10^{-8} = 5.23 \times 10^{-8} \text{ atm-cc./sec. of helium}$$

This the minimum true helium leak rate.

Using Eq. (10-2), the true argon leak rate is:

$$L_{AR} = -\frac{0.5}{5{,}961{,}600}\left[\ln\left(1 - \frac{2569}{9430}\right)\right]$$

$$\begin{aligned} L_{AR} &= -8.387 \times 10^{-8}[\ln(1 - 0.272)] \\ &= -8.387 \times 10^{-8}(\ln 0.72757) \\ &= -8.387 \times 10^{-8}(-0.3180) \end{aligned}$$

$$L_{AR} = 2.67 \times 10^{-8} \text{ atm-cc/sec of argon}$$

$$L_{HE} = \sqrt{\frac{40}{4}}\, L_{AR} = 3.16 \times 2.67 \times 10^{-8} = 8.44 \times 10^{-8} \text{ atm-cc/sec of helium}$$

This is the maximum true helium leak rate.

Figure 10-3 is a plot of the L_{HE}/V versus the time after sealing for 100 ppm of oxygen. Assuming oxygen is only reported for values equal to or greater than 100 ppm, Figure 10-6 can be used to determine the maximum true helium leak rate (L_{HE}). If a package has no oxygen after 40 days, for example, the L_{HE}/V curve shows a value of 4×10^{-10}. If the package volume was 2.0 cc, then $L_{HE} = 4 \times 10^{-10} \times 2 = 8 \times 10^{-10}$ atm-cc/sec. This is the maximum true helium leak rate based on the RGA. If the package volume was 0.1 cc, then $L_{HE} = 4 \times 10^{-10} \times 0.1 = 4 \times 10^{-11}$ atm-cc/sec. Depending on the shape of the leak channel, this value could indicate that there is no leak. If the cross-section of the channel was rectangular with a 10:1 ratio, Fig. 10-6 would indicate no leak, because the value is less than the minimum theoretical leak rate. If the leak channel was cylindrical, the leak rate would be real, because it is greater than the minimum theoretical leak rate. Table 10-6 is a RGA report for a hybrid with a volume of 1.94 cc. This hybrid was sealed on June 19 with 90% nitrogen and 10% helium. It was burned in for a week at an ambient of 125°C and fine leak tested after bombing.

The measured leak value was less than 5.0×10^{-9} which corresponds to a nominal true helium leak rate of less than 5.0×10^{-8} atm-cc/sec. The RGA was performed on November 3. The time between seal and RGA was 137 days = 137 × 24 × 3600 = 11,836,800 sec. The RGA shows 17.5% helium, so the maximum true helium leak rate based on the measured value of:

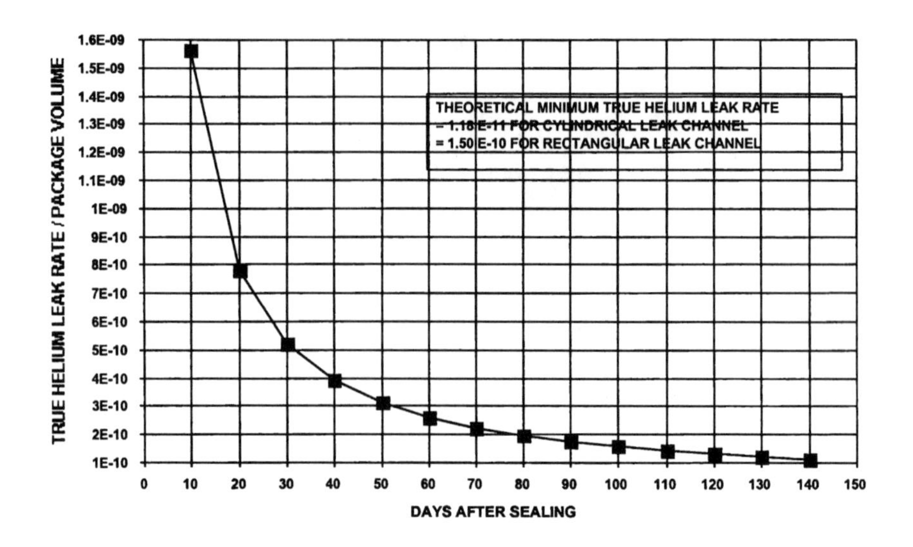

Figure 10-3. The true helium leak rate divided by the package volume, to allow 100 ppm of oxygen to enter the package.

Table 10-6. Residual Gas Analysis of a Hybrid Showing Conflicting Data

Gas	Percent/PPM	Hybrid
Ammonia	percent	1.35
Argon	ppm	ND
Carbon dioxide	percent	2.41
Fluorocarbons	ppm	ND
Helium	percent	17.5
Hydrocarbons	ppm	345
Hydrogen	ppm	ND
Methanol	ppm	ND
Moisture	ppm	1765
Nitrogen	percent	78.6
Oxygen	PPM	ND

$$5 \times 10^{-9} \text{ is } \frac{5\times10^{-9}}{0.175} = 2.86 \times 10^{-8} \text{ atm-cc/sec}$$

No oxygen was reported in the RGA, so that Fig. 10-3 can be used to estimate the true helium leak rate based on the RGA. Figure 10-3 shows a L_{HE}/V value after 137 days of approximately 1.2×10^{-10}. Multiplying by the package volume of 1.94 cc, the true helium leak rate = 3.9×10^{-10} atm-cc/sec.

4.0 INTERPRETATION OF RGA DATA

Example 1. Table 10-7 is a RGA of a pinless package used to verify the atmosphere in the sealing chamber. This 2.5 cc package was sealed with approximately 10% helium and leak tested without bombing on July 13. It passed the fine leak measured limit of 5×10^{-9} atm-cc/sec. On July 15, it passed the gross leak test. The package was sent for a RGA which was performed on July 25. The time from seal to test was 13 days = $13 \times 24 \times 3600$ = 1,123,200 sec. The RGA raises the question, was the oxygen in the sealing chamber or did it leak in?

Table 10-7. Residual Gas Analysis of an Empty Package Used to Verify the Atmosphere of the Sealing Chamber

Gas	Percent/PPM	Hybrid
Ammonia	ppm	ND
Argon	ppm	177
Carbon dioxide	ppm	<100
Fluorocarbons	ppm	ND
Helium	percent	11.9
Hydrocarbons	ppm	ND
Hydrogen	ppm	ND
Methanol	ppm	ND
Moisture	ppm	313
Nitrogen	percent	87.8
Oxygen	PPM	2590

The oxygen to argon ratio is 14.63. Calculating L_{HE} from L_{O_2}:

$$L_{O_2} = -\frac{2.5}{1{,}123{,}200}\left[\ln\left(1 - \frac{0.2590}{20.95}\right)\right]$$
$$= -2.23 \times 10^{-6}\,[\ln(1 - 0.01236)]$$
$$= -2.23 \times 10^{-6}\,[\ln 0.98764]$$
$$= -2.23 \times 10^{-6}\,(-0.01244)$$

$$L_{O_2} = 2.77 \times 10^{-8} \text{ atm-cc/sec of oxygen}$$

$$L_{HE} = 2.83 \times 2.77 \times 10^{-8} = 7.83 \times 10^{-8} \text{ atm-cc/sec of helium}$$

Calculating L_{HE} from L_{AR}:

$$L_{AR} = -2.23 \times 10^{-6}\left[\ln\left(1 - \frac{177}{9430}\right)\right]$$
$$= -2.23 \times 10^{-6}\,[\ln(1 - 0.01877)]$$
$$= -2.23 \times 10^{-6}\,(\ln 0.9812)$$
$$= -2.23 \times 10^{-6}\,(-0.01895)$$

$$L_{AR} = 4.23 \times 10^{-8} \text{ atm-cc/sec of argon}$$

$$L_{HE} = 3.16 \times 4.23 \times 10^{-8} = 1.33 \times 10^{-7} \text{ atm-cc/sec of helium}$$

Although the exact measured leak rate coming out of the sealer is unknown, the maximum value is 5×10^{-9} atm-cc/sec/helium. The true helium leak rate based on this assumed measured value is 5×10^{-9}/atm of helium in the package and is equal to $5 \times 10^{-9}/0.119$ atm of helium which equals 4.2×10^{-8} atm-cc/sec of helium. This is the maximum helium leak rate based on the fine leak measurement.

Using this maximum leak rate, the amount of oxygen that leaked in during this 13 days can be calculated using Eq. (10-1). The true leak rate in this equation is now the oxygen leak rate based on the true helium leak rate of 4.2×10^{-8}.

$$L_{O_2} = L_{HE}/2.83 = 4.2 \times 10^{-8}/2.83 = 1.48 \times 10^{-8}$$

$$Q_{O_2} = 0.2095\left[1 - e^{-\frac{(1.48\times10^{-8})(1,123,200)}{2.5}}\right]$$

$$= 0.2095\left(1 - e^{-0.006649}\right)$$

$$Q_{O_2} = 0.2095(1 - 0.99337) = 0.2095(0.006627)$$

$$Q_{O_2} = 0.001388 \text{ atm of oxygen} = 1,388 \text{ ppm}$$

Because this value is less than the amount shown in the RGA, the oxygen was in the sealing chamber.

Using this maximum leak rate, the amount of argon that leaked in during this 13 days can be calculated using Eq. (1). The true leak rate in this equation is now the argon leak rate based on the true helium leak rate.

$$L_{AR} = 4.2 \times 10^{-8}/3.16 = 1.329 \times 10^{-8}$$

$$Q_{AR} = 9340\left[1 - e^{-\frac{(1.329\times10^{-8})(1,123,200)}{2.5}}\right]$$

$$= 9340\left(1 - e^{-0.005971}\right) = 9340(1 - 0.994046)$$

$$= 9340(0.005954)$$

The quantity $Q_{AR} = 55.6 = 56$ ppm, which would not be reported as it is less than 100 ppm. The argon reported as 177 ppm indicates that air or argon was in the sealing chamber.

The amount of oxygen leaking into a package, when the package is in air, is plotted in Figs. 10-4–10-6. Each curve in the figures is for a different value of the true helium leak rate divided by the package volume (L_{HE}/V). The curves start out linearly, the smaller the L_{HE}/V, the longer they remain linear (as shown in Figs. 10-4 and 10-5).

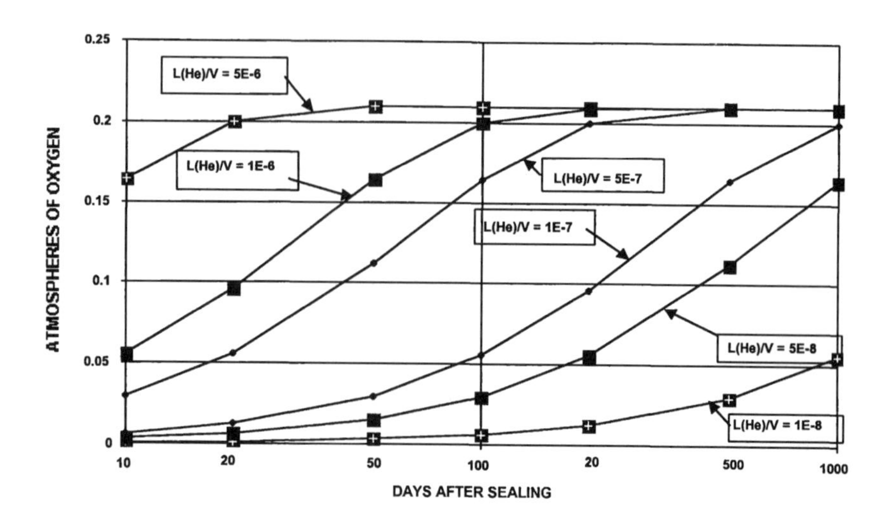

Figure 10-4. The amount of oxygen leaking into a package when in ambient air, versus the days after sealing. L_{HE}/V curves are from 1×10^{-8} to 5×10^{-6}.

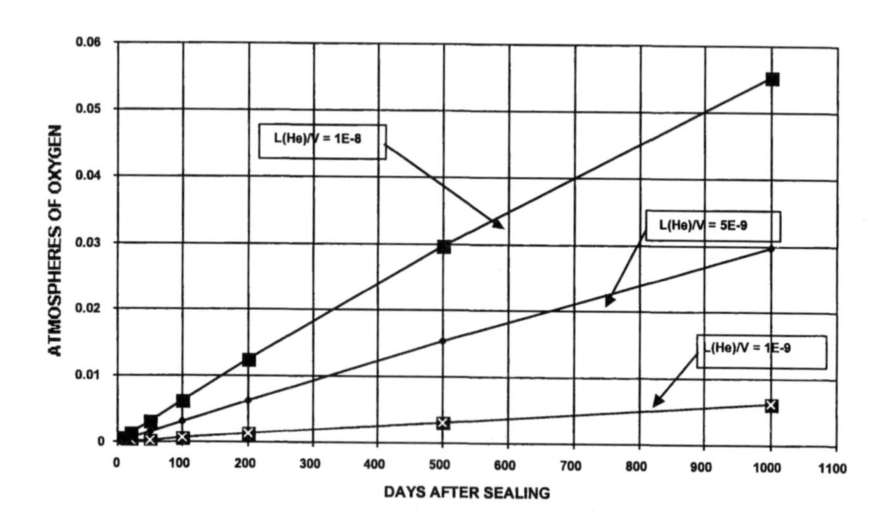

Figure 10-5. The amount of oxygen leaking into a package when in ambient air, versus the days after sealing. L_{HE}/V curves are from 1×10^{-9} to 1×10^{-8}.

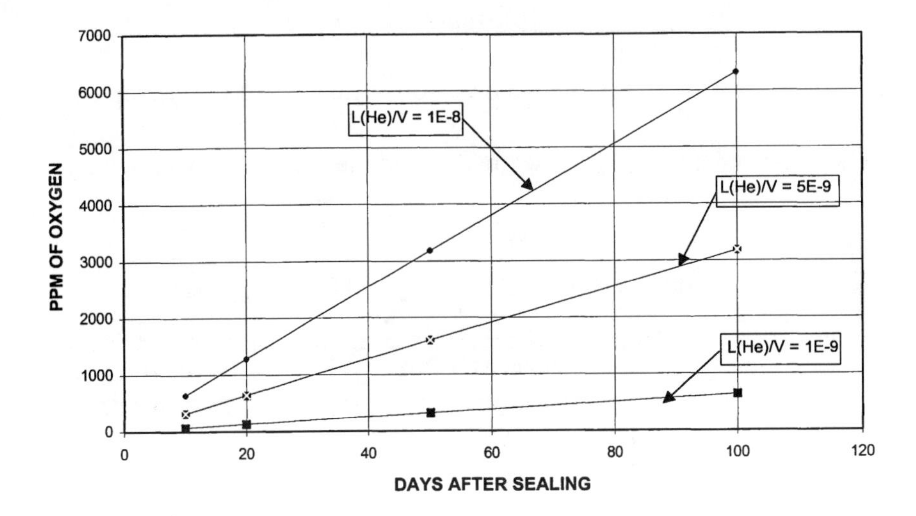

Figure 10-6. PPM of oxygen leaking into a package, when in ambient air, for the first 100 days after sealing. L_{HE}/V are from 1×10^{-9} to 1×10^{-8}.

Example 2. Table 10-8 shows a RGA which disagrees with the previous hermeticity tests. This hybrid had a package volume of 2.5 cc. and was sealed in 90% nitrogen and 10% helium on August 24. It was fine leak tested on September 7 and the measured value was 2×10^{-9} atm-cc/sec. The RGA was performed on September 12. The time between seal and RGA was 19 days which equals $19 \times 24 \times 3600 = 1{,}641{,}600$ sec. The true helium leak rate equals the measured value divided by the atmospheres of helium in the package $= 2 \times 10^{-9}/0.152 = 1.32 \times 10^{-8}$ atm-cc/sec. The L_{HE}/V value is $1.32 \times 10^{-8}/2.5$ cc $= 5.28 \times 10^{-9}$ atm-cc/sec. Figure 6 shows an oxygen content of 600 ppm for this value of L_{HE}/V after 19 days. This is magnitudes different from the 4.51% in the RGA.

The leak rate of oxygen can be calculated using this value:

$$L_{O_2} = -\frac{2.5}{1641600}\left[\ln\left(1 - \frac{4.51}{20.95}\right)\right]$$

$$= -1.52 \times 10^{-6}[\ln(1 - 0.2153)]$$

$$= -1.52 \times 10^{-6}(\ln 0.7840)$$

$$= -1.52 \times 10^{-6}(-0.2424)$$

$$L_{O_2} = 3.68 \times 10^{-7} \text{ atm-cc/sec}$$

$$L_{HE} = 2.83 \times L_{O_2} = 2.83 \times 3.68 \times 10^{-7} = 1.04 \times 10^{-6} \text{ atm-cc/sec}$$

This is two orders of magnitude greater than the 1.32×10^{-8} atm-cc/sec.

Table 10-8. Residual Gas Analysis of a Hybrid Showing 4.51% Oxygen

Gas	Percent/PPM	Hybrid
Ammonia	percent	2.32
Argon	ppm	2180
Carbon dioxide	ppm	6110
Fluorocarbons	ppm	ND
Helium	percent	15.2
Hydrocarbons	ppm	272
Hydrogen	ppm	ND
Methanol	ppm	ND
Moisture	percent	1.26
Nitrogen	percent	75.9
Oxygen	percent	4.51

The true helium leak rate based on the argon content is:

$$L_{AR} = -\frac{2.5}{1641600}\left[\ln\left(1 - \frac{2180}{9340}\right)\right]$$

$$L_{AR} = -1.52 \times 10^{-6}[\ln(1 - 0.234)]$$

$$= 11.52 \times 10^{-6}(\ln 0.766595)$$

$$= -1.52 \times 10^{-6}(-0.265796)$$

$$= 4.00 \times 10^{-7} \text{ atm-cc/sec}$$

$$L_{HE} = 3.16 \times L_{AR} = 3.16 \times 4 \times 10^{-7} = 1.26 \times 10^{-6} \text{ atm-cc/sec}$$

This value is also about 100 times greater than the 1.32×10^{-8} atm-cc/sec.

The RGA shows no fluorocarbons. The amount of fluorocarbons expected in the RGA if the true helium leak rate was 1.3×10^{-6}, can be estimated using the values in Table 10-2. The table has $R_{DET} = 3.87 \times 10^{-8}$ for $L_{HE} = 1.41 \times 10^{-6}$, and $R_{DET} = 1.18 \times 10^{-8}$ for $L_{HE} = 5.06 \times 10^{-7}$. By a linear extrapolation, $R_{DET} = 2.36 \times 10^{-8}$ for $L_{HE} = 1.3 \times 10^{-6}$. This is the leak rate of the detector liquid into the package. The package was bombed in detector liquid for 4 hr at 3 atm absolute. The atmospheres of detector liquid forced into the package is calculated using Eq. (8-6):

$$Q_{DET+} = \Delta P_i \left(1 - e^{-\frac{R_{DET} t}{V \Delta P_i}}\right)$$

where: ΔP_i = the difference in total pressure from outside to inside the package = 3 – 1 = 2 atm

R_{DET} = 2.36×10^{-8}

t = 4 hr = 14,400 sec

V = 2.5 cc

$$Q_{DET} = 2\left[1 - e^{-\frac{(2.36\times10^{-8})(14400)}{2.5\times2}}\right] = 2\left(1 - e^{-0.000067968}\right)$$

$$Q_{DET} = 2(1 - 0.999932) = 2(6.80 \times 10^{-5}) = 1.36 \times 10^{-4} \text{ atm}$$

Volume of detector liquid = Q_{DET} × package volume =$1.36 \times 10^{-4} \times 2.5$ = 3.4×10^{-4} ml = 0.34 μl. This would be detected in the gross leak test. It would be reported in the RGA. Using Eq. (8-14):

$$\text{atmospheres fluorocarbon gas} = \frac{0.607 \text{ microliters detector liquid}}{\text{package volume}}$$

$$\text{atmospheres fluorocarbon gas} = \frac{0.607 \times 0.34}{2.5} = 0.083$$

Using Eq. (8-15):

$$\%\ \text{fluorocarbons} = \frac{\text{atmospheres fluorcarbon gas} \times 100}{1.26 + \text{atmospheres fluorocarbon gas}}$$

$$\%\ \text{fluorocarbons} = \frac{0.083 \times 100}{1.26 + 0.083} = 6.18$$

The package was leak tested through the hole after the RGA was performed. The true helium leak rate through the hole was 3×10^{-9} atm-cc/sec. The true helium leak rate after bombing (1.32×10^{-8}) and the true helium leak rate through the hole (3×10^{-9}), although not in agreement, are orders of magnitude smaller than a value to allow 4.51% oxygen. If the 4.51% oxygen leaked in, the leak channel would allow detector liquid to enter the package corresponding to 6.18% in the RGA. Leak rate measurements do not support the large amount of oxygen reported in the RGA.

The sealing chamber could have had a leak and some air could have been sealed in the package. The oxygen to argon ratio in the RGA equals 20.7. This is low compared to the 22.4 ratio in air. Argon could have been an impurity in the nitrogen. If no oxygen leaked in, then the composition in the sealing chamber would have been:

$$\%\ \text{air} = \frac{\%\ \text{oxygen in RGA} \times 100}{\%\ \text{oxygen in air}} = \frac{4.51 \times 100}{20.9} = 21.6$$

Consisting of 4.51 % oxygen

$$\%\ \text{argon} = \frac{\%\ \text{oxygen in RGA}}{\text{oxygen to argon ratio in air}} = \frac{4.51}{22.4} = 0.2013 = 2{,}013\ \text{ppm}$$

The additional argon needed from the nitrogen equals

$$2180 \text{ ppm} - 2013 \text{ ppm} = 167 \text{ ppm}$$

Assuming the air had 2% water, the water = .02 × 21.6% air = 0.43% water. Additional water from outgassing in the package = 1.26% – 0.43% = 0.83%.

The discrepancy between the RGA and the leak rate measurements lead to the following possible explanations:

1. The fine leak measurements must have been correct, or the package would have failed gross leak, and fluorocarbons would have been reported on the RGA. Therefore air was in the sealing chamber.
2. The RGA was incorrect or belonged to a different package.

Further investigation into the condition of the sealer on the day the package was sealed, and into the RGA would be required to determine which is correct.

Residual Gas Analysis can also distinguish between epoxies. Each column in Table 10-9 is the average of four 1000 hr life test hybrids having the same package size, same circuit, and same components. The old epoxy column contains an amine type film epoxy, a conductive paste epoxy, and an insulative paste epoxy. These epoxies were not 5011 qualified, but were used for several years in a high reliability missile application. The film epoxy was cured at 170°C and the pastes at 155°C. The hybrid was outgassed in vacuum at 150°C for 24 hr just prior to sealing. The cure temperatures for the hybrids in the last four columns were changed to 180°C for the film and 170°C for the pastes. Vacuum outgassing remained the same.

The hybrid in column "New Conductive #1" had the same film and insulative epoxy as the "Old Epoxies" column, but had a new 5011 type conductive epoxy. The RGA showed no significant difference when the "New Conductive #1" replaced the old conductive epoxy.

The hybrid in column "New Conductive #2" had the same film and insulative epoxy but had a another new 5011 type conductive epoxy. The RGA shows methanol which is not in the previous two columns and had no hydrocarbons. The carbon dioxide is also significantly less.

The hybrid in column "New Insulative #1" had the same film and conductive epoxy as in the old epoxy column but had a new 5011 insulative epoxy paste. The RGA shows isopropyl alcohol and methanol which is new and a decrease in carbon dioxide. There are no hydrocarbons.

The hybrid in column "New Insulative #2" had the same film and conductive epoxy as in the old epoxy column but had another new 5011 insulative epoxy paste. The RGA shows a decrease in carbon dioxide and hydrocarbons.

The relative constant amount of ammonia in all the hybrids indicate that raising the cure temperature of the film had no effect on the outgassing of ammonia.

Table 10-9. Residual Gas Analysis of Different Epoxies

Gas	%/PPM	Old Epoxies	New Conductive #1	New Conductive #2	New Insulative #1	New Insulative #2
Ammonia	%	6.78	7.12	6.93	6.33	5.90
Argon	ppm	370	310	103	153	133
Carbon dioxide	%	3.88	4.03	2.93	2.09	2.28
Fluorocarbons	ppm	ND	ND	ND	ND	ND
Helium	%	11.24	11.2	11.2	13.65	11.75
Hydrocarbons	ppm	630	406	ND	ND	94
Hydrogen	ppm	617	567	476	666	196
Isopropyl alcohol	ppm	ND	ND	ND	617	ND
Methanol	ppm	ND	ND	179	227	ND
Moisture	ppm	3341	3179	3100	1984	2500
Nitrogen	%	78.25	77.2	78.55	77.53	79.80
Oxygen	ppm	ND	ND	ND	ND	ND

5.0 THE QUALIFICATION OF SMALL PACKAGES USING RGA

A small and large package having the same size leak channel will allow the same amount of oxygen, water, and argon to leak into the package. However, the percentages of the gases that leaked in, relative to the other gases in the package, will not be the same. The percentage in the smaller package will be greater than that in the larger package. For this reason the hermeticity requirement of small packages is often tighter than for a larger package. The typical hermeticity requirement guaranteed by the package vender is a true helium leak rate of 1×10^{-8} atm-cc/sec. Small packages may require 1×10^{-9} or less, bringing into question the accuracy of the bombed leak rate measurement.

The leak rate measurement value for a true helium leak rate of 1×10^{-9} for 0.1 cc package can be calculated. Assuming a helium bombing pressure of 3 atm absolute for 4 hr, the amount of helium forced into the package can be calculated using Eq. (10-1):

$$Q_{HE} = 3\left[1 - e^{-\frac{(1\times10^{-9})(14{,}400)}{0.1}}\right]$$

$$Q_{HE} = 3(1 - e^{-0.000144}) = 3(1 - 0.999856) = 3(0.000144)$$

$$Q_{HE} = 0.00043 \text{ atm of helium}$$

The measured leak rate equals the true leak rate multiplied by the atmospheres of helium in the package = $1 \times 10^{-9} \times 0.00043 = 4.32 \times 10^{-13}$ atm-cc/sec. This can not be measured.

If the package is backfilled with 10% helium, then the measured limit is $1 \times 10^{-9} \times 0.1 = 1 \times 10^{-10}$ atm-cc/sec. This is very difficult to measure and has measurement errors greater than 1×10^{-10} (see Ch. 7).

RGA can be used to calculate very small leak rates if the time between seal and RGA test is known.

Example 3. A package having a volume of 0.1 cc had a true helium leak rate limit of 1×10^{-9} atm-cc/sec. The package was sealed with 10% helium so the measured limit was 1×10^{-10}. The initial measured leak rate was less than 1×10^{-9} and therefore unreliable. The package was subjected to thermal and mechanical stress and then sent out for RGA. The time between seal and RGA test was 20 days. From Eq. (10-1):

$$Q_{O_2} = \Delta p_{O_2}\left(1 - e^{-\frac{L_{O_2} t}{V}}\right)$$

where Q_{O_2} = the oxygen that has leaked in 20 days

Δp_{O_2} = the oxygen in the air = 20.96%

L_{O_2} = the true oxygen leak rate = $\sqrt{\frac{4}{32}} \times 1 \times 10^{-9}$ =3.53 × 10^{-10}

V = 0.1 cc

t = 20 × 24 × 3600 = 1,728,000 sec

$$Q_{O_2} = 20.96\left[1 - e^{-\frac{(3.53\times10^{-10})(1,728,000)}{0.1}}\right]$$

$$Q_{O_2} = 20.96(1 - e^{-0.0060998}) = 20.96(1 - 0.993919) = 20.96\ (0.00608)$$

$$Q_{O_2} = 0.1274\% = 1275 \text{ ppm}$$

If the RGA showed less than 1275 ppm oxygen, the true helium leak rate would be less than 1 × 10^{-9} atm-cc/sec. There is a possible error here because some oxygen that leaked in could have reacted with some component within the package. The amount of argon in the package could also be used to calculate the leak rate, but if there was an argon impurity in the nitrogen, there would also be an error.

Example 4. Table 10-10 shows the RGA of five packages that have gone through mechanical and thermal qualification testing. The volume of the packages was 0.071 cc and they were backfilled with a nominal 10% helium The time between seal and RGA test was 54 days.

No argon or oxygen is reported in the RGA, so that the maximum argon and oxygen is 100 ppm. Figure 10-3 gives a true helium leak rate divided by the package volume for 100 ppm of oxygen. At 54 days the L_{HE}/V value = 2.8 × 10^{-10} and L_{HE} (the true helium leak rate) = 2.8 × 10^{-10} × 0.071 cc = 2.0 × 10^{-11} atm-cc/sec. For any leak channel other than cylindrical in shape, these packages have zero leaks. The minimum theoretical leak rate for a cylindrical leak channel 0.1 cm long is 1.18 × 10^{-11} atm-cc/sec. For a rectangular leak channel 0.1 cm long, the theoretical minimum leak is 1.50 × 10^{-10} atm-cc/sec.

Table 10-10. RGA for the Qualification of Small Packages

Gas	%/PPM	Package # 1	Package # 2	Package # 3	Package # 4	Package # 5
Ammonia	ppm	ND	ND	ND	ND	ND
Argon	ppm	ND	ND	ND	ND	ND
Carbon dioxide	ppm	<100	<100	<100	<100	<100
Fluorocarbons	ppm	ND	ND	ND	ND	ND
Helium	%	13.0	13.2	13,5	13.3	13.6
Hydrocarbons	ppm	ND	ND	ND	ND	ND
Hydrogen	ppm	4049	1949	4417	3314	1779
Isopropyl alcohol	ppm	ND	ND	ND	ND	ND
Methanol	ppm	ND	ND	ND	ND	ND
Moisture	ppm	<100	829	645	763	565
Nitrogen	%	86.6	86.5	86.0	86.3	86.2
Oxygen	ppm	ND	ND	ND	ND	ND

6.0 PROBLEMS AND THEIR SOLUTIONS

Problem 1. A 20 cc package is sealed with 100% nitrogen. A RGA is performed 20 days after sealing. It shows neither oxygen nor argon. What is the true helium leak rate (L_{HE}), assuming no reaction with any oxygen?

Solution. When no oxygen is indicated, the maximum oxygen is less than 100 ppm. Figure 10-3 is a plot of the true helium leak rate divided by the package volume, versus days after sealing. The curve shows a value of 7.9×10^{-10} for L_{HE}/V after 20 days. The true maximum helium leak rate = $7.9 \times 10^{-10} \times 20 = 1.6 \times 10^{-8}$ atm-cc/sec or less.

Problem 2. A package with a volume of 10.16 cc is tested for residual gases 20 days after it was sealed. The RGA is repeated below. What is the true helium leak rate?

Ammonia	ppm	ND
Argon	ppm	330
Carbon dioxide	ppm	130
Fluorocarbons	ppm	ND
Helium	ppm	ND
Hydrocarbons	ppm	ND
Hydrogen	ppm	190
Isopropyl alcohol	ppm	ND
Methanol	ppm	ND
Moisture	ppm	ND
Nitrogen	%	99+
Oxygen	ppm	2750

Solution. The true oxygen leak rate can be calculated from the amount of oxygen in the package. The true argon leak rate can be calculated from the amount of argon in the package. These leak rates can be converted to the equivalent true helium leak rate by the square root of the inverse of their molecular weights or by using Table 3-1.

Using Eq. 10-2:

$$L_{O_2} = -\frac{V}{t}\left[\ln\left(1 - \frac{Q}{\Delta p_i}\right)\right]$$

where: L_{O_2} = the true oxygen leak rate

V = 10.16 cc.

t = 20 days = 20 24 × 3600 = 1,728,000 sec

Q = the amount of gas leaked in

Δp_i = the percentage of the gas in the air

$$L_{O_2} = -\frac{10.16}{1{,}728{,}000}\left[\ln\left(1 - \frac{0.2750}{20.96}\right)\right]$$

$$L_{O_2} = -5.88 \times 10^{-6}[\ln(1 - 0.1312)] = -5.88 \times 10^{-6}(\ln 0.98688)$$

$$L_{O_2} = -5.88 \times 10^{-6}(-0.01312) = 7.77 \times 10^{-8} \text{ atm-cc/sec}$$

$$L_{HE} = \sqrt{\frac{32}{4}}\, L_{O_2} = 2.83 \times 7.77 \times 10^{-8} = 2.20 \times 10^{-7} \text{ atm-cc/sec}$$

This is the minimum true helium leak. Some of the oxygen that leaked in could have reacted with something within the package and therefore was not reported.

$$L_{AR} = -5.88 \times 10^{-6}\left[\ln\left(1 - \frac{330}{9430}\right)\right]$$

$$L_{AR} = -5.88 \times 10^{-6}[\ln(1 - 0.03499)] = -5.88 \times 10^{-6}(\ln 0.9650)$$

$$L_{AR} = -5.88 \times 10^{-6}(-0.03562) = 2.09 \times 10^{-7} \text{ atm-cc/sec}$$

$$L_{HE} = \sqrt{\frac{40}{4}}\, L_{AR} = 3.16 \times 2.09 \times 10^{-7} = 6.61 \times 10^{-7} \text{ atm-cc/sec}$$

This is the maximum true helium leak rate because some of the reported argon may be due to an impurity in the nitrogen. The true helium leak rate is between 2.20×10^{-7} and 6.61×10^{-7} atm-cc/sec.

Problem 3. A 2.0 cc MCM is backfilled with 90% nitrogen and 10% helium. Immediately after sealing, it measures 2×10^{-8} atm-cc/sec. RGA testing is performed 1,000 hr after sealing. Assuming no oxygen reaction in the package and no argon impurity in the nitrogen, what is the expected oxygen to argon ratio?

Solution. Figure 10-1 gives the oxygen to argon ratio as a function of L_{HE}/V and hours after sealing. L_{HE} is equal to the measured value divided by the atmospheres of helium in the package $= 2 \times 10^{-8}/0.1 = 2 \times 10^{-7}$ atm-cc/sec.

$$L_{HE}/V = 2 \times 10^{-7}/2 = 1 \times 10^{-7} \text{ atm/sec}$$

The oxygen to argon ratio at 1,000 hr when $L_{HE}/V = 1 \times 10^{-7}$ is 24.7.

Problem 4. A 2.5 cc package is sealed with 100% nitrogen. The RGA shows 2,500 ppm of oxygen 50 days after sealing. What is the minimum true helium leak rate?

Solution 4. Equation (10-2) will solve for the true oxygen leak rate:

$$L_{O_2} = -\frac{V}{t}\left[\ln\left(1-\frac{Q_{O_2}}{\Delta pi}\right)\right]$$

$$L_{O_2} = -\frac{2.5}{4{,}320{,}000}\left[\ln\left(1-\frac{0.2500}{20.96}\right)\right]$$
$$= -5.79 \times (\ln 0.98807)$$

$$L_{O_2} = -5.79 \times 10^{-7}(-0.0120) = 6.95 \times 10^{-9} \text{ atm-cc/sec}$$

$$L_{HE} = L_{O_2} \times \sqrt{\frac{32}{4}} = 6.95 \times 10^{-9} \times 2.83 = 1.97 \times 10^{-8} \text{ atm-cc/sec}$$

This is the minimum true helium leak rate because more than 2,500 ppm oxygen could have entered the package, but the excess over 2,500 could have reacted with something in the package.

Problem 5. The RGA of a 10.0 cc package shows no argon. The time between sealing and RGA testing is 50 days. What is the maximum true helium leak rate?

Solution. When no argon is reported, the maximum amount of argon present is 100 ppm. Using Eq. (10-2):

$$L_{AR} = -\frac{V}{t}\left[\ln\left(1-\frac{Q_{AR}}{\Delta p_i}\right)\right]$$

where V = 10 cc

$t = 50 \times 24 \times 3600 = 4{,}320{,}000$ sec

$Q_{AR} = 100$ ppm

$\Delta p_i = 9430$ ppm

$$L_{AR} = -\frac{10}{4{,}320{,}000}\left[\ln\left(1-\frac{100}{9430}\right)\right]$$

$$L_{AR} = -2.31 \times 10^{-6}[\ln(1-0.0106)] = -2.31 \times 10^{-6}(-0.0107)$$

$$L_{AR} = 2.46 \times 10^{-8} \text{ atm-cc/sec}$$

$$L_{HE} \max = L_{AR} = 2.46 \times 10^{-8} \times 3.16 = 7.78 \times 10^{-8} \text{ atm-cc/sec}$$

Problem 6. The RGA of a 0.1 cc package shows no oxygen and 300 ppm of argon. The RGA is performed 40 days after sealing. What is the true helium leak rate?

Solution. The true helium leak rate can be calculated on the basis of there being no oxygen, and also can be calculated on the basis of 300 ppm of argon. Some oxygen might have leaked in but reacted with material within the package. Therefore, the true helium leak rate based on the oxygen will be a minimum value. The calculation of the true helium leak rate based on the 300 ppm of argon will be a maximum value, because some argon may have been sealed in the package as an impurity in the nitrogen.

The true helium leak rate (L_{HE}) based on no oxygen reported can be determined from Fig. 10-3. No oxygen reported means less than 100 ppm. Figure 10-3 shows the true helium leak rate divided by the package volume (L_{HE}/V) to be 3.9×10^{-10} atm-cc/sec.

$$L_{HE} = 3.9 \times 10^{-10} \times 0.1 \text{ cc} = 3.9 \times 10^{-11} \text{ atm-cc/sec}$$

This is the minimum true helium leak rate.

The true helium leak rate based on the argon is calculated using Eq. (10-2):

$$40 \text{ days} = 40 \times 24 \times 3600 = 3{,}456{,}000 \text{ sec}$$

$$L_{AR} = \frac{0.1}{3{,}456{,}000}\left[\ln\left(1-\frac{300}{9430}\right)\right]$$

$$L_{AR} = -2.89 \times 10^{-8}[\ln(1-0.031810] = -2.89 \times 10^{-8} (\ln 0.968187)$$

$$L_{AR} = -2.89 \times 10^{-8}(-0.03233) = 9.34 \times 10^{-10} \text{ atm-cc/sec}$$

$$L_{HE} = L_{AR} \times \sqrt{\frac{40}{4}} = 9.34 \times 10^{-10} \times 3.16 = 2.95 \times 10^{-9} \text{ atm-cc/sec}$$

This is the maximum true helium leak rate. The true helium leak rate is between 3.9×10^{-11} and 2.95×10^{-9} atm-cc/sec.

Problem 7. A small package (0.05 cc) is being qualified by means of the RGA. If the RGA is performed 60 days after sealing, what is the oxygen and argon limits for a true helium leak rate of 1×10^{-9} atm-cc/sec?

Solution. To calculate the amount of oxygen, the true oxygen leak rate corresponding to the true helium leak rate must be used in Eq. (10-1):

$$L_{O_2} = L_{HE} \times \sqrt{\frac{4}{32}} = 1 \times 10^{-9} \times 0.354 = 3.53 \times 10^{-10} \text{ atm-cc/sec}$$

$$Q_{O_2} = \Delta p_i \left(1 - e^{-\frac{L_{O_2} t}{V}}\right)$$

$$t = 60 \text{ days} = 60 \times 24 \times 3600 = 5{,}184{,}000 \text{ sec}$$

$$Q_{O_2} = 20.96\% \left[1 - e^{-\frac{(3.53 \times 10^{-10})(5{,}184{,}000)}{0.05}}\right]$$

$$Q_{O_2} = 20.96\%(1 - e^{-0.036599}) = 20.96\%(1 - 0.96406)$$

$$Q_{O_2} = 20.96\%(0.03594)$$

$$Q_{O_2} = 0.753\%$$

To calculate the maximum allowed amount of argon, the true argon leak rate (L_{AR}) is used.

$$L_{AR} = L_{HE} \times \sqrt{\frac{4}{40}} = 1 \times 10^{-9} \times 0.316 = 3.16 \times 10^{-10} \text{ atm-cc/sec}$$

$$Q_{AR} = 9430 \text{ ppm}\left[1 - e^{-\frac{(3.16\times10^{-10})(5,184,000)}{0.05}}\right]$$

$$Q_{AR} = 9430 \text{ ppm}(1 - e^{-0.03276}] = 9430 \text{ ppm}(1 - 0.96777)$$

$$Q_{AR} = 9430 \text{ ppm} \times 0.03223 = 304 \text{ ppm}$$

Problem 8. What is the true helium leak rate of a 2.0 cc package that contains 2,000 ppm oxygen and 150 ppm argon? The time between sealing and RGA is 30 days and the package was in air.

Solution. Using Eq. (10-2):

$$L_{O_2} = -\frac{V}{t}\left[\ln\left(1 - \frac{Q_{O_2}}{\Delta pi}\right)\right]$$

$$t = 30 \text{ days} = 30 \times 24 \times 3600 = 2,592,000 \text{ sec}$$

$$L_{O_2} = -\frac{2.0}{2,592,000}\left[\ln\left(1 - \frac{0.2\%}{20.96\%}\right)\right]$$

$$L_{O_2} = -7.72 \times 10^{-7}(\ln 0.990458) = -7.72 \times 10^{-7}(-0.0095878)$$

$$L_{O_2} = 7.40 \times 10^{-9} \text{ atm-cc/sec}$$

$$L_{HE} = L_{O_2} \times \sqrt{\frac{32}{4}} = 7.40 \times 10^{-9} \times 2.83 = 2.09 \times 10^{-8} \text{ atm-cc/sec}$$

$$L_{AR} = -\frac{2.0}{2{,}592{,}000}\left[\ln\left(1-\frac{150}{9430}\right)\right]$$

$$L_{AR} = -7.72 \times 10^{-7}(\ln 0.984093) = -7.72 \times 10^{-7}(-0.01603)$$

$$L_{AR} = 1.24 \times 10^{-8} \text{ atm-cc/sec}$$

$$L_{HE} = L_{AR} \times \sqrt{\frac{40}{4}} = 1.24 \times 10^{-8} \times 3.16 = 3.92 \times 10^{-8} \text{ atm-cc/sec}$$

The true helium leak rate (L_{HE}) is between 2.09×10^{-8} and 3.92×10^{-8} atm-cc/sec.

REFERENCES

1. Greaves, C., *Vacuum*, 20(2):69–70, Pergamon Press (1970)
2. Roberts, R. W., and Vanderslice, T. A., *Ultrahigh Vacuum And Its Application*, pp. 48–49, Prentice Hall (1963)

Appendix

1.0 LIST OF SYMBOLS AND DIMENSIONS

A = cross-sectional area or area

a = inside package cavity dimension

ATM_{MIN} = minimum atmospheric change that can be detected optically

D = diameter

D = the diffusion of a gas through a solid per second

d = thickness

E = modulus of elasticity

F = conductance F_{DET} = conductance of detector liquid

F_m = molecular conductance

F_{mc} = molecular conductance of a uniform cylinder

F_{mr} = molecular conductance of a uniform rectangular tube

F_{ms} = molecular conductance of a uniform slit

F_O = conductance of an orifice

F_t = conductance of a tube

F_V = viscous conductance

F_{vc} = viscous conductance of a uniform cylinder

F_{VCHE} = viscous conduction of helium in a cylindrical leak channel

F_{HE} = viscous conductance of helium

F_{vr} = viscous conductance of a uniform rectangular tube

F_{VRHE} = viscous conductance of helium in a rectangular leak channel

F_{VR} = viscous conductance of a uniform rectangular tube

G = correction factor for conduction through a rectangular leak channel

HeD = helium at the detector

K = permeability of a gas-solid combination

k = Boltzman constant = 1.38×10^{-16} dyne-cm/degree

L = true leak rate or equivalent standard air leak rate

L_1 = equivalent standard air leak rate

L_{AIR} = equivalent standard air leak rate

L_{AR} = true argon leak rate

L_{HE} = true helium leak rate

L_{H_2O} = true water leak rate

L_N = true nitrogen leak rate

L_O = true oxygen leak rate

ℓ = length

M = molecular weight

MDL = minimum detectable leak

MDS = minimum detectable signal

m = mass of one molecule

m = reciprocal of Poisons Ratio

mfp = mean free path

ml = milliliters

N = number of molecules in volume V

N_A = Avogadro's Number = number of molecules in a molecular weight = 6.02×10^{23}

P = pressure

P_{atm} = pressure in atmospheres

P_{AVE} = average pressure for viscous flow

P_i = initial pressure

P_{mm} = pressure in mm of mercury = P_T

$P\mu$ = pressure in microns of mercury

$P_{\mu b}$ = pressure in microbars

P_T = pressure in Torr = P_{mm}

P_t = pressure at time t

ppm = parts per million

Q = quantity of gas or liquid

Q_{DET} = quantity of detector liquid

R = measured helium leak rate = R_1 = $R1$

R_O = gas constant = k/m

R_{DET} = calculated leak rate of detector liquid

S = solubility of a gas in a solid

T = absolute temperature or time

T = lid thickness

t = time

V = volume

w = pounds per square inch

Y = correction factor for the conductance through a cylindrical leak channel

y = lid deflection

Z = vacuum impedance

γ = mass of gas in volume V

η = viscosity

$\nu_a = \nu_{AVE}$ = average molecular velocity

ν_S = root mean square of velocity = $\sqrt{\nu_{ave}^{2}}$

ρ = density

μl = microliters

σ = molecular diameter

2.0 DIMENSIONS

m = mass

l = length

t = time

T = temperature

Conductance is $[l^3 t^{-1}]$

Leak rate is $[ml^2 t^{-3}]$

Pressure is $[ml^{-1} t^{-2}]$

R_o is $[ml^2 t^{-2} T^{-1}]$

Viscosity is $[ml^{-1} t^{-1}]$

dyne = g-cm/sec is $[ml\, t^{-2}]$

erg = dyne-cm is $[ml^2 t^{-2}]$

3.0 CONVERSION FACTORS FOR PRESSURE/VACUUM

To Convert A	To B	Multiply A By
atmospheres	bars	1.0133
atmospheres	dynes/cm^2	$1{,}010 \times 10^3$
atmospheres	microns ($\propto$ of mercury)	760×10^3
atmospheres	millibars	1,013.3
atmospheres	mm of mercury	760
atmospheres	pascals (Pa)	101×10^3
atmospheres	torr	760
$\frac{\text{atmospheres-cc}}{\text{second}}$	$\frac{\text{micron-liters}}{\text{hour}}$	2.736×10^6
$\frac{\text{atmospheres-cc}}{\text{second}}$	$\frac{\text{torr-liters}}{\text{second}}$	0.760
bars	atmospheres	0.987
bars	dynes/cm^2	1×10^6
bars	microns	750×10^3
bars	millibars	1,000
bars	mm of mercury	750
bars	pascals	1×10^5
bars	torr	750
dynes/cm^2	atmospheres	9.901×10^{-7}
dynes/cm^2	bars	1×10^{-6}
dynes/cm^2	microns	0.75
dynes/cm^2	millibars	1×10^{-3}
dynes/cm^2	mm of mercury	7.5×10^{-4}
dynes/cm^2	pascals	0.1
dynes/cm^2	torr	7.5×10^{-4}

To Convert A	To B	Multiply A By
microns	atmospheres	1.316×10^{-6}
microns	bars	1.333×10^{-6}
microns	dynes/cm^2	1.333×10^{6}
microns	millibars	1.333×10^{-3}
microns	mm of mercury	1,000
microns	pascals	0.1333
microns	torr	1,000
$\frac{\text{micron-liters}}{\text{hour}}$	$\frac{\text{atmospheres-cc}}{\text{second}}$	3.65497×10^{-7}
millibars	atmospheres	9.87×10^{-4}
millibars	bars	1×10^{-3}
millibars	dynes/cm^2	1,000
millibars	microns	750
millibars	mm of mercury	0.750
millibars	pascals	100
millibars	torr	0.750
pascals	atmospheres	9.87×10^{-6}
pascals	bars	1×10^{-5}
pascals	dynes/cm^2	10
pascals	microns	7.50
pascals	millibars	0.01
pascals	mm of mercury	7.5×10^{-3}
pascals	newtons/meter^2	1
pascals	torr	7.5×10^{-3}
torr	atmospheres	1.32×10^{-3}
torr	bars	1.33×10^{-3}
torr	dynes/cm^2	1.33×10^{3}
torr	microns	1×10^{3}
torr	millibars	1.33

To Convert A	To B	Multiply A By
torr	mm of mercury	1.0
torr	pascals	133
$\frac{\text{torr-liters}}{\text{second}}$	$\frac{\text{atm-cc}}{\text{second}}$	1.315
$\frac{\text{torr-liters}}{\text{second}}$	$\frac{\text{Pa-meters}}{\text{second}}$	0.1

Acknowledgment

Thanks to my colleagues of the many years who have encouraged me to write this book, as there was no book addressing this specialized topic. I would like to thank the Communications Systems Division of the AlliedSignal Corporation for the use of the several hundred Residual Gas Analysis Reports. A special thanks to Danita Eichenlaub for her review of several chapters, for her technical discussions, and for her long time encouragement.

Index